THE BOOK ®

Kawasaki ZX600 & 750 Fours

Service and Repair Manual

by Bob Henderson
and John H Haynes Member of the Guild of Motoring Writers

Models covered

(1780-260-3Y3)

ZX600A (GPZ600R). 592cc. UK 1985 through 1990
ZX600A (Ninja 600R). 592cc. US 1985 through 1987
ZX600B alumininum frame model (Ninja 600RX). 592cc. US 1987
ZX600C (GPX600R). 592cc. UK 1988 through 1996
ZX600C (Ninja 600R). 592cc. US 1988 through 1997
ZX750F (GPX750R). 748cc. UK 1987 through 1991
ZX750F (Ninja 750R). 748cc. US 1987 through 1990

ABCDE
FGHIJ

© Haynes Publishing 1999

A book in the **Haynes Service and Repair Manual Series**

All rights reserved. No part of this book may be reproduced or transmitted in any form or by any means, electronic or mechanical, including photocopying, recording or by any information storage or retrieval system, without permission in writing from the copyright holder.

ISBN **1 85960 230 4**

Library of Congress Catalog Card Number 97-72743

British Library Cataloguing in Publication Data
A catalogue record for this book is available from the British Library.

Printed in the USA

Haynes Publishing
Sparkford, Yeovil, Somerset BA22 7JJ, England

Haynes North America, Inc
861 Lawrence Drive, Newbury Park, California 91320, USA

Editions Haynes
4, Rue de l'Abreuvoir
92415 COURBEVOIE CEDEX, France

Haynes Publishing Nordiska AB
Box 1504, 751 45 UPPSALA, Sweden

Contents

LIVING WITH YOUR KAWASAKI

Introduction

Daily (pre-ride) checks

MAINTENANCE

Routine maintenance and servicing

Contents

REPAIRS AND OVERHAUL

Kawasaki
The Green Meanies

by Julian Ryder

Kawasaki Heavy Industries

Kawasaki is a company of contradictions. It is the smallest of the big four Japanese manufacturers but the biggest company, it was the last of the four to make and market motorcycles yet it owns the oldest name in the Japanese industry, and it was the first to set up a factory in the USA. Kawasaki Heavy Industries, of which the motorcycle operation is but a small component, is a massive company with its heritage firmly in the old heavy industries like shipbuilding and railways; nowadays it is as much involved in aerospace as in motorcycles.

In fact it may be because of this that Kawasaki's motorcycles have always been quirky, you get the impression that they are designed by a small group of enthusiasts who are given an admirably free hand. More

realistically, it may be that Kawasaki's designers have experience with techniques and materials from other engineering disciplines. Either way, Kawasaki have managed to be the factory who surprise us more than the rest. Quite often, they do this by totally ignoring a market segment the others are scrabbling over, but more often they hit us with pure, undiluted performance.

The origins of the company, and its name, go back to 1878 when Shozo Kawasaki set up a dockyard in Tokyo. By the late 1930s, the company was making its own steel in massive steelworks and manufacturing railway locos and rolling stock. In the run up to war, the Kawasaki Aircraft Company was set up in 1937 and it was this arm of the now giant operation that would look to motorcycle engine manufacture in post-war Japan.

They bought their high-technology

experience to bear first on engines which were sold on to a number of manufacturers as original equipment. Both two- and four-stroke units were made, a 58 cc and 148 cc OHC unit. One of the customer companies was Meihatsu Heavy Industries, another company within the Kawasaki group, which in 1961 was shaken up and renamed Kawasaki Auto Sales. At the same time, the Akashi factory which was to be Kawasaki's main production facility until the Kobe earthquake of 1995, was opened. Shortly afterwards, Kawasaki took over the ailing Meguro company, Japan's oldest motorcycle maker, thus instantly obtaining a range of bigger bikes which were marketed as Kawasaki-Meguros. The following year, the first bike to be made and sold as a Kawasaki was produced, a 125 cc single called the B8 and in 1963 a motocross version, the B8M appeared.

Model development

Kawasaki's first appearance on a road-race circuit came in 1965 with a batch of disc-valve 125 twins. They were no match for the opposition from Japan in the shape of Suzuki and Yamaha or for the fading force of the factory MZs from East Germany. Only after the other Japanese factories had pulled out of the class did Kawasaki win, with British rider Dave Simmonds becoming World 125 GP Champion in 1969 on a bike that looked astonishingly similar to the original racer. That same year Kawasaki reorganised once again, this time merging three

The H1 three cylinder two-stroke 500

companies to form Kawasaki Heavy Industries. One of the new organisation's objectives was to take motorcycle production forward and exploit markets outside Japan.

KHI achieved that target immediately and set out their stall for the future with the astonishing and frightening H1. This three-cylinder air-cooled 500 cc two-stroke was arguably the first modern pure performance bike to hit the market. It hypnotised a whole generation of motorcyclists who'd never before encountered such a ferocious, wheelie inducing power band or such shattering straight-line speed allied to questionable handling. And as for the 750 cc version ...

The triples perfectly suited the late '60s, fitting in well with the student demonstrations of 1968 and the anti-establishment ethos of the Summer of Love. Unfortunately, the oil crisis would put an end to the thirsty strokers but Kawasaki had another high-performance ace up their corporate sleeve. Or rather they thought they did.

The 1968 Tokyo Show saw probably the single most significant new motorcycle ever made unveiled: the Honda CB750. At Kawasaki it caused a major shock, for they also had a 750 cc four, code-named New York Steak, almost ready to roll and it was a double, rather than single, overhead cam motor. Bravely, they took the decision to go ahead - but with the motor taken out to 900 cc. The result was the Z1, unveiled at the 1972 Cologne Show. It was a bike straight out of the same mould as the H1, scare stories spread about unmanageable power, dubious straight-line stability and frightening handling, none of which stopped the sales graph rocketing upwards and led to the coining of the term 'superbike'. While rising fuel prices cut short development of the big two-strokes,

The first Superbike, Kawasaki's 900 cc Z1

the Z1 went on to found a dynasty, indeed its genes can still be detected in Kawasaki's latest products like the ZZ-R1100 (Ninja ZX-11).

This is another characteristic of the way Kawasaki operates. Models quite often have very long lives, or gradually evolve. There is no major difference between that first Z1 and the air-cooled GPz range. Add water-cooling and you have the GPZ900, which in turn metamorphosed into the GPZ1000RX and then the ZX-10 and the ZZ-R1100. Indeed, the last three models share the same 58 mm

stroke. The bikes are obviously very different but it's difficult to put your finger on exactly why.

Other models have remained effectively untouched for over a decade: the KH and KE single-cylinder air-cooled two-stroke learner bikes, the GT550 and 750 shaft-drive hacks favoured by big city despatch riders and the GPz305 being prime examples. It's only when they step outside the performance field that Kawasakis seems less sure. Their first factory customs were dire, you simply got the impression that the team that designed them didn't have their heart in the job. Only when the Classic range appeared in 1995 did they get it right.

Racing success

Kawasaki also have a more focused approach to racing than the other factories. The policy has always been to race the road bikes and with just a couple of exceptions that's what they've done. Even Simmonds' championship winner bore a strong resemblance to the twins they were selling in the late '60s and racing versions of the 500 and 750 cc triples were also sold as over-the-counter racers, the H1R and H2R. The 500 was in the forefront of the two-stroke assault on MV Agusta but wasn't a Grand Prix winner. It was the 750 that made the impact and carried the factory's image in F750 racing against the Suzuki triples and Yamaha fours.

The factory's decision to use green, usually regarded as an unlucky colour in sport, meant its bikes and personnel stood out and the phrase 'Green Meanies' fitted them perfectly. The Z1 motor soon became a full 1000 cc and powered Kawasaki's assault in F1 racing, notably in endurance which Kawasaki saw as

One of the two-stroke engined KH and KE range - the KH125EX

The GT750 - a favourite hack for despatch riders

being most closely related to its road bikes.

That didn't stop them dominating 250 and 350 cc GPs with a tandem twin two-stroke in the late '70s and early '80s, but their path-breaking monocoque 500 while a race winner never won a world title. When Superbike arrived, Kawasaki's road 750s weren't as track-friendly as the opposition's out-and-out race replicas. This makes Scott Russell's World title on the ZXR750 in 1993 even more praiseworthy, for the homologation bike, the ZXR750RR, was much heavier and much more of a road bike than the Italian and Japanese competition.

The company's Supersport 600 contenders have similarly been more sports-tourers than race-replicas, yet they too have been competitive on the track. Indeed, the flagship bike, the ZZ-R1100, is most definitely a sports tourer capable of carrying two people and their luggage at high speed in comfort all day and then doing it again the next day. Try that on one of the race replicas and you'll be in need of a course of treatment from a chiropractor.

Through doing it their way Kawasaki developed a brand loyalty for their performance bikes that kept the Z1's derivatives in production until the mid-'80s

and turned the bike into a classic in its model life. You could even argue that the Z1 lives on in the shape of the 1100 Zephyr's GPz1100-derived motor. And that's another Kawasaki invention, the retro bike. But when you look at what many commentators refer to as the retro boom, especially in

Japan, you find that it is no such thing. It is the Zephyr boom. Just another example of Japan's most surprising motorcycle manufacturer getting it right again.

The high-performance ZXR750

The ZX600 and 750 Fours

For a first go, the ZX600A (GPZ600R in the UK, Ninja 600R in the US) wasn't bad. In fact it was marvelous. Up until its introduction in 1985, Kawasaki had never made a water-cooled bike. Their air-cooled ZX models (note the lower-case 'zed' in the UK GPz designation denoting air cooling, a capital means liquid cooled) middleweights had ruled the roost since the start of the decade, advancing through the usual Kawasaki progression of naked twin-shock bike, naked monoshocker, then growing a bikini fairing, a half fairing and a full fairing in successive years.

The ZX600A, though, hit the ground running as a fully-enclosed water-cooled bike with the full set of mid '80s trickery including a 16-inch front wheel and suspension equipped with anti-dive and adjustable air pressure and damping. Critically, it was a year ahead of the Honda CBR600 and swept all before it in the middleweight sales stakes. Nevertheless, there were two sets of modifications; the first was a new set of disc brakes for the A4 model of 1988, the second a major reworking of engine internals for the A5 of 1989 which increased power by 9% and peak torque by 13%. This didn't really do much for sales as by then the modern 600s had arrived.

The ZX600A's chassis package was all new but the motor was based around the old 52.4 mm stroke of the GPz550 with the bore taken out 2 mm to 60 mm, another typical piece of Kawasaki evolution. As was the ZX600C model (GPX600R in the UK, Ninja 600R in the US) which arrived in 1988. The engine - at the bottom end at least - was the same as the ZX600A but up top the drive for inlet efficiency led the valves to be operated by short rocker arms pivoting in ball and pillow mountings as opposed to the cam lobes operating on

Kawasaki ZX600C

buckets over the top of the valve stem. The new design allowed the inlet tracts to be straightened out.

Externally, the ZX600C was slimmer and a lot easier on the eyes than the ZX600A, although that pearlescent white paintwork does look a little dated in the late '90s! Naturally, Kawasaki added loads of lightness, bring it down a whopping 25 kg to 180 kg but didn't go with the trend back to bigger wheels seen on the competition, giving it a 16-incher at the back as well as the front. It had the full set of '80s high-tech but the electronic anti-dive system (ESCS) was quietly dropped on later models. The character of the bikes were very similar; both needed to be revved and worked hard to get the best out of them, although there is no doubt that the ZX600C doesn't just look more modern, it goes that way, too.

The ZX600A now looks like a soft all rounder but in its day it was a cutting edge sportster. The ZX600C was much harsher (especially the rear suspension) and more

focused, but by the time it hit the market it had the CBR600 to deal with.

The bigger ZX750F (GPX750R in the UK, Ninja 750R in the US) had similar problems as it tried to make up for the flop that was the ZX750G (GPZ750R), only more of them; the FZ, VFR and GSX-R 750s. This time the new bike wasn't a derivation of a previous model as the brand new 68 x 51.5 mm bore and stroke show, dimensions which, incidentally, were carried over to the ZXR750. Externally it looked so like the 600 that it was difficult to know which was which without looking at the decals. The 18-inch rear wheel on the 750 was a visual clue, though. It rode like the smaller bikes, too, with a pronounced reluctance to do anything in the low or mid-range followed by a manic rush at the top end. The chassis seemed to tighten up when you hit the power and what seemed like a soft, gutless bike in town turned into a rev hound when you hit the open road; all the classic characteristics of a sporty Kawasaki.

Acknowledgements

Our thanks are due to Kawasaki Motors (UK) Ltd for permission to reproduce certain illustrations used in this manual. We would also like to thank the Avon Rubber Company, who kindly supplied information and technical assistance on tyre fitting, and NGK Spark plugs (UK) Ltd for information on spark plug maintenance and electrode conditions.

Thanks are also due to the Kawasaki Information Service and Kel Edge for supplying colour transparencies. The main front cover photograph was taken by Phil Flowers. The introduction, "Kawasaki - The Green Meanies" was written by Julian Ryder.

About this Manual

The aim of this manual is to help you get the best value from your motorcycle. It can do so in several ways. It can help you decide what work must be done, even if you choose to have it done by a dealer; it provides information and procedures for routine maintenance and servicing; and it offers diagnostic and repair procedures to follow when trouble occurs.

We hope you use the manual to tackle the work yourself. For many simpler jobs, doing it yourself may be quicker than arranging an appointment to get the motorcycle into a dealer and making the trips to leave it and pick it up. More importantly, a lot of money can be saved by avoiding the expense the shop must pass on to you to cover its labour and overhead costs. An added benefit is the sense of satisfaction and accomplishment that you feel after doing the job yourself.

References to the left or right side of the motorcycle assume you are sitting on the seat, facing forward.

We take great pride in the accuracy of information given in this manual, but motorcycle manufacturers make alterations and design changes during the production run of a particular motorcycle of which they do not inform us. No liability can be accepted by the authors or publishers for loss, damage or injury caused by any errors in, or omissions from, the information given.

Professional mechanics are trained in safe working procedures. However enthusiastic you may be about getting on with the job at hand, take the time to ensure that your safety is not put at risk. A moment's lack of attention can result in an accident, as can failure to observe simple precautions.

There will always be new ways of having accidents, and the following is not a comprehensive list of all dangers; it is intended rather to make you aware of the risks and to encourage a safe approach to all work you carry out on your bike.

Asbestos

● Certain friction, insulating, sealing and other products - such as brake pads, clutch linings, gaskets, etc. - contain asbestos. Extreme care must be taken to avoid inhalation of dust from such products since it is hazardous to health. If in doubt, assume that they do contain asbestos.

Fire

● Remember at all times that petrol is highly flammable. Never smoke or have any kind of naked flame around, when working on the vehicle. But the risk does not end there - a spark caused by an electrical short-circuit, by two metal surfaces contacting each other, by careless use of tools, or even by static electricity built up in your body under certain conditions, can ignite petrol vapour, which in a confined space is highly explosive. Never use petrol as a cleaning solvent. Use an approved safety solvent.

● Always disconnect the battery earth terminal before working on any part of the fuel or electrical system, and never risk spilling fuel on to a hot engine or exhaust.

● It is recommended that a fire extinguisher of a type suitable for fuel and electrical fires is kept handy in the garage or workplace at all times. Never try to extinguish a fuel or electrical fire with water.

Fumes

● Certain fumes are highly toxic and can quickly cause unconsciousness and even death if inhaled to any extent. Petrol vapour comes into this category, as do the vapours from certain solvents such as trichloroethylene. Any draining or pouring of such volatile fluids should be done in a well ventilated area.

● When using cleaning fluids and solvents, read the instructions carefully. Never use materials from unmarked containers - they may give off poisonous vapours.

● Never run the engine of a motor vehicle in an enclosed space such as a garage. Exhaust fumes contain carbon monoxide which is extremely poisonous; if you need to run the engine, always do so in the open air or at least have the rear of the vehicle outside the workplace.

The battery

● Never cause a spark, or allow a naked light near the vehicle's battery. It will normally be giving off a certain amount of hydrogen gas, which is highly explosive.

● Always disconnect the battery ground (earth) terminal before working on the fuel or electrical systems (except where noted).

● If possible, loosen the filler plugs or cover when charging the battery from an external source. Do not charge at an excessive rate or the battery may burst.

● Take care when topping up, cleaning or carrying the battery. The acid electrolyte, evenwhen diluted, is very corrosive and should not be allowed to contact the eyes or skin. Always wear rubber gloves and goggles or a face shield. If you ever need to prepare electrolyte yourself, always add the acid slowly to the water; never add the water to the acid.

Electricity

● When using an electric power tool, inspection light etc., always ensure that the appliance is correctly connected to its plug and that, where necessary, it is properly grounded (earthed). Do not use such appliances in damp conditions and, again, beware of creating a spark or applying excessive heat in the vicinity of fuel or fuel vapour. Also ensure that the appliances meet national safety standards.

● A severe electric shock can result from touching certain parts of the electrical system, such as the spark plug wires (HT leads), when the engine is running or being cranked, particularly if components are damp or the insulation is defective. Where an electronic ignition system is used, the secondary (HT) voltage is much higher and could prove fatal.

Remember...

✗ **Don't** start the engine without first ascertaining that the transmission is in neutral.

✗ **Don't** suddenly remove the pressure cap from a hot cooling system - cover it with a cloth and release the pressure gradually first, or you may get scalded by escaping coolant.

✗ **Don't** attempt to drain oil until you are sure it has cooled sufficiently to avoid scalding you.

✗ **Don't** grasp any part of the engine or exhaust system without first ascertaining that it is cool enough not to burn you.

✗ **Don't** allow brake fluid or antifreeze to contact the machine's paintwork or plastic components.

✗ **Don't** siphon toxic liquids such as fuel, hydraulic fluid or antifreeze by mouth, or allow them to remain on your skin.

✗ **Don't** inhale dust - it may be injurious to health (see Asbestos heading).

✗ **Don't** allow any spilled oil or grease to remain on the floor - wipe it up right away, before someone slips on it.

✗ **Don't** use ill-fitting spanners or other tools which may slip and cause injury.

✗ **Don't** lift a heavy component which may be beyond your capability - get assistance.

✗ **Don't** rush to finish a job or take unverified short cuts.

✗ **Don't** allow children or animals in or around an unattended vehicle.

✗ **Don't** inflate a tyre above the recommended pressure. Apart from overstressing the carcass, in extreme cases the tyre may blow off forcibly.

✔ **Do** ensure that the machine is supported securely at all times. This is especially important when the machine is blocked up to aid wheel or fork removal.

✔ **Do** take care when attempting to loosen a stubborn nut or bolt. It is generally better to pull on a spanner, rather than push, so that if you slip, you fall away from the machine rather than onto it.

✔ **Do** wear eye protection when using power tools such as drill, sander, bench grinder etc.

✔ **Do** use a barrier cream on your hands prior to undertaking dirty jobs - it will protect your skin from infection as well as making the dirt easier to remove afterwards; but make sure your hands aren't left slippery. Note that long-term contact with used engine oil can be a health hazard.

✔ **Do** keep loose clothing (cuffs, ties etc. and long hair) well out of the way of moving mechanical parts.

✔ **Do** remove rings, wristwatch etc., before working on the vehicle - especially the electrical system.

✔ **Do** keep your work area tidy - it is only too easy to fall over articles left lying around.

✔ **Do** exercise caution when compressing springs for removal or installation. Ensure that the tension is applied and released in a controlled manner, using suitable tools which preclude the possibility of the spring escaping violently.

✔ **Do** ensure that any lifting tackle used has a safe working load rating adequate for the job.

✔ **Do** get someone to check periodically that all is well, when working alone on the vehicle.

✔ **Do** carry out work in a logical sequence and check that everything is correctly assembled and tightened afterwards.

✔ **Do** remember that your vehicle's safety affects that of yourself and others. If in doubt on any point, get professional advice.

● **If** in spite of following these precautions, you are unfortunate enough to injure yourself, seek medical attention as soon as possible.

Frame and engine numbers

The frame serial number is stamped into the right side of the steering head and the engine serial number is stamped into the right engine case. Both of these numbers should be recorded and kept in a safe place so they can be furnished to law enforcement officials in the event of theft.

The frame serial number, engine serial number and carburetor identification number should also be kept in a handy place (such as with your driver's licence) so they are always available when purchasing or ordering parts for your machine.

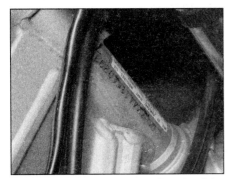

The frame number is stamped on the right side of the steering head

The engine number (arrow) is stamped into the top of the crankcase, just inboard of the clutch cover

US models
US models covered by this manual are as follows:

ZX600A (Ninja 600R)	A1 (1985), A2 (1986), A3 (1987)
ZX600B - Aluminum frame model	B1 (1987)
ZX600C (Ninja 600R)	C1 (1988), C2 (1989), C3 (1990), C4 (1991), C5 (1992), C6 (1993), C7 (1994), C8 (1995), C9 (1996), C10 (1997)
ZX750F (Ninja 750R)	F1 (1987), F2 (1988), F3 (1989), F4 (1990)

UK ZX600 A models (GPZ600R):

Year	Model Code	Initial frame number	Initial engine number
1985	ZX600 A1	ZX600A-000001 on	ZX600AE000001 on
1986	ZX600 A2	ZX600A-025001 on	ZX600AE025001 on
1987	ZX600 A3	ZX600A-046001 on	ZX600AE052040 on
1988	ZX600 A4/A4A	ZX600A-054001 on	ZX600ZE052040 on
1989/90	ZX600 A5/A5A	ZX600A-055801 on	ZX600AE069501 on

UK ZX600 C models (GPX600R):

Year	Model Code	Initial frame number	Initial engine number
1988	ZX600 C1	ZX600C-000001 on	ZX600AE052040 on
1989	ZX600 C2	ZX600C-011501 on	ZX600AE069501 on
1990-92	ZX600 C3	ZX600C-019001 on	ZX600AE069501 on
1993	ZX600 C6	ZX600C-600001 on	ZX600AE069501 on
1994-96	ZX600 C7	ZX600C-601551 on	ZX600AE069501 on

UK ZX750F models (GPX750R):

Year	Model Code	Initial frame number	Initial engine number
1987	ZX750 F1	ZX750F-000001 to 016200	ZX750FE000001 to 018000
1988	ZX750 F2	ZX750F-016201 to020200	ZX750FE018001 to 024000
1989-91	ZX750 F3	ZX750F-020201 on	ZX750FE024001 on

Buying spare parts

Once you have found all the identification numbers, record them for reference when buying parts. Since the manufacturers change specifications, parts and vendors (companies that manufacture various components on the machine), providing the ID numbers is the only way to be reasonably sure that you are buying the correct parts.

Whenever possible, take the worn part to the dealer so direct comparison with the new component can be made. Along the trail from the manufacturer to the parts shelf, there are numerous places that the part can end up with the wrong number or be listed incorrectly.

The two places to purchase new parts for your motorcycle - the accessory store and the franchised dealer - differ in the type of parts they carry. While dealers can obtain virtually every part for your motorcycle, the accessory dealer is usually limited to normal high wear items such as shock absorbers, tune-up parts, various engine gaskets, cables, chains, brake parts, etc. Rarely will an accessory outlet have major suspension components, cylinders, transmission gears, or cases.

Used parts can be obtained for roughly half the price of new ones, but you can't always be sure of what you're getting. Once again, take your worn part to the breaker (wrecking yard) for direct comparison.

Whether buying new, used or rebuilt parts, the best course is to deal directly with someone who specialises in parts for your particular make.

1 Engine/transmission oil level

Before you start:
✔ Take the motorcycle on a short run to allow it to reach normal operating temperature.

Caution: Do not run the engine in an enclosed space such as a garage or shop.

✔ Stop the engine and support the motorcycle in an upright position on level ground. Position the motorcycle on its centrestand. Allow it to stand undisturbed for five minutes to allow the oil level to stabilise.

The correct oil
● Modern, high-revving engines place great demands on their oil. It is very important that the correct oil for your bike is used.
● Always top up with a good quality oil of the specified type and viscosity and do not overfill the engine.

Oil type	API grade SG
Oil viscosity	
Cold climates	SAE 10W/40 or 10W/50
Warm climates	SAE 20W/40 or 20W/50

✔ The oil level is viewed through the window in the clutch cover on the right-hand side of the engine. Wipe the glass clean before inspection to make the check easier.

Bike care:
● If you have to add oil frequently, you should check whether you have any oil leaks. If there is no sign of oil leakage from the joints and gaskets the engine could be burning oil (see *Fault Finding*).

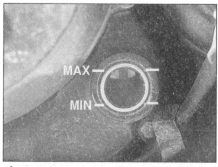

1 The oil level should lie between the MAX and MIN level lines on the window.

2 If the level is below the MIN line, remove the filler cap from the top of the clutch cover . . .

3 . . . and top up with the recommended grade and type of oil, to bring the level up to the MAX line on the window.

2 Clutch fluid level (750 models)

 Warning: Brake and clutch hydraulic fluid can harm your eyes and damage painted surfaces, so use extreme caution when handling and pouring it. Do not use fluid that has been standing open for some time, as it absorbs moisture from the air which can cause a loss of clutch effectiveness.

Before you start:
✔ Position the motorcycle on its centrestand and turn the handlebars until the top of the master cylinder is as level as possible.

✔ Make sure you have the correct hydraulic fluid. DOT 4 is recommended.

Bike care:
● If the fluid reservoir requires repeated topping-up this is an indication of an hydraulic leak somewhere in the system, which should be investigated immediately.
● Check for signs of fluid leakage from the hydraulic hoses and components - if found, rectify immediately.
● Check the operation of the clutch; if there is evidence of air in the system (spongy feel to lever), it must be bled (Chapter 2B).

1 Clutch fluid level is checked via sightglass - it must be above LOWER level mark(arrow).

2 Remove the two screws to free the reservoir cap.

3 Top up with new clean hydraulic fluid of the recommended type so that the level is above the LOWER level mark. Take care to avoid spills (see **Warning** above).

4 Ensure that the diaphragm is correctly folded before installing the cover.

3 Brake fluid levels

Warning: Brake and clutch hydraulic fluid can harm your eyes and damage painted surfaces, so use extreme caution when handling and pouring it and cover surrounding surfaces with rag. Do not use fluid that has been standing open for some time, as it absorbs moisture from the air which can cause a dangerous loss of braking effectiveness.

Before you start:

✔ Position the motorcycle on its centrestand, and turn the handlebars until the top of the front brake master cylinder is as level as possible. Remove the right side cover (see Chapter 8) to access the rear brake fluid reservoir.

✔ Make sure you have the correct hydraulic fluid - DOT 4 is recommended. Wrap a rag around the reservoir being worked on to ensure that any spillage does not come into contact with painted surfaces.

Bike care:

● The fluid in the front and rear brake master cylinder reservoirs will drop slightly as the brake pads wear down. Remove the right side cover for access to the rear brake fluid reservoir.

● If either fluid reservoir requires repeated topping-up this is an indication of an hydraulic leak somewhere in the system, which should be investigated immediately.

● Check for signs of fluid leakage from the hydraulic hoses and components - if found, rectify immediately.

● Check the operation of both brakes before taking the machine on the road; if there is evidence of air in the system (spongy feel to lever or pedal), it must be bled as described in Chapter 7.

1 The front brake fluid level is checked via the sightglass in the reservoir - it must be above the LOWER level mark (arrow).

2 If the level is below the LOWER level mark, remove the two screws (arrows) to free the reservoir cover, then remove the cover, and diaphragm.

3 Top up with new clean hydraulic fluid of the recommended type, until the level is above the LOWER mark. Take care to avoid spills (see **Warning** above).

4 Ensure that the diaphragm is correctly seated before installing the cover.

5 The rear brake fluid level can be seen through the translucent body of the reservoir. The fluid must lie between the LOWER and UPPER level marks. Top up as described for the front reservoir.

4 Suspension, steering and drive chain

Suspension and Steering:

● Check that the front and rear suspension operates smoothly without binding.
● Check that the suspension is adjusted as required.
● Check that the steering moves smoothly from lock-to-lock.

Drive chain:

● Check that the drive chain slack isn't excessive, and adjust if necessary (see Chapter 1).
● If the chain looks dry, lubricate it (see Chapter 1).

5 Coolant level

Warning: DO NOT remove the radiator pressure cap to add coolant. Topping up is done via the coolant reservoir tank filler. DO NOT leave open containers of coolant about, as it is poisonous.

Before you start:

✔ Make sure you have a supply of coolant available (a mixture of 50% soft water and 50% corrosion inhibited ethylene glycol antifreeze is needed).

✔ Place the motorcycle on its centrestand whilst checking the level. Make sure the motorcycle is on level ground.

✔ Make sure the engine is cold. Do not perform this check just after the engine has been run.

Bike care:

● Use only the specified coolant mixture. It is important that antifreeze is used in the cooling system all year round, not just during the winter months. Don't top-up with water alone, as the antifreeze will become too diluted.

● Do not overfill the reservoir tank. If the coolant is significantly above the upper line at any time, the surplus coolant should be siphoned off to prevent it from being expelled out of the breather hose when the engine is running.

● If the coolant level falls steadily, check the system for leaks as described in Chapter 1. If no leaks are found and the level still continues to fall, it is recommended that the machine be taken to a Kawasaki dealer who will pressure test the system.

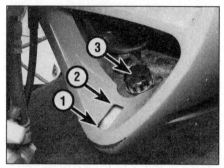

1 On ZX600A and B models check coolant level through the slot in the lower fairing. Level should lie between lower mark (1) and full mark (2). Top up via filler cap (3).

2 On ZX600C models turn the handlebars to the right and look into the left side of the upper fairing to check the coolant level. View level through window (1). Level should lie between full (2) and lower (3) level marks.

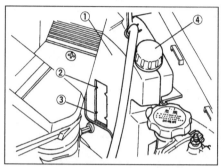

3 On ZX750F models check coolant level through window in the right inner fairing. Level of coolant in tank (1) should lie between full (2) and lower (3) marks. Top up via filler cap (4).

6 Legal and safety checks

Lighting and signalling:

● Take a minute to check that the headlight, tail light, brake light, instrument lights and turn signals all work correctly.
● Check that the horn sounds when the switch is operated.
● A working speedometer is a statutory requirement in the UK.

Safety:

● Check that the throttle grip rotates smoothly and snaps shut when released, in all steering positions. Also check for the correct amount of freeplay (see Chapter 1).
● On all ZX600 models, check that the clutch lever operates smoothly and with the correct amount of freeplay (see Chapter 1).
● Check that sidestand return spring holds the stand securely up when retracted. The same applies to the centrestand.
● Check that the engine STOP switch works correctly.

Fuel:

● This may seem obvious, but check that you have enough fuel to complete your journey. If you notice signs of fuel leakage - rectify the cause immediately.
● Ensure you use the correct grade unleaded or low-lead fuel - see Chapter 4 Specifications.

7 Tyres

Tyre care:

● Check the tyres carefully for cuts, tears, embedded nails or other sharp objects and excessive wear. Operation of the motorcycle with excessively worn tyres is extremely hazardous, as traction and handling are directly affected.
● Check the condition of the tyre valve and ensure the dust cap is in place.
● Pick out any stones or nails which may have become embedded in the tyre tread. If left, they will eventually penetrate through the casing and cause a puncture.
● If tyre damage is apparent, or unexplained loss of pressure is experienced, seek the advice of a tyre fitting specialist without delay.

Tyre tread depth:

● At the time of writing UK law requires that tread depth must be at least 1 mm over 3/4 of the tread breadth all the way around the tyre, with no bald patches. Many riders, however, consider 2 mm tread depth minimum to be a safer limit. Kawasaki recommend the following minimum tread depths.

Regular speed	Front	Rear
Up to 80 mph (130 kmh)	1 mm	2 mm
Above 80 mph (130 kmh)	1 mm	3 mm

● Many tyres now incorporate wear indicators in the tread. Identify the triangular pointer, or TWI marking, on the tyre sidewall to locate the indicator bar and replace the tyre if the tread has worn down to the bar.

1 Check the tyre pressures when the tyres are **cold** and keep them properly inflated.

2 Measure tread depth at the centre of the tyre using a tread depth gauge.

3 Tyre tread wear indicator bar and its location marking (usually either an arrow, a triangle or the letters TWI) on the sidewall (arrows).

The correct pressures:

● The tyres must be checked when **cold**, not immediately after riding. Note that low tyre pressures may cause the tyre to slip on the rim or come off. High tyre pressures will cause abnormal tread wear and unsafe handling.
● Use an accurate pressure gauge.
● Proper air pressure will increase tyre life and provide maximum stability and ride comfort.

Model	Front	Rear
ZX600 - all models	32 psi	36 psi
ZX750F		
Up to 98 kg (215 lbs) load	32 psi	36 psi
Above 98 kg (215 lbs) load, or high speed	36 psi	41 psi

Notes

Chapter 1
Routine maintenance and servicing

Contents

Degrees of difficulty

Easy, suitable for novice with little experience	**Fairly easy,** suitable for beginner with some experience	**Fairly difficult,** suitable for competent DIY mechanic 	**Difficult,** suitable for experienced DIY mechanic	**Very difficult,** suitable for expert DIY or professional 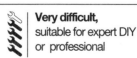

Engine

Spark plugs - 600 models
   Type
      US models . . . . NGK D9EA or ND X27ES-U
      UK and Canadian models . . . . NGK DR8ES or ND X27ESR-U
   Gap . . . . 0.6 to 0.7 mm (0.024 to 0.028 in)
Spark plugs - 750 models
   Type - US
      Standard plug . . . . NGK D8EA or ND X24ES-U
      For racing . . . . NGK D9EA or ND X27ES-U
      For cold operation . . . . NGK D7EA or ND X22ES-U
   Type - Canada
      Standard plug . . . . NGK DR8ES-L or ND X24ESR-U
      For racing . . . . NGK DR8ES or ND X27ESR-U
      For cold operation . . . . NGK DR7ES or ND X22ESR-U
   Type - UK
      Standard and racing plug . . . . NGK DR8ES or ND X27ESR-U
      For cold operation . . . . NGK DR8ES-L or DR7ES, ND X24ESR-U or X22ESR-U
   Gap . . . . 0.6 to 0.7 mm (0.024 to 0.028 in)
Engine idle speed - 600 models
   All except California models . . . . 1050 ± 50 rpm
   California models . . . . 1300 ± 50 rpm
Engine idle speed - 750 models
   All except California models . . . . 950 to 1050 rpm
   California models . . . . 1200 to 1300 rpm
Valve clearances (COLD engine) - 600 models
   Intake . . . . 0.13 to 0.18 mm (0.005 to 0.007 in)
   Exhaust . . . . 0.18 to 0.23 mm (0.007 to 0.009 in)
Valve clearances (COLD engine) - 750 models
   Intake . . . . 0.08 to 0.13 mm (0.0031 to 0.0051 in)
   Exhaust . . . . 0.12 to 0.17 mm (0.0047 to 0.0066 in)
Cylinder compression pressure - 600 models
   Acceptable range . . . . 109 to 171 psi
   Maximum difference between cylinders . . . . 14 psi
Cylinder compression pressure - 750 models
   Acceptable range . . . . 139 to 213 psi
   Maximum difference between cylinders . . . . Not specified
Carburettor synchronization (vacuum difference between cylinders) . . Less than 2 cm (0.391 in) Hg
Cylinder numbering (from left side to right side of bike) 1-2-3-4
Firing order 1-2-4-3

Miscellaneous

Brake pad minimum thickness . . . . 1.0 mm (0.040 in)
Freeplay adjustments
   Throttle grip . . . . 2 to 3 mm (0.08 to 0.12 in)
   Clutch lever (gap between lever and lever bracket when freeplay
   is taken up) - 600 models . . . . 2 to 3 mm (0.08 to 0.12 in)
Drive chain
   Slack . . . . 35 to 40 mm (1.38 to 1.57 in)
   20-link length . . . . 323 mm (12.73 in) maximum
Battery electrolyte specific gravity . . . . 1.280 at 68°F (20°C)
Minimum tyre tread depth
   Front . . . . 1.0 mm (0.040 in)
   Rear
      Up to 80 mph (130 kmh) . . . . 2.0 mm (0.079 in)
      Above 80 mph (130 kmh) . . . . 3.0 mm (0.118 in)
Tyre pressures (cold) - 600 models
   Front . . . . 32 psi
   Rear . . . . 36 psi

Tyre pressures (cold) - 750 models	**Front**	**Rear**
Up to 215 lbs (98 kg) load	32 psi	36 psi
215 to 401 lbs (98 to 182 kg) load, or high speed	36 psi	41 psi

Suspension air pressures
   Forks - 600 A and B models . . . . 7 to 10 psi
   Rear shock absorber
      600 A and B models . . . . 0 to 50 psi
      600 C models . . . . 0 to 28 psi
      750 model . . . . 0 to 21 psi

Torque specifications

Oil drain plug	20 Nm (14.5 ft-lbs)
Oil filter mounting bolt	20 Nm (14.5 ft-lbs)
Coolant drain bolt	8 Nm (69 in-lbs)
Spark plugs	13.3 Nm (120 in-lbs)
Valve cover bolts	8 Nm (87 in-lbs)
Alternator mounting bolts - 750 models	39 Nm (29 ft-lbs)

Recommended lubricants and fluids

Engine/transmission oil

Type	API grade SG multigrade and fuel efficient oil
Viscosity	
In cold climates	SAE 10W40 or 10W50
In warm climates	SAE 20W40 or 20W50
Capacity - 600 models	
With filter change	3.0 lit (3.2 US qt, 5.3 Imp pt)
Oil change only	2.6 lit (2.7 US qt, 4.6 Imp pt)
Capacity - 750 models	
With filter change	3.2 lit (3.4 US qt, 5.6 Imp pt)
Oil change only	3.0 lit (3.2 US qt, 5.3 Imp pt)
Coolant	
Type	50/50 mixture of ethylene glycol based antifreeze and soft water
Capacity	
600 A and B models and 750 models	2.0 lit (2.1 US qt, 3.5 Imp pt)
600 C models	2.3 lit (2.4 US qt, 4.0 Imp pt)
Brake and clutch fluid	DOT 4
Fork oil	
Type	SAE 10W20 fork oil
Amount - 600 A and B models	
Dry fill	321 ± 4 cc
At oil change	Approximately 273 cc
Oil level - 600 A and B models (forks fully extended - no spring)	334 ± 2 mm (13.16 ± 0.008 in)
Amount - C1 to C5 (UK), C1 to C6 (US) models	
Dry fill	
Left fork tube	311 ± 4 cc
Right fork tube	356 ± 4 cc
At oil change	
Left fork tube	Approximately 265 cc
Right fork tube	Approximately 305 cc
Oil level - C1 to C5 (UK), C1 to C6 (US) models (forks fully compressed - no spring)	
Left fork tube	182 ± 2 mm (7.2 ± 0.008 in)
Right fork tube	154 ± 2 mm (6.07 ± 0.008 in)
Amount - C6 and C7 (UK), C7 to C10 (US) models	
Dry fill	349 ± 4 cc
At oil change	Approximately 300 cc
Oil level - C6 and C7 (UK), C7 to C10 (US) models (forks fully compressed - no spring)	126 ± 2 mm (4.71 ± 0.008 in)
Amount - 750 models	
Dry fill	380 ± 4 cc
At oil change	Approximately 325 cc
Oil level - 750 models (forks fully compressed - no spring)	
Left fork tube	120 ± 2 mm (4.724 ± 0.079 in)
Right fork tube	149 ± 2 mm (5.866 ± 0.079 in)
Miscellaneous	
Wheel bearings	Medium weight, lithium-based multi-purpose grease
Swingarm pivot bearings	Medium weight, lithium-based multi-purpose grease
Cables and lever pivots	Chain and cable lubricant or 10W30 motor oil
Sidestand/centerstand pivots	Chain and cable lubricant or 10W30 motor oil
Brake pedal/shift lever pivots	Chain and cable lubricant or 10W30 motor oil
Throttle grip	Multi-purpose grease or dry film lubricant

Component locations on right side

1 Rear brake fluid reservoir
2 Battery
3 Steering head bearings
4 Front brake fluid reservoir

5 Coolant reservoir - 750 models
6 Fork oil drain plug
7 Brake pads
8 Engine oil window

9 Clutch cable lower adjuster - 600 models
10 Engine oil filler cap
11 Rear brake pedal height adjuster

Component locations on left side

1 Coolant reservoir - 600 C models
2 Clutch cable upper adjuster -
  600 models, clutch fluid reservoir -
  750 models
3 Fuel tap filter
4 Idle speed adjuster
5 Air filter
6 Drive chain
7 Engine oil drain bolt and filter
8 Alternator belt - 750 models
9 Coolant reservoir - 600 A and B models
10 Brake pads

Note: *The daily (pre-ride) checks outlined in the owner's manual covers those items which should be inspected on a daily basis. Always perform the pre-ride inspection at every maintenance interval (in addition to the procedures listed). The intervals listed below are the intervals recommended by the manufacturer for each particular operation during the model years covered in this manual. Your owner's manual may have different intervals for your model.*

Daily or before riding

See *'Daily (pre-ride) checks'* at the beginning of this manual.

After the initial 500 miles (800 km)

Note: *This check is usually performed by a Kawasaki dealer after the first 500 miles (800 km) from new. Thereafter, maintenance is carried out according to the following intervals of the schedule.*

Every 200 miles (300 km)

☐ Lubricate the drive chain (Sec 1).

Every 500 miles (800 km)

☐ Check/adjust the drive chain slack (Sec 2).

Every 3000 miles (5000 km)

☐ Clean and gap the spark plugs (Sec 3).
☐ Check the operation of the air suction valve - US models (Sec 4).
☐ Check the idle speed (Sec 5).
☐ Check the carburettor synchronisation (Sec 6).
☐ Check the evaporative emission control system - California models (Sec 7).
☐ Adjust the clutch freeplay - 600 models (Sec 8).
☐ Check the drive chain and sprockets for wear (Sec 9).
☐ Check the brake pads (Sec 10).
☐ Check the brake system (Sec 11).
☐ Lubricate all cables, levers, pedal and stand pivots (Sec 12).
☐ Change the engine oil and oil filter (Sec 13).
☐ Clean the air filter element (Sec 14).
☐ Check the steering head bearing adjustment (Sec 15).
☐ Check the tyres and wheels (Sec 16).
☐ Check the battery electrolyte level (Sec 17).
☐ Check the exhaust system for leaks (Sec 18).
☐ Check the tightness of all fasteners (Sec 19).

Every 6000 miles (10,000 km)

All of the items above plus:
☐ Check the valve clearances (Sec 20)
☐ Check the throttle cable freeplay (Sec 21).
☐ Check the cleanliness of the fuel system and the condition of the fuel and vacuum hoses (Sec 22).
☐ Lubricate the swingarm needle bearings and Uni-trak linkage (Sec 23).
☐ Replace the spark plugs (Sec 24).
☐ Check and adjust the alternator drive belt - 750 models (Sec 25).
☐ Check the cooling system (Sec 26).

Every 18,000 miles (30,000 km)

☐ Change the coolant (Sec 27).
☐ Change the fork oil (Sec 28).

Every year

☐ Check the carburettor warmer system components - UK models only (Sec 29).

Every two years

☐ Change the brake and clutch fluid (Sec 30).
☐ Replace the anti-dive seals and metal pipe - 600 A and B models (Sec 31).
☐ Replace the seals in the brake calipers and master cylinders (Sec 32).
☐ Replace the seals in the clutch slave cylinder and master cylinder - 750 models (Sec 33).
☐ Lubricate the steering head bearings (Sec 34).
☐ Check the wheel bearings (Sec 35).
☐ Replace the fuel hoses (Sec 36).
☐ Replace the brake hoses (Sec 37).

Non-scheduled maintenance

☐ Check the cylinder compression (Sec 38).
☐ Check the suspension (Sec 39).

Introduction

1 This Chapter is designed to help the home mechanic maintain his/her motorcycle for safety, economy, long life and peak performance.

2 Deciding where to start or plug into the routine maintenance schedule depends on several factors. If the warranty period on your motorcycle has just expired, and if it has been maintained according to the warranty standards, you may want to pick up routine maintenance as it coincides with the next mileage or calendar interval. If you have owned the machine for some time but have never performed any maintenance on it, then you may want to start at the nearest interval and include some additional procedures to ensure that nothing important is overlooked. If you have just had a major engine overhaul, then you may want to start the maintenance routine from the beginning. If you have a used machine and have no knowledge of its history or maintenance record, you may desire to combine all the checks into one large service initially and then settle into the maintenance schedule prescribed.

3 Before beginning any maintenance or repair, the machine should be cleaned thoroughly, especially around the oil filter, spark plugs, valve cover, side panels, carburettors, etc. Cleaning will help ensure that dirt does not contaminate the engine and will allow you to detect wear and damage that could otherwise easily go unnoticed.

4 Certain maintenance information is sometimes printed on decals attached to the motorcycle. If the information on the decals differs from that included here, use the information on the decal.

Every 200 miles (300 km)

1 Drive chain - lubrication

Note: *If the chain is extremely dirty, it should be removed and cleaned before it's lubricated (see Chapter 6).*

1 The best time to lubricate the chain is after the motorcycle has been ridden. When the chain is warm, the lubricant will penetrate the joints between the side plates, pins, bushings and rollers to provide lubrication of the internal load bearing areas.

2 Use a good quality chain lubricant and apply it to the area where the side plates overlap - not the middle of the rollers **(see illustration)**. After applying the lubricant, let it soak in a few minutes before wiping off any excess.

Caution: If one of the commercial aerosol chain lubricants is used, make sure it is marked as being suitable for O-ring chains.

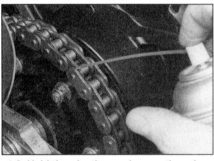

1.2 Hold the plastic nozzle near the edge of the chain and turn the wheel by hand - as the lubricant sprays out repeat the procedure on the inside edge of the chain

Every 500 miles (800 km)

2 Drive chain - check and adjustment

Check

1 A neglected drive chain won't last long and can quickly damage the sprockets. Routine chain adjustment and lubrication isn't difficult and will ensure maximum chain and sprocket life.

2 To check the chain, place the bike on its centerstand and shift the transmission into Neutral. Make sure the ignition switch is off.

3 Push up on the bottom run of the chain and measure the slack midway between the two sprockets **(see illustration)**, then compare your measurements to the value listed in this Chapter's Specifications. As wear occurs, the chain will actually stretch, which means adjustment usually involves removing some slack from the chain. In some cases where lubrication has been neglected, corrosion and galling may cause the links to bind and kink, which effectively shortens the chain's length. If the chain is tight between the sprockets, rusty or kinked, it's time to replace it with a new one.

Adjustment

4 Rotate the rear wheel until the chain is positioned with the least amount of slack present.

5 Loosen the torque link-to-rear caliper holder bolt.

6 Loosen and back-off the locknuts on the adjuster bolts **(see illustration)**.

7 Remove the cotter pin and loosen the axle nut **(see illustration)**.

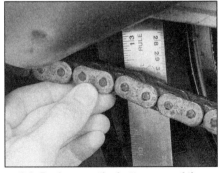

2.3 Push up on the bottom run of the chain and measure how far it deflects

2.6 Loosen the locknuts on the adjusting bolts

2.7 Remove the cotter pin and loosen the axle nut

8 Turn the axle adjusting bolts on both sides of the swingarm until the proper chain tension is obtained. Be sure to turn the adjusting bolts evenly to keep the rear wheel in alignment. If the adjusting bolts reach the end of their travel, the chain is excessively worn and should be replaced with a new one (see Chapter 6).

9 When the chain has the correct amount of slack, make sure the marks on the adjusters correspond to the same relative marks on each side of the swingarm **(see illustration)**. Tighten the axle nut to the torque listed in the Chapter 7 Specifications, then install a **new** cotter pin (split pin). If necessary, turn the nut an additional amount to line up the cotter pin (split pin) hole with the castellations in the nut - don't loosen the nut to do this.

10 Tighten the locknuts and the torque link nut securely.

2.9 The mark on each adjuster should be aligned with the same relative marks on each side of the swingarm

Every 3000 miles (5000 km)

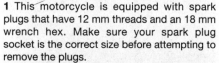

3 Spark plugs - clean and gap

1 This motorcycle is equipped with spark plugs that have 12 mm threads and an 18 mm wrench hex. Make sure your spark plug socket is the correct size before attempting to remove the plugs.

2 Remove the fuel tank (see Chapter 4), then disconnect the spark plug caps from the spark plugs. If available, use compressed air to blow any accumulated debris from around the spark plugs. Remove the plugs **(see illustration)**.

3.2 Use an extension and a deep socket to remove the spark plugs

3.6a Using a wire type gauge to check the plug gap

3 Inspect the electrodes for wear. Both the center and side electrodes should have square edges and the side electrode should be of uniform thickness. Look for excessive deposits and evidence of a cracked or chipped insulator around the center electrode. Compare your spark plugs to the colour spark plug reading chart at the end of this manual. Check the threads, the washer and the porcelain insulator body for cracks and other damage.

4 If the electrodes are not excessively worn, and if the deposits can be easily removed with a wire brush, the plugs can be regapped and reused (if no cracks or chips are visible in the insulator). If in doubt concerning the condition of the plugs, replace them with new ones, as the expense is minimal.

5 Cleaning spark plugs by sandblasting is permitted, provided you clean the plugs with a high flash-point solvent afterwards.

6 Before installing new plugs, make sure they are the correct type and heat range. Check the gap between the electrodes, as they are not preset. For best results, use a wire-type gauge rather than a flat gauge to check the gap **(see illustration)**. If the gap must be adjusted, bend the side electrode only and be very careful not to chip or crack the insulator nose **(see illustration)**. Make sure the washer is in place before installing each plug.

7 Since the cylinder head is made of aluminum, which is soft and easily damaged, thread the plugs into the head by hand.

3.6b Bend the side electrode only, as indicated by the arrows

Since the plugs are quite recessed, slip a short length of hose over the end of the plug to use as a tool to thread it into position. The hose will grip the plug well enough to turn it, but will start to slip if the plug begins to cross-thread in the hole.

 HAYNES HiNT *If a stripped spark plug thread is discovered, note that it can be repaired by installing a thread insert - refer to Tools and Workshop Tips in the Reference section of this manual for details.*

8 Once the plugs are finger tight, the job can be finished with a socket. If a torque wrench is available, tighten the spark plugs to the torque listed in this Chapter's Specifications. If you do not have a torque wrench, tighten the plugs finger tight (until the washers bottom on the cylinder head) then use a wrench to tighten them an additional 1/4 turn. Regardless of the method used, do not over-tighten them.

9 Reconnect the spark plug caps.

4 Air suction valves - check (US models)

1 The air suction valves, installed on US models only, are one-way check valves that allow fresh air to flow into the exhaust ports.

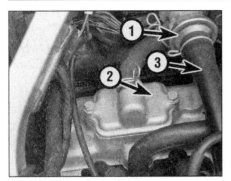

4.4 Details of the air suction valves and hoses

1 *Vacuum switching valve*
2 *Air suction valve*
3 *Hose to air filter housing*

The suction developed by the exhaust pulses pulls the air from the air cleaner, through a hose to the vacuum switch valve, through a pair of hoses and two pairs of reed valves, and finally into the exhaust ports. The introduction of fresh air helps ignite any fuel that may not have been burned by the normal combustion process.

2 Remove the fuel tank (see Chapter 4).

3 Remove the ignition coils (see Chapter 5).

4 Disconnect the hoses from the air suction valves **(see illustration)**. Remove the bolts and lift off the covers.

5 Check the valve for cracks, warping, burning or other damage **(see illustration)**. Check the area where the reeds contact the valve holder for scratches, separation and grooves. If any of these conditions are found, replace the valve.

6 Wash the valves with solvent if carbon has accumulated between the reed and the valve holder.

7 Installation of the valves is the reverse of removal. Be sure to use a new gasket.

5 Idle speed - check and adjustment

1 The idle speed should be checked and adjusted after the carburettors are synchronised and when it is obviously too high or too low. Before adjusting the idle speed, make sure the valve clearances and spark plug gaps are correct. Also, turn the handlebars back-and-forth and see if the idle speed changes as this is done. If it does, the throttle cable may not be adjusted correctly, or it may be worn out. Be sure to correct this problem before proceeding.

2 The engine should be at normal operating temperature, which is usually reached after 10 to 15 minutes of stop and go riding. Place the motorcycle on the centerstand and make sure the transmission is in Neutral.

3 Turn the throttle stop screw, located on the left side of the bike, just forward of the

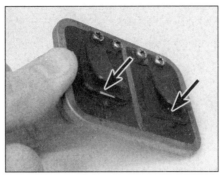

4.5 Check the reeds (arrows) on the air suction valve for damage and carbon build-up

carburettor for cylinder no. 1 **(see illustration)**, until the idle speed listed in this Chapter's Specifications is obtained.

4 Snap the throttle open and shut a few times, then recheck the idle speed. If necessary, repeat the adjustment procedure.

5 If a smooth, steady idle can't be achieved, the fuel/air mixture may be incorrect. Refer to Chapter 4 for additional carburettor information.

6 Carburettor synchronisation - check and adjustment

Warning: Gasoline (petrol) is extremely flammable, so take extra precautions when you work on any part of the fuel system. Don't smoke or allow open flames or bare light bulbs near the work area, and don't work in a garage where a natural gas-type appliance (such as a water heater or clothes dryer) is present. If you spill any fuel on your skin, rinse it off immediately with soap and water. When you perform any kind of work on the fuel system, wear safety glasses and have a class B type fire extinguisher on hand.

1 Carburettor synchronization is simply the process of adjusting the carburettors so they pass the same amount of fuel/air mixture to each cylinder. This is done by measuring the vacuum produced in each cylinder. Carburettors that are out of synchronization will result in decreased fuel mileage, increased engine temperature, less than ideal throttle response and higher vibration levels.

2 To properly synchronise the carburettors, you will need some sort of vacuum gauge setup, preferably with a gauge for each cylinder, or a mercury manometer, which is a calibrated tube arrangement that utilizes columns of mercury to indicate engine vacuum.

3 A manometer can be purchased from a motorcycle dealer or accessory shop and should have the necessary rubber hoses supplied with it for hooking into the vacuum hose fittings on the carburettors.

5.3 Turn the idle speed adjusting screw (arrow) in or out until the specified idle is obtained

4 A vacuum gauge setup can also be purchased from a dealer or fabricated from commonly available hardware and automotive vacuum gauges.

5 The manometer is the more reliable and accurate instrument, and for that reason is preferred over the vacuum gauge setup; however, since the mercury used in the manometer is a liquid, and extremely toxic, extra precautions must be taken during use and storage of the instrument.

6 Because of the nature of the synchronization procedure and the need for special instruments, most owners leave the task to a dealer service department or a reputable motorcycle repair shop.

7 Start the engine and let it run until it reaches normal operating temperature, then shut it off.

8 Remove the fuel tank (see Chapter 4).

9 Detach the vacuum hoses, or blanking caps, from the fittings on the carburettors **(see illustration)**, then hook up the vacuum gauge set or the manometer according to the manufacturer's instructions. Make sure there are no leaks in the setup, as false readings will result.

10 Reconnect the fuel line to the fuel tank (it's not necessary to hook-up the vacuum line to the fuel tap). Have an assistant hold the fuel tank out of the way, but in such a position that fuel can still be delivered and access to the carburettors is unobstructed. Place the fuel tap lever in the Prime position on 600 models. On 750 models, the tap knob must be

6.9 Detach the vacuum hoses from the fittings on the front of the carburetors

6.13 Synchronising screw for cylinders 1 and 2 (the carburetors for cylinders 3 and 4 also have a screw like this between them)

6.15 Turn this screw to synchronise carburettors 1 and 2 to carburettors 3 and 4

depressed to allow fuel to flow through the tap without the vacuum pipe connected.

11 Start the engine and make sure the idle speed is correct.

12 The vacuum readings for all of the cylinders should be the same, or at least within the tolerance listed in this Chapter's Specifications. If the vacuum readings vary, adjust as necessary.

13 To perform the adjustment, synchronise the carburettors for cylinders 1 and 2 by turning the butterfly valve adjusting

screw between those two carburettors, as needed, until the vacuum is identical or nearly identical for those two cylinders **(see illustration)**.

14 Next, synchronise the carburettors for cylinders 3 and 4, using the butterfly valve adjusting screw situated between those two carburettors.

15 Finally, synchronise the carburettors for cylinders 1 and 2 to the carburettors for cylinders 3 and 4 by turning the center adjusting screw **(see illustration)**.

16 When the adjustment is complete, recheck the vacuum readings and idle speed, then stop the engine. Remove the vacuum gauge or manometer and attach the hoses to the fittings on the carburettors. Reinstall the fuel tank and seat.

7 Evaporative emission control system (California models) - check

1 This system, installed on California models to conform to stringent emission control standards, routes fuel vapors from the fuel system into the engine to be burned, instead of letting them evaporate into the atmosphere. When the engine isn't running, vapors are stored in a carbon canister.

Hoses

2 To begin the inspection of the system, remove the seat and side covers (see Chapter 8 if necessary). Inspect the hoses from the fuel tank, carburettor and liquid/vapour separator to the canister for cracking, kinks or other signs of deterioration **(see illustrations)**.

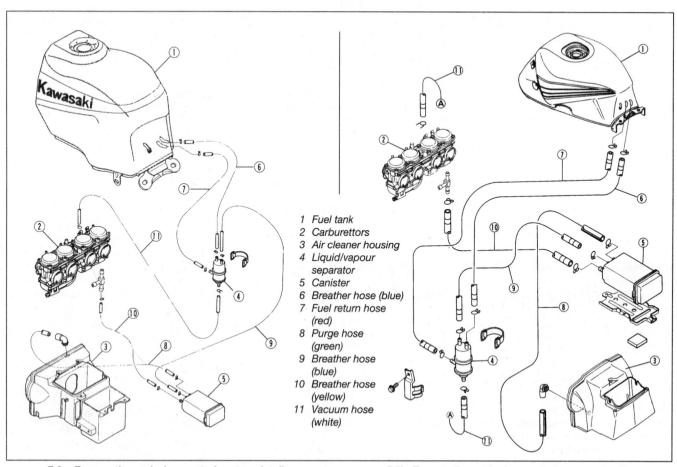

1 Fuel tank
2 Carburettors
3 Air cleaner housing
4 Liquid/vapour separator
5 Canister
6 Breather hose (blue)
7 Fuel return hose (red)
8 Purge hose (green)
9 Breather hose (blue)
10 Breather hose (yellow)
11 Vacuum hose (white)

7.2a Evaporative emission control system details (600 A and B models)

7.2b Evaporative emission control system details (600 C models)

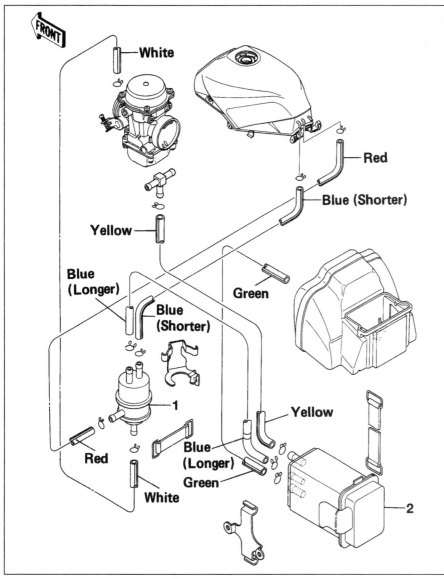

7.2c Evaporative emission control system details (750 models)

1 Liquid-vapor separator   2 Canister

7.3 The liquid/vapour separator (arrow) is retained by a strap

8.2 Pull the clutch lever in until resistance is felt then measure clearance between the lever and bracket

1 Clearance   2 Lock wheel   3 Adjuster

Liquid/vapour separator

3 To check the liquid/vapour separator, label and disconnect the hoses from it **(see illustration)**, then remove it from the machine. Check it closely for cracks or other signs of damage. Reinstall the separator and connect the hoses, except for the breather hose. Using a syringe, inject approximately 20 cc of gasoline (petrol) into the separator.

4 Disconnect the fuel return hose from the fuel tank and direct the end of the hose into an approved gasoline (petrol) container. Hold the container level with the top of the fuel tank.

5 Start the engine and allow it to idle. If the fuel that was squirted into the separator comes out of the hose, it's working properly. If fuel doesn't come out of the hose, replace the separator.

Canister

6 Remove the canister from under the passenger's seat and inspect it for cracks or other signs of damage. Tip the canister so the nozzles point down. If fuel runs out of the canister, the liquid/vapor separator is probably bad - check it as described above. The fuel inside the canister has probably caused damage, so it would be a good idea to replace it also.

---

8 Clutch - check and adjustment (600 models)

1 Correct clutch freeplay is necessary to ensure proper clutch operation and reasonable clutch service life. Freeplay normally changes because of cable stretch and clutch wear, so it should be checked and adjusted periodically.

2 Clutch cable freeplay is checked at the lever on the handlebar. Slowly pull in on the lever until resistance is felt, then note how far the lever has moved away from its bracket at the pivot end **(see illustration)**. Compare this distance with the value listed in this Chapter's Specifications. Too little freeplay may result in the clutch not engaging completely. If there is too much freeplay, the clutch might not release fully.

3 Freeplay adjustments can be made at the clutch lever by loosening the lock wheel and turning the adjuster until the desired freeplay is obtained. Always retighten the lock wheel once the adjustment is complete. If the lever adjuster reaches the end of its travel, try adjusting the cable at its bracket on the engine.

4 On A models, remove the right-side fairing stay (see Chapter 8).

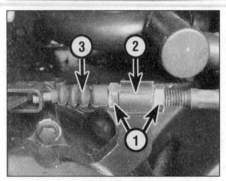

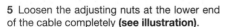

8.5 Details of the lower end of the clutch cable

1 Adjusting nuts   2 Bracket   3 Dust cover

5 Loosen the adjusting nuts at the lower end of the cable completely **(see illustration)**.
6 Loosen the knurled lock wheel at the clutch lever and turn the adjuster in or out until the gap between the adjuster and lock wheel is approximately 5 or 6 mm **(see illustration)**.
7 Pull the clutch cable tight to remove all slack, then tighten the adjusting nuts against the bracket at the lower end of the cable.
8 Turn the adjuster at the clutch lever until the correct freeplay is obtained. When the cable is properly adjusted, the angle between the cable and the release lever should be 80 to 90° **(see illustration)**.
9 If the proper amount of freeplay still can't be obtained, the cable must be replaced (see Chapter 2A).
10 Install the right-side fairing stay if you're working on an A model.

9 Drive chain and sprockets - wear check

Drive chain

1 Remove the chain guard (it's held on by two bolts). Check the entire length of the chain for damaged rollers, loose links and pins.

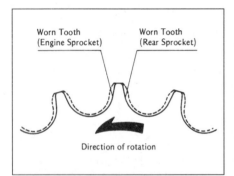

9.3 Check the sprockets in the areas indicated to see if they are worn excessively

---

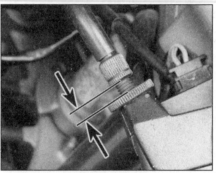

8.6 Turn the adjuster out until the gap between the shoulder of the adjuster and the locknut is 5 to 6 mm, then tighten the lock wheel

2 Hang a 20-lb weight on the bottom run of the chain and measure the length of 20 links along the top run. Rotate the wheel and repeat this check at several places on the chain, since it may wear unevenly. Compare your measurements with the maximum 20-link length listed in this Chapter's Specifications. If any of your measurements exceed the maximum, replace the chain. **Note:** *Never install a new chain on old sprockets, and never use the old chain if you install new sprockets - replace the chain and sprockets as a set.*

Sprockets

3 Remove the shift lever and engine sprocket cover (see Chapter 6, Section 18). Check the teeth on the engine sprocket and the rear sprocket for wear **(see illustration)**.
4 Refer to Chapter 6 for the sprocket diameter measurement procedure if the sprockets appear to be worn excessively.

10 Brake pads - wear check

1 The brake pads should be checked at the recommended intervals and replaced with new ones when worn beyond the limit listed in this Chapter's Specifications.

10.2 The front brake pads (arrows) are visible from the front of the bike – these pads look like they're ready for replacement

---

8.8 When the cable adjustment is correct, the angle between the cable and the release lever should be 80 to 90°

2 To check the front brake pads, position the front wheel so you can see clearly into the front of the brake caliper. The brake pads are visible from this angle and should have at least the specified minimum amount of lining material remaining on the metal backing plate **(see illustration)**. Be sure to check the pads in both calipers.
3 Check the rear brake pads by looking into the caliper from the rear of the machine **(see illustration)**.
4 If the pads are worn excessively, they must be replaced with new ones (see Chapter 7).

11 Brake system - general check

General

1 Check the brake lever and pedal for loose connections, excessive play, bends, and other damage. Replace any damaged parts with new ones (see Chapter 7).
2 Make sure all brake fasteners are tight and that the fluid level in the reservoir is correct (see *Daily (pre-ride) checks*). Look for leaks at the hose connections and check for cracks in the hoses. If the lever is spongy, bleed the brakes as described in Chapter 7.

10.3 Check the rear pads (arrows) by looking between the caliper and disc

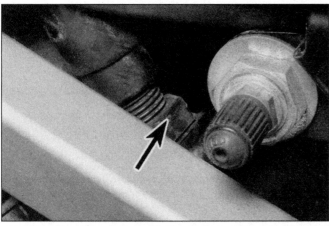

11.4 To adjust the rear brake light switch, turn the adjusting nut (arrow) to move the switch in or out

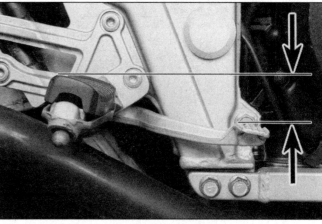

11.5 The top of the brake pedal should be about 40 mm below the top of the footpeg

Brake light switches

3 Make sure the brake light operates when the brake lever is depressed. The front brake light switch is not adjustable. If it fails to operate properly, replace it with a new one (see Chapter 9).

4 Make sure the brake light is activated when the rear brake pedal is depressed approximately 11 mm (7/16 in). If adjustment is necessary, hold the switch and turn the adjusting nut on the switch body **(see illustration)** until the brake light is activated when required. Turning the switch out will cause the brake light to come on sooner, while turning it in will cause it to come on later. If the switch doesn't operate the brake lights, check it as described in Chapter 9.

Brake pedal position

5 Rear brake pedal position is largely a matter of personal preference. Locate the pedal so that the rear brake can be engaged quickly and easily without excessive foot movement. The recommended factory setting is approximately 40 mm below the top of the footpeg **(see illustration)**.

6 To adjust the position of the pedal, loosen the locknut on the master cylinder clevis, then

remove the master cylinder (see Chapter 7). To lower the brake pedal, turn the clevis clockwise. To raise the position of the brake pedal, turn the clevis counterclockwise.

7 Install the master cylinder, fill the rear brake master cylinder with the recommended brake fluid, then bleed the air from the system (see Chapter 7). If necessary, adjust the brake light switch (see previous sub-section).

12 Lubrication - general

1 Since the controls, cables and various other components of a motorcycle are exposed to the elements, they should be lubricated periodically to ensure safe and trouble-free operation.

2 The footpegs, clutch and brake lever, brake pedal, shift lever and side and centerstand pivots should be lubricated frequently. In order for the lubricant to be applied where it will do the most good, the component should be disassembled. However, if chain and cable lubricant is being used, it can be applied to the pivot joint gaps and will usually work its way into the areas where friction occurs. If motor oil or light grease is being used, apply it

sparingly as it may attract dirt (which could cause the controls to bind or wear at an accelerated rate). **Note:** *One of the best lubricants for the control lever pivots is a dry-film lubricant (available from many sources by different names).*

3 On 600 models, the clutch cable should be separated from the handlebar lever and bracket before it is lubricated **(see illustration)**. It should be treated with motor oil or a commercially available cable lubricant which is specially formulated for use on motorcycle control cables. Small adapters for pressure lubricating the cables with spray can lubricants are available and ensure that the cable is lubricated along its entire length **(see illustration)**. If motor oil is being used, tape a funnel-shaped piece of heavy paper or plastic to the end of the cable, then pour oil into the funnel and suspend the end of the cable upright **(see illustration)**. Leave it until the oil runs down into the cable and out the other

12.3a To disconnect the clutch cable from the lever and bracket, line-up the slots in the bracket, lock wheel and adjuster, then pull the cable in the direction of the arrow and slide it through the slots

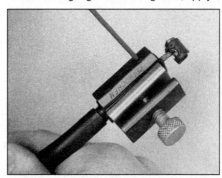

12.3b Lubricating a cable with a pressure lube adapter (make sure the tool seals around the inner cable)

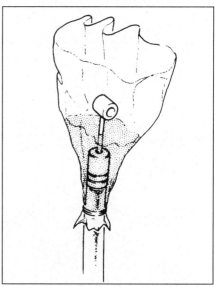

12.3c Lubricating a control cable with a makeshift funnel and motor oil

13.4 The oil drain plug is located in the center of the oil pan, between the exhaust pipes

13.5a Unscrew the oil filter mounting bolt . . .

13.5b . . . then lower the filter from the crankcase

end. When attaching the cable to the lever, be sure to lubricate the barrel-shaped fitting at the end with multi-purpose grease.

4 To lubricate the throttle and choke cables, disconnect the cable(s) at the lower end, then lubricate the cable with a pressure lube adapter **(see illustration 12.3b)**. See Chapter 4 for the choke cable removal procedure.

5 Speedometer and tachometer cables should be removed from their housings and lubricated with motor oil or cable lubricant.

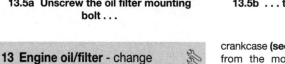

13 Engine oil/filter - change

1 Consistent routine oil and filter changes are the single most important maintenance procedure you can perform on a motorcycle. The oil not only lubricates the internal parts of the engine, transmission and clutch, but it also acts as a coolant, a cleaner, a sealant, and a protectant. Because of these demands, the oil takes a terrific amount of abuse and should be replaced often with new oil of the recommended grade and type. Saving a little money on the difference in cost between a good oil and a cheap oil won't pay off if the engine is damaged.

2 Before changing the oil and filter, warm up the engine so the oil will drain easily. Be careful when draining the oil, as the exhaust pipes, the engine, and the oil itself can cause severe burns.

3 Put the motorcycle on the centerstand over a clean drain pan. If you're working on a 600 A or B model, remove the lower fairing (see Chapter 8). Remove the oil filler cap to vent the crankcase and act as a reminder that there is no oil in the engine.

4 Next, remove the drain plug from the engine **(see illustration)** and allow the oil to drain into the pan. Do not lose the sealing washer on the drain plug.

5 As the oil is draining, remove the oil filter mounting bolt and lower the filter out of the

crankcase **(see illustrations)**. Separate the filter from the mounting bolt, then remove the washer, spring and oil fence **(see illustration)**. If additional maintenance is planned for this time period, check or service another component while the oil is allowed to drain completely.

6 Clean the filter cover and housing with solvent or clean rags. Wipe any remaining oil off the filter cover sealing area of the crankcase.

7 Remove the mounting bolt from the filter cover. The oil filter bypass valve is located inside the mounting bolt. Wash the mounting bolt in solvent and check the bypass valve for damage. If the valve is full of sludge, drive out the retaining pin and remove the spring and steel ball **(see illustrations)**.

8 Clean the components and check them for damage - especially, be sure to check the spring for distortion. If any damage is found, replace the mounting bolt/bypass valve assembly. If the components are okay, reassemble the valve and install the retaining pin.

9 Check the condition of the drain plug threads and the sealing washer. Use a new O-ring on the filter housing when it is installed.

10 Install a new O-ring on the mounting bolt, lubricate the mounting bolt with clean engine oil and insert the bolt through the filter cover. Place the oil fence over the mounting bolt.

11 Install a new O-ring on the filter cover, then install the spring and washer. Twist the new oil filter down the mounting bolt, making sure the rubber grommets on the filter don't slip out of place **(see illustration)**.

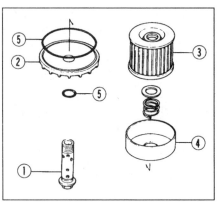

13.5c Oil filter details

1 *Mounting bolt/bypass valve assembly*
2 *Filter cover*
3 *Oil filter*
4 *Oil fence*
5 *O-ring*

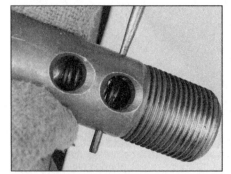

13.7a To disassemble the bypass valve, drive out the retaining pin . . .

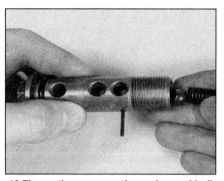

13.7b . . . then remove the spring and ball from the mounting bolt

13.11 Lubricate the mounting bolt and twist the new filter onto it, making sure the grommets aren't dislodged from the filter

12 Guide the oil filter assembly up into the crankcase. Tighten the mounting bolt with your fingers until the filter cover contacts the crankcase making sure the O-ring on the filter cover stays in its groove and seals properly. Tighten the mounting bolt to the torque listed in this Chapter's Specifications.

13 Slip the sealing washer over the drain plug, then install and tighten the plug. Tighten the drain plug to the torque listed in this Chapter's Specifications. Avoid overtightening, as damage to the engine case will result.

> **HAYNES HINT** *Check the old oil carefully - if it is very metallic coloured, then the engine is experiencing wear from break-in (new engine) or from insufficient lubrication. If there are flakes or chips of metal in the oil, then something is drastically wrong internally and the engine will have to be disassembled for inspection and repair. If there are pieces of fibre-like material in the oil, the clutch is experiencing excessive wear and should be checked.*

14 Refill the crankcase to the proper level with the recommended oil and install the filler cap. Start the engine and let it run for two or three minutes. Shut it off, wait a few minutes, then check the oil level. If necessary, add more oil to bring the level up to the Maximum mark. Check around the drain plug and filter housing for leaks.

15 The old oil drained from the engine cannot be reused in its present state and should be disposed of. Oil reclamation centers, auto repair shops and gas stations will normally accept the oil, which can be refined and used again (be sure to check with the repair shop or gas station first). After the oil has cooled, it can be drained into a suitable container (capped plastic jugs, topped bottles, milk cartons, etc.) for transport to one of these disposal sites.

Note: *It is antisocial and illegal to dump oil down the drain. To find the location of your local oil recycling bank, call this number free.*

0800 66 33 66

In the USA, note that any oil supplier must accept used oil for recycling.

14 Air filter element - clean

Note: *Replace the air filter element every five cleanings (or more frequently, if the bike is operated in dusty conditions).*

600 A and B models

1 Remove the seat and the fuel tank (see Chapter 4).
2 Remove the fuel tank bracket **(see illustration 17.1).**
3 Remove the screws that secure the air filter housing cover, then remove the cover **(see illustration).**
4 Remove the air filter element **(see illustration).** Wipe out the housing with a clean rag.

5 Tap the filter element on a solid surface to dislodge dirt and dust from the paper. If compressed air is available, use it to clean the element by blowing from the inside out. If the paper is extremely dirty or torn, replace the element with a new one.

6 Reinstall the filter by reversing the removal procedure. Make sure the element is seated properly in the filter housing before installing the cover. Reinstall the fuel tank bracket, fuel tank and seat.

600 C and 750 models

7 Remove the seat.
8 Remove the screws that secure the air filter housing cover, then remove the cover **(see illustration).**
9 Pull the air filter element out of the housing and separate it from the element frame. Wipe out the housing with a clean rag.
10 Wash the filter element with solvent and wring it out. Blow dry it with compressed air, if available.
11 Check the element for holes and tears and replace it if necessary.
12 Install the filter element on the element frame and install it in the filter housing, making sure the pins on the element frame engage with the grooves on the side of the housing **(see illustration).**
13 Install the seat.

15 Steering head bearings - check and adjustment

1 This vehicle is equipped with tapered-roller type steering head bearings which can become dented, rough or loose during normal use of the machine. In extreme cases, worn or loose steering head bearings can cause steering wobble that is potentially dangerous.

Check

2 To check the bearings, place the motorcycle on the centerstand and block the machine so the front wheel is in the air.

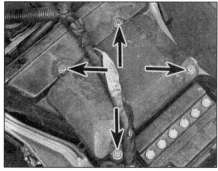

14.3 Remove the four screws (arrows) and lift off the air filter housing cover (600 A and B models)

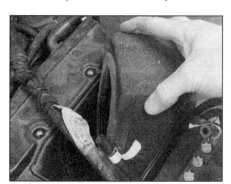

14.4 Detach the air filter element from the front of the filter housing, then lift it out (600 A and B models)

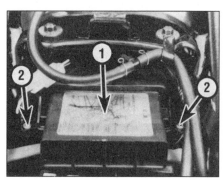

14.8 The filter housing cover on 600 C and 750 models is retained by two screws
1 Cover   2 Screws

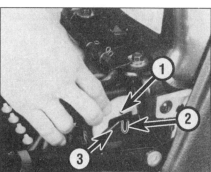

14.12 When installing the filter element on a 600 C or 750 model, make sure the pins on the filter frame engage with the grooves in the sides of the housing

1 Frame end   3 Air cleaner
2 Groove   housing

15.4 If play is felt when attempting to move the forks back-and-forth, adjust the steering head bearings

3 Point the wheel straight ahead and slowly move the handlebars from side-to-side. Dents or roughness in the bearing races will be felt and the bars will not move smoothly.
4 Next, grasp the fork legs and try to move the wheel forward and backward **(see illustration)**. Any looseness in the steering head bearings will be felt. If play is felt in the bearings, adjust the steering head as follows.

 Freeplay in the fork bushes can be misinterpreted for steering head play - do not confuse the two.

Adjustment

5 Remove the fuel tank (see Chapter 4). On 600 C models it will be necessary to remove the lower and upper fairings and knee grip covers (see Chapter 8). On 600 A and B models and 750 models, loosen the fork lower pinch bolts and the steering head nut **(see illustration)**. On 600 C models, remove the steering head nut, the handlebars and the steering head (see Chapter 6). Use a spanner wrench to loosen the steering stem locknut.

17.1 To gain access to the battery, unbolt the fuel tank from the rear mount, then remove the bolts (arrows) securing the mount to the frame (fuel tank removed for clarity)

15.5 Steering head and fork details (600 A and B models)

1 Fork lower pinch bolts
2 Steering head nut

6 Carefully tighten the steering stem locknut until the steering head is tight but does not bind when the forks are turned from side-to-side **(see illustration)**.
7 Retighten the steering head nut and the fork pinch bolts, in that order, to the torque values listed in the Chapter 6 Specifications. On 600 C models, make sure the top ends of the fork tubes protrude 16.0 to 17.5 mm from the upper surface of the fork clamps. On 750 models the forks should protrude 15 mm above the upper surface of the fork clamps.
8 Recheck the steering head bearings for play as described above. If necessary, repeat the adjustment procedure. Reinstall all parts previously removed.
9 Refer to Chapter 6 for steering head bearing lubrication and replacement procedures.

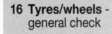

16 Tyres/wheels - general check

Tyres

1 Refer to *Daily (pre-ride) checks*.

17.2 To remove the battery, unscrew the bolts securing the battery cables to the terminals (negative first, positive last), then pull the battery straight up

1 Negative cable
2 Positive cable and boot

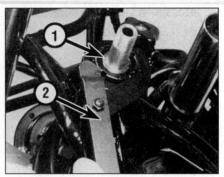

15.6 Tighten the steering stem locknut with a special spanner wrench (600 C model shown)

1 Stem locknut
2 Spanner wrench (Kawasaki tool no. 57001–1100)

Wheels

2 The cast wheels used on this machine are virtually maintenance free, but they should be kept clean and checked periodically for cracks and other damage. Never attempt to repair damaged cast wheels; they must be replaced with new ones.
3 Check the valve stem locknuts to make sure they are tight. Also, make sure the valve stem cap is in place and tight. If it is missing, install a new one made of metal or hard plastic.

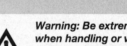

17 Battery electrolyte level/specific gravity - check

⚠️ *Warning: Be extremely careful when handling or working around the battery. The electrolyte is very caustic and an explosive gas (hydrogen) is given off when the battery is charging.*

1 To check and replenish the battery electrolyte, it will be necessary to remove the seat and the fuel tank bracket **(see illustration)**.
2 Remove the bolts securing the battery cables to the battery terminals (remove the negative cable first, positive cable last) **(see illustration)**. Pull the battery straight up to remove it. The electrolyte level will now be visible through the opaque battery case - it should be between the Upper and Lower level marks.
3 If it is low, remove the cell caps and fill each cell to the upper level mark with distilled water. Do not use tap water (except in an emergency), and do not overfill. The cell holes are quite small, so it may help to use a plastic squeeze bottle with a small spout to add the water. If the level is within the marks on the case, additional water is not necessary.
4 Next, check the specific gravity of the electrolyte in each cell with a small

hydrometer made especially for motorcycle batteries (if the electrolyte level is known to be sufficient it won't be necessary to remove the battery. These are available from most dealer parts departments or motorcycle accessory stores.

5 Remove the caps, draw some electrolyte from the first cell into the hydrometer **(see illustration)** and note the specific gravity. Compare the reading to the Specifications listed in this Chapter. **Note:** *Add 0.007 to the reading for every 10°C above 20°C, and subtract 0.007 for every 10°C below 20°C. Add 0.004 points to the reading for every 10°F above 68°F - subtract 0.004 points from the reading for every 10° below 68°F.* Return the electrolyte to the appropriate cell and repeat the check for the remaining cells. When the check is complete, rinse the hydrometer thoroughly with clean water.

6 If the specific gravity of the electrolyte in each cell is as specified, the battery is in good condition and is apparently being charged by the machine's charging system.

7 If the specific gravity is low, the battery is not fully charged. This may be due to corroded battery terminals, a dirty battery case, a malfunctioning charging system, or loose or corroded wiring connections. On the other hand, it may be that the battery is worn out, especially if the machine is old, or that infrequent use of the motorcycle prevents normal charging from taking place.

8 Be sure to correct any problems and charge the battery if necessary. Refer to Chapter 9 for additional battery maintenance and charging procedures.

9 Install the battery cell caps, tightening them securely. Reconnect the cables to the battery, attaching the positive cable first and the negative cable last. Make sure to install the insulating boot over the positive terminal.

17.5 Check the specific gravity with an hydrometer

Install the fuel tank mount and the seat. Be very careful not to pinch or otherwise restrict the battery vent tube (if equipped), as the battery may build up enough internal pressure during normal charging system operation to explode.

18 Exhaust system - check
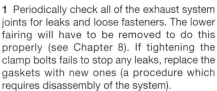

1 Periodically check all of the exhaust system joints for leaks and loose fasteners. The lower fairing will have to be removed to do this properly (see Chapter 8). If tightening the clamp bolts fails to stop any leaks, replace the gaskets with new ones (a procedure which requires disassembly of the system).

2 The exhaust pipe flange nuts at the cylinder heads **(see illustration)** are especially prone to loosening, which could cause damage to the head. Check them frequently and keep them tight.

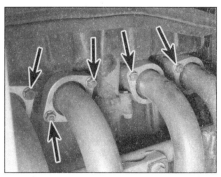

18.2 The exhaust pipe flange bolts (arrows) should be checked frequently and tightened if necessary

19 Fasteners - check

1 Since vibration of the machine tends to loosen fasteners, all nuts, bolts, screws, etc. should be periodically checked for proper tightness.

2 Pay particular attention to the following:

Spark plugs
Engine oil drain plug
Oil filter cover bolt
Gearshift lever
Footpegs and sidestand
Engine mount bolts
Shock absorber mount bolts
Uni-trak linkage bolts
Front axle and clamp bolt
Rear axle nut

3 If a torque wrench is available, use it along with the Torque specifications at the beginning of this, or other, Chapters.

Every 6000 miles (10,000 km)

20 Valve clearances - check and adjustment

1 The engine must be completely cool for this maintenance procedure, so let the machine sit overnight before beginning.

2 Disconnect the cable from the negative terminal of the battery.

3 Refer to Chapter 4 and remove the fuel tank.

4 Remove the valve cover (see Chapter 2).

5 Remove the pickup coil cover (see Chapter 5).

6 Position the number 1 piston (on the left side of the engine) at Top Dead Centre (TDC) on the compression stroke. Do this by turning the crankshaft, with a wrench placed on the crankshaft nut, until the TDC mark on the rotor is aligned with the timing mark on the

crankcase **(see illustration)**. Now, check the position of the camshaft sprockets - the IN and EX marks should be pointing towards each other **(see illustration)**. If they aren't,

20.6a Turn the crankshaft by the nut and align the TDC mark on the rotor with the timing mark on the crankcase (arrows)

turn the crankshaft one complete revolution and realign the mark. Piston number 1 is now at TDC compression, which can also be verified by looking at the cam lobes for that

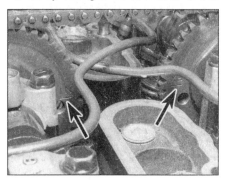

20.6b The IN and EX marks on the camshaft sprockets will be pointing toward each other if piston no. 1 is positioned at TDC on the compression stroke

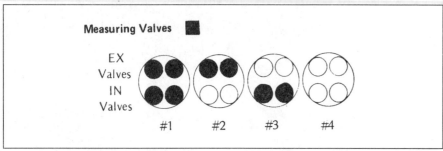

20.7 With cylinder no. 1 at TDC compression, the shaded valves can be adjusted

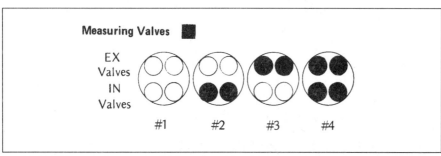

20.12 With cylinder no. 4 at TDC compression, the shaded valves can be adjusted

20.8 Checking the valve clearance

cylinder - they should not be depressing the rocker arms for either the intake valves or the exhaust valves.

7 With the engine in this position, all of the valves for cylinder no. 1 can be checked, as well as the exhaust valves for cylinder no. 2 and the intake valves for cylinder no. 3 **(see illustration)**.

8 Start with the no. 1 intake valve clearance. Insert a feeler gauge of the thickness listed in this Chapter's Specifications between the valve stem and adjuster screw on 600 models **(see illustration)**, or between the valve stem and rocker arm on 750 models. Pull the feeler gauge out slowly - you should feel a slight drag. If there's no drag or a heavy drag, loosen the adjuster screw locknut and turn the adjuster screw in or out, as needed, until you can feel a slight drag on the feeler gauge as you withdraw it. If required, the Kawasaki valve adjusting tool can be used (Pt. No. 57001-1217 for 600 models or Pt. No. 57001-1232 for 750 models).

9 Hold the adjuster screw (to keep it from turning) and tighten the locknut. Recheck the clearance to make sure it hasn't changed.

10 Now adjust the no. 1 exhaust valves, following the same procedure you used for the intake valves. Make sure to use a feeler gauge of the specified thickness.

11 Proceed to adjust the no. 2 exhaust valves and the no. 3 intake valves.

12 Rotate the crankshaft one complete revolution and align the TDC mark on the rotor with the timing mark on the crankcase, which will position piston no. 4 at TDC compression. Adjust all four valves on cylinder no. 4, followed by the no. 3 exhaust valves and the no. 2 intake valves **(see illustration)**.

13 Install the valve cover and all of the components that had to be removed to get it off.

14 Install the fuel tank and reconnect the cable to the negative terminal of the battery.

21 Throttle cables - freeplay check and adjustment

Check

1 Make sure the throttle grip rotates easily from fully closed to fully open with the front wheel turned at various angles. The grip should return automatically from fully open to fully closed when released. If the throttle sticks, check the throttle cables for cracks or kinks in the housings. Also, make sure the inner cables are clean and well-lubricated.

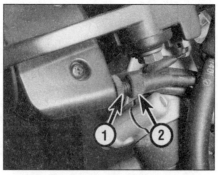

21.3 Loosen the accelerator cable lockwheel (1) and turn the adjuster (2) in or out to obtain the correct throttle freeplay

2 Check for a small amount of freeplay at the grip and compare the freeplay to the value listed in this Chapter's Specifications.

Adjustment

Note: *These motorcycles use two throttle cables - an accelerator cable and a decelerator cable.*

3 Freeplay adjustments can be made at the throttle end of the cable. Loosen the lockwheel on the cable **(see illustration)** and turn the adjuster until the desired freeplay is obtained, then retighten the lockwheel.

4 If the cables can't be adjusted at the grip end, adjust them at the lower ends. To do this, first remove the fuel tank (see Chapter 4).

600 A and B models

5 Loosen the locknuts on both throttle cables **(see illustration)**, then turn both adjusting nuts in completely.

6 Turn out the adjusting nut of the decelerator cable until the inner cable becomes tight, then tighten the locknut.

7 Turn the accelerator adjusting nut until the desired freeplay is obtained, then tighten the locknut. Make sure the throttle linkage lever contacts the idle adjusting screw when the throttle grip is at rest **(see illustration)**.

600 C and 750 models

8 Loosen the locknuts and screw the adjuster in completely at the upper end of the

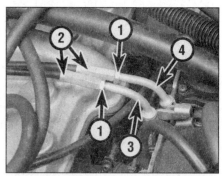

21.5 Throttle cable details (600 A and B models)

1 Locknuts   3 Decelerator cable
2 Adjusting nuts   4 Accelerator cable

21.7 The throttle linkage lever must contact the idle adjusting screw (arrow) when the throttle is closed

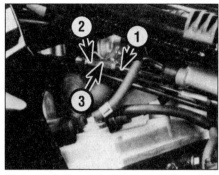

21.9a Throttle cable details (600 C models)

1 Accelerator cable   3 Adjusting nut
2 Locknut

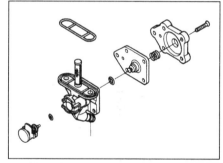

21.9b Throttle cable lower adjusting nuts (750 models)

A Accelerator cable   C Adjusting nuts
B Decelerator cable   D Lock nuts

accelerator cable. Tighten the locknut at the upper end of the cable.

9 Loosen the locknut at the lower part of the cable **(see illustrations)**. Turn the adjusting nut until the desired freeplay is obtained. Tighten the locknut at the lower part of the cable.

10 If the freeplay is still not adequate, try using the adjuster at the upper end of the cable again.

22 Fuel system - check and filter cleaning

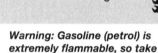

⚠️ **Warning: Gasoline (petrol) is extremely flammable, so take extra precautions when you work on any part of the fuel system. Don't smoke or allow open flames or bare light bulbs near the work area, and don't work in a garage where a natural gas-type appliance (such as a water heater or clothes dryer) is present. If you spill any fuel on your skin, rinse it off immediately with soap and water. When you perform any kind of work on the fuel system, wear safety glasses and**

have a class B type fire extinguisher on hand.

1 Check the fuel tank, the fuel tap, the lines and the carburettors for leaks and evidence of damage.

2 If carburettor gaskets are leaking, the carburettors should be disassembled and rebuilt by referring to Chapter 4.

3 If the fuel tap is leaking, tightening the screws may help. If leakage persists, the tap should be disassembled and repaired or replaced with a new one.

4 If the fuel lines are cracked or otherwise deteriorated, replace them with new ones.

5 Check the vacuum hose connected to the fuel tap. If it is cracked or otherwise damaged, replace it with a new one.

6 The fuel filter, which is attached to the fuel tap, may become clogged and should be removed and cleaned periodically. In order to clean the filter, the fuel tank must be drained and the fuel tap removed.

7 Remove the fuel tank (see Chapter 4). Attach a length of hose to the fuel outlet stub (not the vacuum pipe stub) and drain the fuel into an approved fuel container. On 600 models, turn the tap to Pri to drain the fuel; on 750 models, depress the knob on the tap body to drain the fuel.

8 Once the tank is emptied, loosen and remove the screws that attach the fuel tap to the tank. Remove the tap and filter.

9 Clean the filter **(see illustrations)** with solvent and blow it dry with compressed air. If the filter is torn or otherwise damaged, replace the entire fuel tap with a new one. Check the mounting flange O-ring and the gaskets on the screws. If they are damaged, replace them with new ones.

10 Install the O-ring, filter and fuel tap on the tank, then install the tank. Refill the tank and check carefully for leaks around the mounting flange and screws.

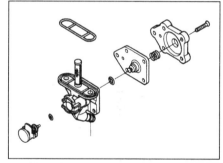

22.9a Fuel tap details - 750 models

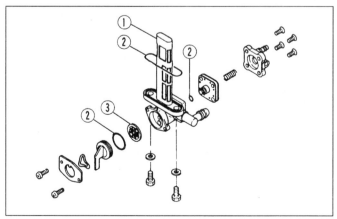

22.9b Fuel tap details – 600 A and B models

1 Filter   2 O-ring   3 Gasket

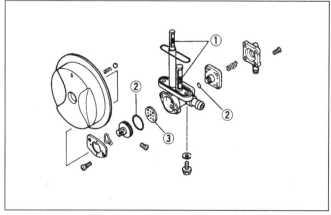

22.9c Fuel tap details – 600 C models

1 Filter   2 O-ring   3 Gasket

25.2 Note which of the marks on the scale (A) aligns with the pointer (B)

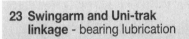

23 Swingarm and Uni-trak linkage - bearing lubrication

1 Over a period of time the grease will harden or dirt will penetrate the bearings.
2 The suspension components are not equipped with grease nipples. Remove the swingarm and the suspension linkage as described in Chapter 6 for greasing of the bearings.

24 Spark plugs - replacement

1 Remove the old spark plugs as described in Section 3 and install new ones.

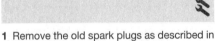

25 Alternator drive belt - check and adjustment (750 models)

1 The belt must be adjusted with the engine cold (at room temperature). The procedure requires a special tool and two torque wrenches, so it may be more practical to have a Kawasaki dealer adjust the belt for you.
2 Note which of the adjustment marks on the alternator scale aligns with the pointer (see illustration). Write this down.
3 Remove the clutch slave cylinder (see Chapter 2B). Remove the alternator bracket, engine sprocket cover and the upper rear engine mounting bolt (see illustration).
4 Insert the alternator tension wrench (Kawasaki tool no. 57001-1235) into the engine mounting bolt hole and place the tool lever against the alternator belt (see illustrations).
5 Loosen the alternator mounting bolts. Push the alternator down against the crankcase, then retighten the alternator mounting bolts with an Allen bolt bit. To get the correct torque for this stage of the procedure, turn the Allen bolt bit with a thumb and two fingers (don't use a ratchet handle).

25.3 The slave cylinder is mounted on the left side of the engine

A Slave cylinder   C Engine sprocket cover
B Bracket   D Engine mounting bolt

6 Attach a torque wrench to the special tool and apply 16 Nm (11.5 ft-lbs) torque.
7 While holding the first torque wrench at the setting specified in Step 6, tighten the alternator mounting bolts to the torque listed in this Chapter's Specifications.
8 Check the position of the pointer relative to the adjustment marks on the scale. It should be at the same position or higher. If it's lower, readjust the belt tension.

26 Cooling system - check

> **Warning: The engine must be cool before beginning this procedure.**

1 The entire cooling system should be checked carefully at the recommended intervals. Look for evidence of leaks, check the condition of the coolant, check the radiator for clogged fins and damage and make sure the fan operates when required.
2 Examine each of the rubber coolant hoses along its entire length. Look for cracks, abrasions and other damage. Squeeze each hose at various points. They should feel firm, yet pliable, and return to their original shape when released. If they are dried out or hard, replace them with new ones.
3 Check for evidence of leaks at each cooling system joint. Tighten the hose clamps carefully to prevent future leaks.

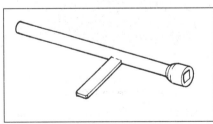

25.4b Alternator tension wrench

25.4a Use the alternator tension wrench (A) to set belt tension

4 Check the radiator for evidence of leaks and other damage (remove the fairings if necessary - see Chapter 8). Leaks in the radiator leave tell-tale scale deposits or coolant stains on the outside of the core below the leak. If leaks are noted, remove the radiator (refer to Chapter 3) and have it repaired at a radiator shop or replace it with a new one.
Caution: Do not use a liquid leak stopping compound to try to repair leaks.
5 Check the radiator fins for mud, dirt and insects, which may impede the flow of air through the radiator. If the fins are dirty, force water or low pressure compressed air through the fins from the backside. If the fins are bent or distorted, straighten them carefully with a screwdriver.
6 Remove the radiator cap by turning it counterclockwise until it reaches a stop. If you hear a hissing sound (indicating there is still pressure in the system), wait until it stops. Now, press down on the cap with the palm of your hand and continue turning the cap counterclockwise until it can be removed (see illustration). Check the condition of the coolant in the radiator. If it is rust colored or if accumulations of scale are visible in the radiator, drain, flush and refill the system with new coolant. Check the cap gaskets for cracks and other damage. Have the cap tested by a dealer service department or replace it with a new one. Install the cap by

26.6 Remove the screw at the rear of the fairing inner panel to gain access to the radiator cap

turning it clockwise until it reaches the first stop, then push down on the cap and continue turning until it can turn no further.

7 Check the antifreeze content of the coolant with an antifreeze hydrometer **(see illustration)**. Sometimes coolant may look like it's in good condition, but might be too weak to offer adequate protection. If the hydrometer indicates a weak mixture, drain, flush and refill the cooling system (see Section 27).

8 Start the engine and let it reach normal operating temperature, then check for leaks again. As the coolant temperature increases, the fan should come on automatically and the temperature should begin to drop. If it does not, refer to Chapter 3 and check the fan and fan circuit carefully.

9 If the coolant level is consistently low, and no evidence of leaks can be found, have the entire system pressure checked by a Kawasaki dealer service department, motorcycle repair shop or service station.

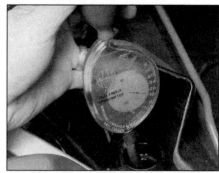

26.7 An antifreeze hydrometer is helpful in determining the condition of the coolant

Every 18,000 miles (30,000 km)

27 Cooling system - draining, flushing and refilling

Warning: Allow the engine to cool completely before performing this maintenance operation. Also, don't allow antifreeze to come into contact with your skin or painted surfaces of the vehicle. Rinse off spills immediately with plenty of water. Antifreeze is highly toxic if ingested. Never leave antifreeze lying around in an open container or in puddles on the floor; children and pets are attracted by its sweet smell and may drink it. Check with local authorities about disposing of used antifreeze. Many communities have collection centers which will see that antifreeze is disposed of safely. Antifreeze is also combustible, so don't store or use it near open flames.

Draining

1 Loosen the radiator cap **(see illustration 26.6)**. Place a large, clean drain pan under the left side of the engine.

2 Remove the lower fairing (see Chapter 8).

3 Remove the drain bolt from the side of the water pump cover on 600 models **(see illustration)**, or from the bottom of the water pump inlet tube on 750 models **(see illustration)**, and allow the coolant to drain into the pan. **Note:** *The coolant will rush out*

with considerable force, so position the drain pan accordingly. Remove the radiator cap completely to ensure that all of the coolant can drain.

4 Drain the coolant reservoir. On 600 A and B models, remove the cap from the reservoir (it's located in the lower fairing) and pour the coolant into the container. On 600 C models, refer to Chapter 3 for the reservoir removal procedure. On 750 models the reservoir tank is drained by blowing air into its overflow tube. Wash the reservoir out with water.

Flushing

5 Flush the system with clean tap water by inserting a garden hose in the radiator filler neck. Allow the water to run through the system until it is clear when it exits the drain bolt hole. If the radiator is extremely corroded, remove it by referring to Chapter 3 and have it cleaned at a radiator shop.

7 Check the drain bolt gasket. Replace it with a new one if necessary.

8 Clean the hole, then install the drain bolt and tighten it to the torque listed in this Chapter's Specifications.

9 Fill the cooling system with clean water mixed with a flushing compound. Make sure the flushing compound is compatible with aluminum components, and follow the manufacturer's instructions carefully.

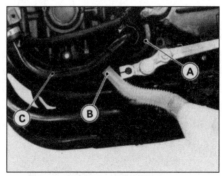

27.3b Remove the drain bolt from the coolant tube on 750 models

A   *Water pump cover*
B   *Drain hole with tube attached*
C   *Coolant tube*

10 Start the engine and allow it to reach normal operating temperature. Let it run for about ten minutes.

11 Stop the engine. Let the machine cool for a while, then cover the radiator cap with a heavy rag and turn it counterclockwise (anti-clockwise) to the first stop, releasing any pressure that may be present in the system. Once the hissing stops, push down on the cap and remove it completely.

12 Drain the system once again.

13 Fill the system with clean water, then repeat Steps 10, 11 and 12.

Refilling

14 Fill the system with the proper coolant mixture (see this Chapter's Specifications). When the system is full (all the way up to the top of the radiator cap filler neck), install the cap and start the engine. Allow the engine to reach normal operating temperature, then shut it off.

15 Let the engine cool off for a while, cover the radiator cap with a heavy rag and loosen it to the first stop to allow any pressure in the system to bleed off before the cap is removed completely. Recheck the coolant level in the radiator filler neck. If it's low, add more coolant until it reaches the top of the filler neck. Reinstall the cap.

16 Allow the engine to cool, then check the coolant level in the reservoir (see *Daily (pre-ride) checks*). If the coolant level is low, add the specified mixture until it reaches the Full mark in the reservoir.

17 Check the system for leaks.

18 Do not dispose of the old coolant by pouring it down a drain. Instead, pour it into a heavy plastic container, cap it tightly and take it to an authorised disposal site or a service station.

28 Fork oil - replacement

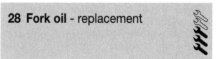

1 Place the motorcycle on the centerstand. Remove the lower fairing and position a jack with a block of wood on the jack head under the engine to support the motorcycle when the fork cap bolts are removed.

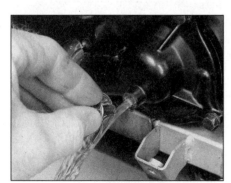

27.3a The drain bolt is located on the water pump on 600 models

28.4 After the handlebar is removed and the fork upper pinch bolts have been loosened, the top plug can be removed

28.5 Fork drain screw location

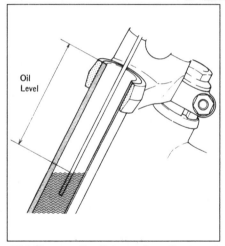

28.9 Measure the distance from the top of the fork tube to the oil and add or drain oil as necessary until the level is correct

2 Release the air pressure in the forks by depressing the air valve core with a small screwdriver.

3 Remove the handlebar from one side (see Chapter 6).

4 Loosen the fork clamp upper pinch bolts and remove the top plug from the exposed fork tube **(see illustration)**. Pull out the fork spring.

5 Place a drain pan under the fork leg and remove the drain screw **(see illustration)**.

 Warning: Do not allow the fork oil to contact the brake discs or pads. If it does, clean the discs with brake system cleaner and replace the pads with new ones before riding the motorcycle.

6 After most of the oil has drained, slowly compress and release the forks to pump out the remaining oil. An assistant will most likely be required to do this procedure.

7 Check the drain screw gasket for damage and replace it if necessary. Apply sealant to the threads of the drain screw, then install the screw and gasket, tightening it securely.

8 Pour the type and amount of fork oil, listed in this Chapter's Specifications, into the fork tube through the opening at the top. Remove the jack from under the engine and slowly pump the forks a few times to purge the air from the upper and lower chambers.

9 If you're working on a 600 A or B model, raise the front wheel off the ground again so that the forks are fully extended. If you're working on a 600 C or 750 model, fully compress the front forks (you may need an assistant to do this). Insert a tape measure into the fork tube and measure the distance from the oil to the top of the fork tube **(see illustration)**. Compare your measurement to the value listed in this Chapter's

Specifications. Drain or add oil, as necessary, until the level is correct.

10 Check the O-rings on the top plug, then coat them with a thin layer of multi-purpose grease. Install the fork spring. Install the top plug and tighten it securely.

11 Tighten the fork tube pinch bolts to the torque listed in the Chapter 6 Specifications. Install the handlebar, tightening the bolts to the torque listed in the Chapter 6 Specifications.

12 Repeat the procedure to the other fork.

13 On 600 A and B models, inject compressed air, a little at a time, until the desired pressure is attained (see this Chapter's Specifications).

14 Install the lower fairing.

Every year

29 Carburettor warmer system components - check (UK models, where fitted)

1 Identify the coolant filter on the engine right side and the check valve on the left side **(see illustrations)**. On 600 C models it may be necessary to remove the right knee grip cover for access (see Chapter 8).

2 Either drain the cooling system as described in Section 27 or clamp the coolant hoses each side of the filter/valve to prevent coolant loss. Release their clamps and pull the hoses free.

 See Tools and Workshop Tips in the Reference section for hose clamping methods.

3 Blow compressed air through the filter and valve, against the normal direction of flow, to clean it. Be sure to install the filter and the

check valve in the proper direction **(see illustration)**.

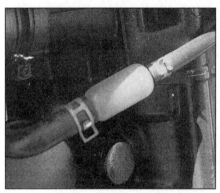

29.1a Carburettor warmer system coolant filter . . .

29.1b . . . and check valve (arrow)

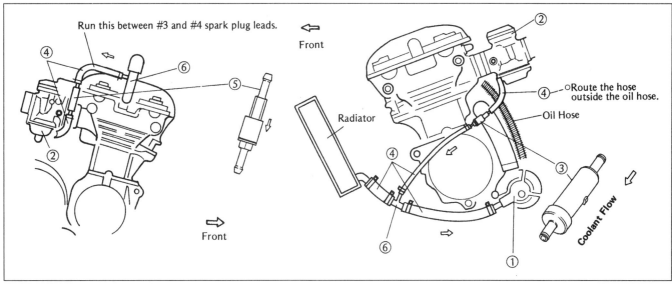

29.3 Carburettor warmer system coolant filter and check valve details (UK models only)

1 Water pump cover   2 Carburettor   3 Check valve   4 Hoses   5 Coolant filter   6 Pipes

Every two years

30 Brake and clutch fluid - change

1 The brake fluid (and clutch fluid on 750 models) should be replaced at the prescribed interval or whenever a master cylinder or caliper overhaul is carried out. Refer to the bleeding section in Chapter 7 for the brakes or Chapter 2 for the clutch, noting that all old fluid must be pumped from the fluid reservoir and hydraulic line before filling with new fluid.

 HAYNES HINT *Old fluid is invariably much darker in colour than new fluid, making it easy to see when all old fluid has been expelled from the system.*

31 Anti-dive seals and metal pipe - replacement (600 A and B models)

1 Refer to Chapter 6, Section 9 and unbolt the plunger from the anti-dive unit on each fork leg. Proceed as described under the overhaul sub-section. Replace the rubber cap and both O-rings on the seal cover with new items.
2 Unbolt the metal pipe linking the plunger unit to the brake hose union and install a new pipe.
3 Take particular care to bleed the system of air when refilling (see Chapter 7, Section 8).

32 Brake caliper and master cylinder seals - replacement

1 Hydraulic seals will deteriorate over a period of time and lose their effectiveness, leading to sticking operation or fluid loss, or allowing the ingress of air and dirt. Refer to Chapter 7 and dismantle the components for seal replacement.

33 Clutch slave cylinder and master cylinder seals - replacement (750 models)

1 Hydraulic seals will deteriorate over a period of time and lose their effectiveness, leading to sticking operation or fluid loss, or allowing the ingress of air and dirt. Refer to Chapter 2B and dismantle the components for seal replacement.

34 Steering head bearings - lubrication

1 Over a period of time the grease will harden or may be washed out of the bearings by incorrect use of jet washes.
2 Disassemble the steering head for re-greasing of the bearings. Refer to Chapter 6 for details.

35 Wheel bearings - check and lubrication

Check

1 Wheel bearings will wear over a period of time and result in handling problems.
2 Place the motorcycle on its centrestand. Check for any play in the bearings by pushing and pulling the wheel against the hub. Also rotate the wheel and check that it rotates smoothly.
3 If any play is detected in the hub, or if the wheel does not rotate smoothly (and this is not due to brake or transmission drag), the wheel bearings must be removed and inspected for wear or damage (see Chapter 7).

Lubrication - 600 A1, A2, A3, A4, B1 models

4 These models use unsealed bearings in the front wheel. Refer to Chapter 7, Section 13 and remove the front wheel bearings for inspection and greasing. Apply grease to the speedometer drive gears on installation.

36 Fuel hoses - replacement

 ⚠ *Warning: Petrol (gasoline) is extremely flammable, so take extra precautions when you*

work on any part of the fuel system. Don't smoke or allow open flames or bare light bulbs near the work area, and don't work in a garage where a natural gas-type appliance is present. If you spill any fuel on your skin, rinse it off immediately with soap and water. When you perform any kind of work on the fuel system, wear safety glasses and have a fire extinguisher suitable for a Class B type fire (flammable liquids) on hand.

1 The fuel delivery and vacuum hoses should be replaced regardless of their condition. On California models, also renew the emission control system hoses.

2 Remove the fuel tank (see Chapter 4). Disconnect the fuel hoses from the fuel tap and from the carburettors, noting the routing of each hose and where it connects (see Chapter 4 if required). It is advisable to make a sketch of the various hoses before removing them to ensure they are correctly installed.

3 Secure each new hose to its unions using new clamps. Run the engine and check for leaks before taking the machine out on the road.

37 Brake and clutch hoses - replacement

1 The hoses will in time deteriorate with age and should be replaced regardless of their apparent condition.

2 Refer to Chapter 7 and disconnect the brake hoses from the master cylinders and calipers. Always replace the banjo union sealing washers with new ones.

3 On 750 models, disconnect the hose from the clutch slave cylinder and master cylinder as described in Chapter 2B. Always replace the banjo union sealing washers with new ones.

Non-scheduled maintenance

38 Cylinder compression - check

1 Among other things, poor engine performance may be caused by leaking valves, incorrect valve clearances, a leaking head gasket, or worn pistons, rings and/or cylinder walls. A cylinder compression check will help pinpoint these conditions and can also indicate the presence of excessive carbon deposits in the cylinder heads.

2 The only tools required are a compression gauge and a spark plug wrench. A compression gauge with a threaded end for the spark plug hole is preferable to the type which requires hand pressure to maintain a tight seal **(see illustration)**. Depending on the outcome of the initial test, a squirt-type oil can may also be needed.

3 Make sure the valve clearances are correctly set (see Section 20).

4 Refer to *Fault Finding Equipment* in the Reference section for details of the compression test.

39 Suspension - check

1 The suspension components must be maintained in top operating condition to ensure rider safety. Loose, worn or damaged suspension parts decrease the vehicle's stability and control.

2 While standing alongside the motorcycle, lock the front brake and push on the handlebars to compress the forks several times. See if they move up-and-down smoothly without binding. If binding is felt, the forks should be disassembled and inspected as described in Chapter 6.

3 Carefully inspect the area around the fork seals for any signs of fork oil leakage **(see illustration)**. If leakage is evident, the seals must be replaced as described in Chapter 6.

4 Check the tightness of all suspension nuts and bolts to be sure none have worked loose.

5 Inspect the shock for fluid leakage and tightness of the mounting nuts. If leakage is found, the shock should be replaced.

6 Set the bike on its centerstand. Grab the swingarm on each side, just ahead of the axle. Rock the swingarm from side to side - there

should be no discernible movement at the rear. If there's a little movement or a slight clicking can be heard, make sure the pivot shaft nut is tight. If the pivot nut is tight but movement is still noticeable, the swingarm will have to be removed and the bearings replaced as described in Chapter 6.

7 Inspect the tightness of the rear suspension nuts and bolts.

8 Check the air pressure in the front forks (600 A and B models only) and the rear shock absorber **(see illustration)**. Note: *The manufacturer recommends against the use of a tyre pressure gauge - a gauge made especially for air suspensions should be used (a tyre pressure gauge may allow too much air to leak out around the valve, thereby giving you an inaccurate reading. Adjust the air pressures to the values listed in this Chapter's Specifications. Add air a little at a time to avoid overpressurising the forks or shock absorber.*

9 The rear shock absorber also has damping adjustment via the knob on the right side of the bike, to the rear of the swingarm pivot. There are four damping positions, position 1 is fully pushed in and provides light damping suitable for light loads, low speed and good road surfaces. Positions 2, 3 and 4 provide progressively greater damping for heavier loads, high speed and poor road surfaces.

38.2 A compression gauge with a threaded fitting for the spark plug hole is required

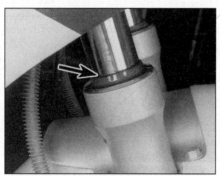

39.3 If oil is leaking past the fork seals (arrow), replace them

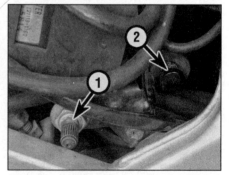

39.8 Rear shock absorber air valve (1) and damping adjuster (2)

Chapter 2 Part A
Engine, clutch and transmission (600 models)

Contents

Degrees of difficulty

| Easy, suitable for novice with little experience | | Fairly easy, suitable for beginner with some experience | | Fairly difficult, suitable for competent DIY mechanic | | Difficult, suitable for experienced DIY mechanic | | Very difficult, suitable for expert DIY or professional | |

Specifications

General

Bore	60.0 mm (2.36 in)
Stroke	52.5 mm (2.06 in)
Displacement	592 cc
Compression ratio	
A and B models	11.0 : 1
C models	11.7 : 1

Camshaft and rocker arms

Lobe height (intake and exhaust)	36.06 mm (1.42 in) minimum
Bearing oil clearance	
1985 and 1986	
Standard	0.078 to 0.121 mm (0.003 to 0.004 in)
Maximum	0.21 mm (0.008 in)
1987 on	
Standard	
Outer journals and centre journal	0.028 to 0.071 mm (0.001 to 0.002 in)
Other two journals	0.078 to 0.121 mm (0.003 to 0.004 in)
Maximum	
Outer journals and centre journal	0.16 mm (0.006 in)
Other two journals	0.21 mm (0.008 in)
Journal diameter	
1985 and 1986	
Standard	22.900 to 22.922 mm (0.902 to 0.903 in)
Minimum	22.87 mm (0.901 in)
1987 on	
Standard	
Outer journals and centre journal	22.950 to 22.972 mm (0.904 to 0.905 in)
Other two journals	22.900 to 22.922 mm (0.902 to 0.903 in)
Minimum	
Outer journals and centre journal	22.92 mm (0.903 in)
Other two journals	22.87 mm (0.901 in)
Bearing journal inside diameter	
Standard	23.000 to 23.021 mm (0.906 to 0.907 in)
Maximum	23.03 mm (0.9073 in)
Camshaft runout (maximum)	0.1 mm (0.003 in)
Camshaft chain 20-link length (maximum)	128.9 mm (5-5/64 in)
Rocker arm inside diameter (maximum)	12.05 mm (0.474 in)
Rocker shaft diameter (minimum)	11.97 mm (0.471 in)

Cylinder head, valves and valve springs

Cylinder head warpage limit	0.05 mm (0.002 in)
Valve stem runout limit	0.05 mm (0.002 in)
Valve stem diameter	
Intake	4.960 to 4.990 mm (0.195 to 0.196 in)
Exhaust	4.940 to 4.970 mm (0.194 to 0.195 in)
Valve guide inside diameter (intake and exhaust)	5.000 to 5.012 mm (0.197 to 0.1974 in)
Valve seat width (intake and exhaust)	0.5 to 1.0 mm (0.020 to 0.040 in)
Valve spring free length (minimum)	
Inner	30.0 mm (1-3/16 in)
Outer	33.4 mm (1-5/16 in)

Pistons

Piston diameter	59.800 to 59.957 mm (2.356 to 2.362 in)
Piston-to-cylinder clearance	0.043 to 0.070 mm (0.001 to 0.002 in)
Oversize pistons and rings	+ 0.5 mm (+0.020 in) (one oversize only)
Ring side clearance	
Top	0.03 to 0.17 mm (0.001 to 0.006 in)
Second	0.02 to 0.16 mm (0.0007 to 0.006 in)
Ring groove width	
Top	1.02 to 1.12 mm (0.040 to 0.044 in)
Second	
1988 and earlier	1.21 to 2.60 mm (0.047 to 0.102 in)
1989 on	1.01 to 1.11 mm (0.040 to 0.043 in)
Oil	
1988 and earlier	2.51 to 2.60 mm (0.100 to 0.102 in)
1989 on	2.01 to 2.11 mm (0.080 to 0.175 in)
Ring thickness	
Top	0.90 to 0.99 mm (0.035 to 0.039 in)
Second	
1988 and earlier	1.10 to 1.19 mm (0.043 to 0.046 in)
1989 on	0.90 to 0.99 mm (0.035 to 0.040 in)
Ring end gap	
Top	0.15 to 0.60 mm (0.006 to 0.023 in)
Second	0.15 to 0.65 mm (0.006 to 0.025 in)

Cylinder block

Bore diameter (maximum)	60.10 mm (141.84 in)
Deck warpage limit	0.05 mm (0.002 in)
Taper limit	0.05 mm (0.002 in)
Out-of-round limit	0.05 mm (0.002 in)

Crankshaft and bearings

Main bearing oil clearance	0.014 to 0.080 mm (0.0005 to 0.003 in)
Main bearing journal diameter	
No mark on crank throw	31.984 to 31.992 mm (1.2601 to 1.2604 in)
"1" mark on crank throw	31.993 to 32.000 mm (1.2605 to 1.2608 in)
Connecting rod side clearance	0.013 to 0.500 mm (0.0005 to 0.020 in)
Connecting rod bearing oil clearance	0.035 to 0.100 mm (0.001 to 0.003 in)
Connecting rod big-end bore diameter	
No mark on side of rod	36.000 to 36.008 mm (1.4184 to 1.4187 in)
"0" mark on side of rod	36.009 to 36.016 mm (1.4187 to 1.4190 in)
Connecting rod journal diameter	
No mark on crank throw	32.984 to 32.992 mm (1.2995 to 1.2998 in)
"0" mark on crank throw	32.993 to 33.000 mm (1.2999 to 1.3002 in)
Primary chain 20-link length (maximum)	193.4 mm (7-39/64 in)

Oil pump and relief valve

Oil pressure (warm)	31 to 40 psi @ 4000 rpm
Relief valve opening pressure	63 to 85 psi

Clutch

Spring free length (minimum)	31.7 mm (1-1/4 in)
Friction plate thickness (minimum)	2.8 mm (0.110 in)
Friction and steel plate warpage limit	0.3 mm (0.019 in)

Transmission

Gear backlash (maximum)	0.25 mm (0.009 in)
Shift fork groove width (maximum)	5.3 mm (0.208 in)
Shift fork ear thickness (minimum)	4.8 mm (0.190 in)
Shift fork guide pin diameter (minimum)	7.8 mm (0.307 in)
Shift drum groove width (maximum)	8.3 mm (0.327 in)

Torque specifications

Valve cover bolts	10 Nm (87 in-lbs)
Camshaft bearing cap bolts	12 Nm (104 in-lbs)
Camshaft gear bolts	15 Nm (11 ft-lbs)
Rocker arm shaft plugs	10 Nm (87 in-lbs)
Oil pipe bolts (on camshaft bearing caps)	12 Nm (104 in-lbs)
Camshaft chain tensioner cap	24 Nm (18 ft-lbs)
Cylinder head nuts	22 Nm (16.5 ft-lbs)
Cylinder block-to-cylinder head bolts	
Initial	8.5 Nm (75 in-lbs)
Final	12 Nm (104 in-lbs)
Cylinder block-to-crankcase nuts	10 Nm (87 in-lbs)
Clutch cover bolts	9 Nm (78 in-lbs)
Clutch spring bolts	9 Nm (78 in-lbs)
Clutch hub nut	136 Nm (100 ft-lbs)
Oil pan bolts	12 Nm (104 in-lbs)
Oil pipe-to-cylinder head union bolts	12 Nm (104 in-lbs)
Oil pipe-to-crankcase union bolt	24 Nm (18 ft-lbs)
Relief valve-to-oil pan	15 Nm (11 ft-lbs)
Engine mounting bolt nuts	34 Nm (25 ft-lbs)
Downtube mounting bolts	24 Nm (18 ft-lbs)
Crankcase bolts	
6 mm bolts	12 Nm (104 in-lbs)
8 mm bolts	27 Nm (20 ft-lbs)
Connecting rod nuts	37 Nm (27 ft-lbs)
Primary chain tensioner bolt	24 Nm (18 ft-lbs)
Chain guide bracket bolts	12 Nm (104 in-lbs)
Secondary shaft nut	73 Nm (54 ft-lbs)
Shift drum guide bolt	24 Nm (18 ft-lbs)
Shift drum positioning bolt	24 Nm (18 ft-lbs)

1 General information

The engine/transmission unit is of the water-cooled, in-line, four cylinder design, installed transversely across the frame. The sixteen valves are operated by double overhead camshafts which are chain driven off the crankshaft. The engine/transmission assembly is constructed from aluminum alloy. The crankcase is divided horizontally.

The crankcase incorporates a wet sump, pressure-fed lubrication system which uses a gear-driven, dual-rotor oil pump, an oil filter and bypass valve assembly, a relief valve and an oil pressure switch. Also contained in the crankcase is the secondary shaft and the starter motor clutch.

Power from the crankshaft is routed to the transmission via the clutch, which is of the wet, multi-plate type and is chain-driven off the crankshaft. The transmission is a six-speed, constant-mesh unit.

2 Operations possible with the engine in the frame

The components and assemblies listed below can be removed without having to remove the engine from the frame. If, however, a number of areas require attention at the same time, removal of the engine is recommended.

Gear selector mechanism external components
Water pump
Starter motor
Alternator
Clutch assembly
Oil pan, oil pump and relief valve
Valve cover, camshafts and rocker arms
Cam chain tensioner
Cylinder head
Cylinder block and pistons

3 Operations requiring engine removal

It is necessary to remove the engine/transmission assembly from the frame and separate the crankcase halves to gain access to the following components:

Crankshaft, connecting rods and bearings
Transmission shafts
Shift drum and forks
Secondary shaft and starter motor clutch
Camshaft chain
Primary chain

4 Major engine repair - general note

1 It is not always easy to determine when or if an engine should be completely overhauled, as a number of factors must be considered.
2 High mileage is not necessarily an indication that an overhaul is needed, while low mileage, on the other hand, does not preclude the need for an overhaul. Frequency of servicing is probably the single most important consideration. An engine that has regular and frequent oil and filter changes, as well as other required maintenance, will most likely give many miles of reliable service. Conversely, a neglected engine, or one which has not been broken in properly, may require an overhaul very early in its life.
3 Exhaust smoke and excessive oil consumption are both indications that piston rings and/or valve guides are in need of attention. Make sure oil leaks are not responsible before deciding that the rings and guides are bad. Refer to Chapter 1 and perform a cylinder compression check to determine for certain the nature and extent of the work required.
4 If the engine is making obvious knocking or rumbling noises, the connecting rod and/or main bearings are probably at fault.
5 Loss of power, rough running, excessive valve train noise and high fuel consumption rates may also point to the need for an overhaul, especially if they are all present at the same time. If a complete tune-up does not remedy the situation, major mechanical work is the only solution.
6 An engine overhaul generally involves restoring the internal parts to the specifications of a new engine. During an overhaul the piston rings are replaced and the cylinder walls are bored and/or honed. If a rebore is done, then new pistons are also required. The main and connecting rod bearings are generally replaced with new ones and, if necessary, the crankshaft is also replaced. Generally the valves are serviced as well, since they are usually in less than perfect condition at this point. While the engine is being overhauled, other components such as the carburettors and the starter motor can be rebuilt also. The end result should be a like-new engine that will give as many trouble free miles as the original.
7 Before beginning the engine overhaul, read through all of the related procedures to familiarize yourself with the scope and requirements of the job. Overhauling an engine is not all that difficult, but it is time consuming. Plan on the motorcycle being tied up for a minimum of two weeks. Check on the availability of parts and make sure that any necessary special tools, equipment and supplies are obtained in advance.

8 Most work can be done with typical shop hand tools, although a number of precision measuring tools are required for inspecting parts to determine if they must be replaced. Often a dealer service department or motorcycle repair shop will handle the inspection of parts and offer advice concerning reconditioning and replacement. As a general rule, time is the primary cost of an overhaul so it doesn't pay to install worn or substandard parts.

 Refer to Tools and Workshop Tips in the Reference section for details of how to use precision measuring tools.

9 As a final note, to ensure maximum life and minimum trouble from a rebuilt engine, everything must be assembled with care in a spotlessly clean environment.

5 Engine - removal and installation

Note: *Engine removal and installation should be done with the aid of an assistant to avoid damage or injury that could occur if the engine is dropped. An hydraulic floor jack should be used to support and lower the engine if possible (they can be rented at low cost).*

Removal

1 Set the bike on its centrestand.
2 Remove the seat and the fuel tank (see Chapter 4).
3 Remove the side covers, knee grip covers (C models), fairing side stays (A models) and the upper and lower fairings (see Chapter 8).
4 Drain the coolant and the engine oil (see Chapter 1).
5 Remove the ignition coils (see Chapter 5).
6 Remove the air suction valve and the vacuum switching valve (see Chapter 1).
7 Remove the carburettors (see Chapter 4) and plug the intake openings with rags.
8 Remove the radiator, radiator hoses and oil cooler (see Chapter 3).
9 Remove the horns (see Chapter 9).
10 Remove the exhaust system (see Chapter 4).
11 Disconnect the lower end of the clutch cable from the lever and bracket (see Chapter 1).
12 Remove the engine sprocket cover, unbolt the engine sprocket and detach the sprocket and chain from the engine (see Chapter 6).

5.14 Remove the bolt (arrow) that holds the ground wire to the engine

5.15 The heat guard (A) is retained by two screws (B) (C models only)

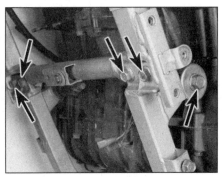

5.16a On A and B models, remove the engine front mounting bolts and nuts and the downtube front mounting bolts (arrows)

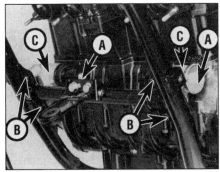

5.16b On C models, remove the engine front mounting bolts (A), the mount bracket bolts (B), then disconnect the bracket (C)

5.16c On A and B models, remove the downtube lower mounting bolts (arrows)

13 Mark and disconnect the wires from the oil pressure switch, neutral switch and the starter motor. Unplug the alternator, sidestand and pickup coil electrical connectors (see Chapters 5 and 9).

14 Remove the bolt securing the ground (earth) wire to the right rear of the engine case **(see illustration)**.

15 If you're working on a C model, remove the heat guard **(see illustration)**.

16 Remove the engine front mounting bolts/nuts and, on A and B models, the bolts holding the downtubes to the frame **(see illustrations)**.

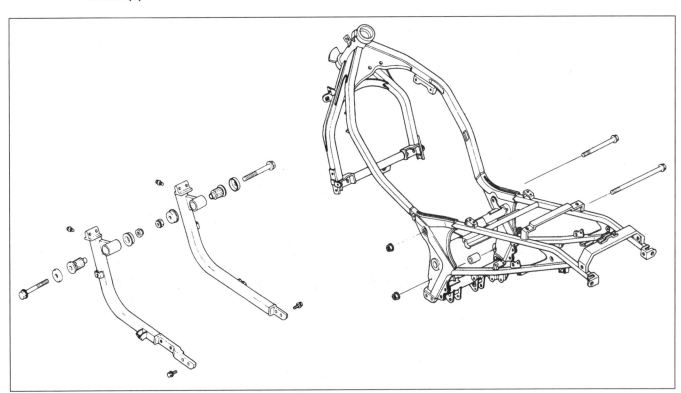

5.16d Engine mounting details - A and B models

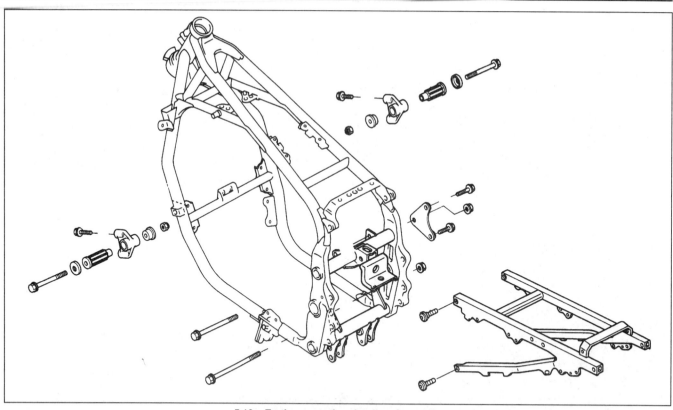

5.16e Engine mounting details - C models

17 Support the engine with a floor jack and a wood block **(see illustration)**.

18 Pry the plugs from the holes in the frame and remove the nuts from the rear mounting bolts **(see illustration)**.

19 With the engine supported, pull the rear mounting bolts out. Make sure no wires or hoses are still attached to the engine assembly.

20 On A and B models, slowly and carefully lower the engine assembly to the floor, then guide it out from under the bike.

21 On C models, raise the engine slightly then, with the help of an assistant, slide the engine out to the right. It would be helpful to have another jack, or a small table or platform that is the same height as the bottom frame tube, which the engine can be slid onto as it is removed.

Installation

22 Installation is basically the reverse of removal. Note the following points:

a) Don't tighten any of the engine mounting bolts until they all have been installed.

b) Use new gaskets at all exhaust pipe connections.

c) Make sure all of the wires on the left side of the engine (under the engine sprocket cover) are routed properly.

d) Tighten the engine mounting bolts and frame downtube bolts to the torque listed in this Chapter's Specifications.

e) Adjust the drive chain, throttle cables, choke cable and clutch cable following the procedures in Chapter 1.

5.17 Support the engine with a floor jack, with a wood block as a cushion

5.18 Remove the nuts from the engine rear mounting bolts, then pull the bolts out

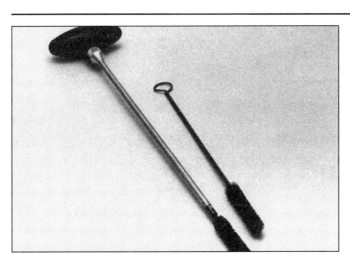

6.2a A selection of brushes is required for cleaning holes and passages in the engine components

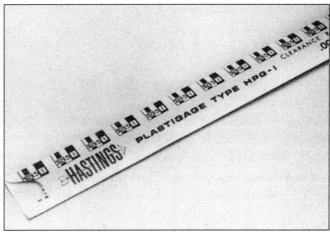

6.2b Type HPG-1 Plastigauge is needed to check the crankshaft, connecting rod and camshaft oil clearances

6 Engine disassembly and reassembly - general information

1 Before disassembling the engine, clean the exterior with a degreaser and rinse it with water. A clean engine will make the job easier and prevent the possibility of getting dirt into the internal areas of the engine.

2 In addition to the precision measuring tools mentioned earlier, you will need a torque wrench, a valve spring compressor, oil galley brushes, a piston ring removal and installation tool, a piston ring compressor, a pin-type spanner wrench and a clutch holder tool (which is described in Section 19). Some new, clean engine oil of the correct grade and type, some engine assembly lube (or moly-based grease), a tube of Kawasaki Bond liquid gasket (part no. 92104-1003) or equivalent. and a tube of RTV (silicone) sealant will also be required. Although it may not be

considered a tool, some Plastigauge (type HPG-1) should also be obtained to use for checking bearing oil clearances **(see illustrations)**.

3 An engine support stand made from short lengths of 2 x 4's bolted together will facilitate the disassembly and reassembly procedures **(see illustration)**. The perimeter of the mount should be just big enough to accommodate the engine oil pan. If you have an automotive-type engine stand, an adapter plate can be made from a piece of plate, some angle iron and some nuts and bolts **(see illustration)**.

4 When disassembling the engine, keep "mated" parts together (including gears, cylinders, pistons, etc. that have been in contact with each other during engine operation) . These "mated" parts must be reused or replaced as an assembly.

5 Engine/transmission disassembly should be done in the following general order with reference to the appropriate Sections.
Remove the cylinder head
Remove the cylinder block

Remove the pistons
Remove the clutch
Remove the oil pan
Remove the external shift mechanism
Remove the alternator rotor/stator coils (see Chapter 9)
Separate the crankcase halves
Remove the secondary sprocket, shaft and starter motor clutch
Remove the crankshaft and connecting rods
Remove the transmission shafts/gears
Remove the shift drum/forks

6 Reassembly is accomplished by reversing the general disassembly sequence.

7 Valve cover - removal and installation

Note: *The valve cover can be removed with the engine in the frame. If the engine has been removed, ignore the steps which don't apply.*

6.3a An engine stand can be made from short lengths of 2 x 4 lumber and lag bolts or nails

6.3b If you have an automotive engine stand, an adapter can be made from a piece of 3/16-inch metal plate, angle iron and some nuts and bolts

7.9 Remove the valve cover bolts (arrows)

7.10 Pry the chain guide out of the cover if it's worn out

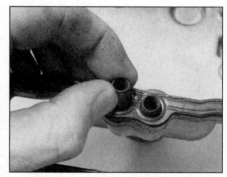

7.11 On US models, be careful not to lose the locating dowels

Removal

1 Set the bike on its centrestand.
2 Drain the engine coolant (see Chapter 1).
3 Remove the fuel tank (see Chapter 4).
4 Remove the upper and lower fairings (see Chapter 8).
5 Remove the air suction valve and the vacuum switching valve (see Chapter 1).
6 Remove the ignition coils and their brackets, along with the spark plug wires (see Chapter 5).
7 Remove the thermostat housing and the upper coolant pipe (see Chapter 3).
8 Remove the baffle plate. On early C models, also remove the reserve lighting device (US and Canadian models only) (see Chapter 9).
9 Remove the valve cover bolts **(see illustration)**.
10 Lift the cover off the cylinder head. If it's stuck, don't attempt to pry it off - tap around the sides of it with a plastic hammer to dislodge it. Check the chain guide in the centre of the cover - if it's excessively worn, pry it out and install a new one **(see illustration)**.

Installation

11 On US models, remove the locating dowels from the valve cover **(see illustration)**.

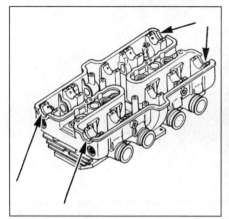

7.12 Apply a thin film of RTV sealant to the half-circle cutouts (arrows)

Peel the rubber gasket from the cover. If it is cracked, hardened, has soft spots or shows signs of general deterioration, replace it with a new one.
12 Clean the mating surfaces of the cylinder head and the valve cover with lacquer thinner, acetone or brake system cleaner. Apply a thin film of RTV sealant to the half-circle cutouts on each side of the head **(see illustration)**.
13 Install the gasket to the cover. Position the cover on the cylinder head, making sure the gasket doesn't slip out of place.
14 Check the rubber seals on the valve cover bolts, replacing them if necessary. Install the bolts, tightening them evenly, to the torque listed in this Chapter's Specifications.
15 The remainder of installation is the reverse of removal. Fill the cooling system with the recommended type and amount of coolant (see Chapter 1).

8 Camshaft chain tensioner - removal and installation

Note: *The camshaft chain tensioner can be removed with the engine in the frame. If the engine has been removed, ignore the steps which don't apply.*

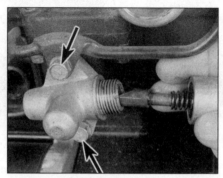

8.3 Remove the tensioner cap, spring and pushrod stop, then remove the tensioner mounting bolts (arrows)

Removal

1 Set the bike on its centrestand.
2 Remove the fuel tank and carburettors (see Chapter 4).
3 Remove the tensioner cap, sealing washer and spring **(see illustration)**.
4 Remove the tensioner mounting bolts and detach it from the cylinder block. If the oil line is in the way, remove the two upper union bolts and carefully guide the tensioner out.
5 Depress the tensioner plunger and pull out the pin. Remove the tensioner components from the tensioner body **(see illustration)** and wash them with solvent.

Installation

6 Lubricate the friction surfaces of the components with moly-based grease. Place the spring on the pushrod and install the pushrod / spring assembly into the tensioner body. Push in on the rod, align the groove in the side of the rod with the pinhole in the side of the tensioner body, then insert the pin into the tensioner body, engaging it

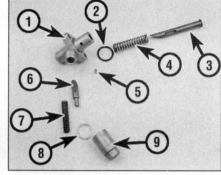

8.5 Components of the camshaft chain tensioner

1	Tensioner body	6	Pushrod stop
2	O-ring	7	Pushrod stop
3	Pushrod		spring
4	Pushrod spring	8	Sealing washer
5	Pin	9	Tensioner cap

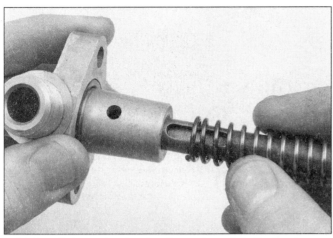

8.6 Install the pushrod and spring part-way into the tensioner body, align the groove in the pushrod with the hole in the tensioner body, depress the pushrod and install the pin

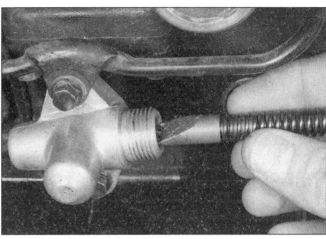

8.9 Install the pushrod stop like this, so the tapered surface will contact the tapered surface of the pushrod

with the groove in the pushrod **(see illustration)**.

7 Check the O-ring on the tensioner body for cracks or hardening. It's a good idea to replace this O-ring as a matter of course.

8 Position the tensioner body on the cylinder block and install the bolts, tightening them securely.

9 Install the pushrod stop into the chain tensioner, so its tapered portion mates with the tapered portion of the pushrod **(see illustration)**. The pushrod stop should protrude approximately 5 mm (3/16-inch) from the end of the tensioner housing. If it sticks out farther than this, the camshaft chain slack hasn't been taken up completely - if this is the case, slowly turn the crankshaft over in the normal direction of rotation (using a wrench placed on the crankshaft bolt under the pickup coil cover - see Chapter 5), pushing the pushrod stop in with your thumb.

10 Slide the spring over the pushrod stop. Install the tensioner cap and sealing washer,

tightening it to the torque listed in this Chapter's Specifications.

11 The remainder of installation is the reverse of removal.

9 Camshafts, rocker arm shafts and rocker arms - removal, inspection and installation

Note: *This procedure can be performed with the engine in the frame.*

Camshafts

Removal

1 Remove the valve cover following the procedure given in Section 7.

2 Remove the camshaft chain tensioner (see Section 8).

3 Remove the pickup coil cover (see Chapter 5).

4 Position the engine at Top Dead Centre (TDC) for cylinders 1 and 4 (see Chapter 1, *Valve clearances - check and adjustment*, for the TDC locating procedure).

5 Remove the rubber caps and lift the oil pipes from the camshaft bearing caps **(see illustration)**. **Note:** *On early models, the oil pipes are secured by banjo bolts and sealing washers.*

6 Unscrew the bearing cap bolts for one of the camshafts, a little at a time, until they are all loose, then unscrew the bearing cap bolts for the other camshaft. Remove the bolts and lift off the bearing caps. Note the numbers on the bearing caps which correspond to the numbers on the cylinder head **(see illustration)**. When you reinstall the caps, be sure to install them in the correct positions.

Caution: If the bearing cap bolts aren't loosened evenly, the camshaft may bind.

7 Pull up on the camshaft chain and carefully guide the camshaft out **(see illustrations)**. With the chain still held taut, remove the other camshaft. Look for marks on the camshafts. The intake camshaft should have an IN mark and the exhaust camshaft should have an EX mark. If you can't find these marks, label the camshafts to ensure they are installed in their original locations. **Note:** *Don't remove the sprockets from the camshafts unless absolutely necessary.*

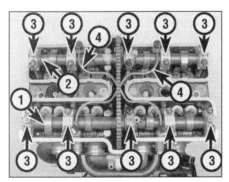

9.5 Details of the camshafts and related components

1 Intake camshaft
2 Exhaust camshaft
3 Camshaft bearing caps
4 Oil pipes

9.6 The camshaft bearing caps have a number cast into them that corresponds to a number cast into the cylinder head (arrows) – make sure the caps are reinstalled in the same positions

9.7a Lift up on the cam chain and carefully guide the camshaft out

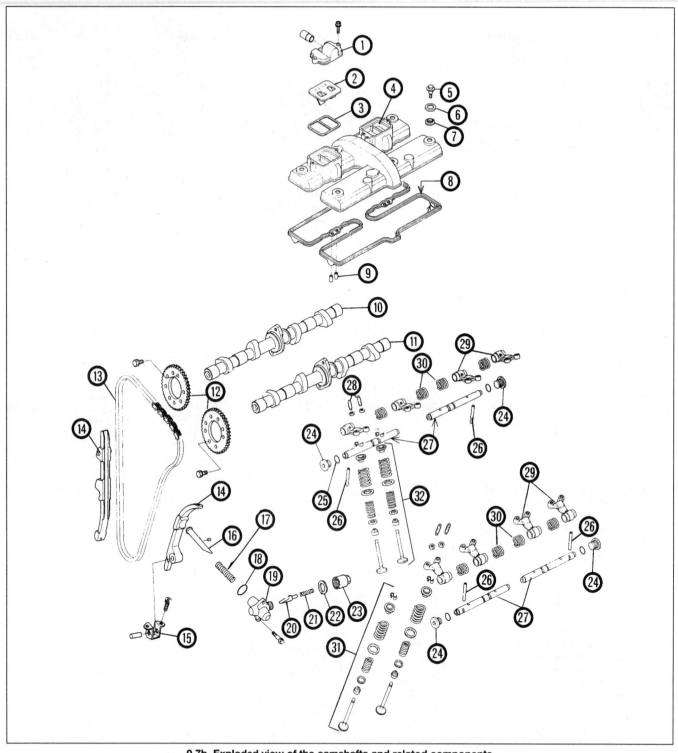

9.7b Exploded view of the camshafts and related components

1 Air suction valve
   cover
2 Air suction valve
3 Gasket
4 Valve cover
5 Valve cover bolt
6 Cap
7 Seal

8 Gasket
9 Locating dowels
10 Exhaust camshaft
11 Intake camshaft
12 Camshaft sprockets
13 Camshaft chain
14 Chain guides

15 Chain guide bracket
16 Tensioner pushrod
17 Tensioner spring
18 O-ring
19 Tensioner body
20 Tensioner pushrod
   stop

21 Spring
22 Sealing washer
23 Tensioner cap
24 Plug
25 O-ring
26 Locating pin
27 Rocker arm shafts

28 Valve adjusting
   screws
29 Rocker arms
30 Retaining springs
31 Intake valve and
   related components
32 Exhaust valve and
   related components

9.8 While the camshafts are out, wire the cam chain to another component to keep tension on it

9.10a Check the lobes of the camshaft for wear – here's a good example of damage which will require replacement (or repair) of the camshaft

9.10b Measure the height of the camshaft lobes with a micrometer

8 While the camshafts are out, don't allow the chain to go slack - if you do, it will become detached from the gear on the crankshaft and may bind between the crankshaft and case, which could cause damage to these components. Wire the chain to another component to prevent it from dropping down **(see illustration)**. Also, cover the top of the cylinder head with a rag to prevent foreign objects from falling into the engine.

Inspection

Before replacing camshafts or the cylinder head and bearing caps because of damage, check with local machine shops specialising in motorcycle engine work. In the case of the camshafts, it may be possible for cam lobes to be welded, reground and hardened, at a cost far lower than that of a new camshaft. If the bearing surfaces in the cylinder head are damaged, it may be possible for them to be bored out to accept bearing inserts. Due to the cost of a new cylinder head it is recommended that all options be explored before condemning it as trash!

9 Inspect the cam bearing surfaces of the head and the bearing caps. Look for score marks, deep scratches and evidence of spalling (a pitted appearance).
10 Check the camshaft lobes for heat discoloration (blue appearance), score marks, chipped areas, flat spots and spalling **(see illustration)**. Measure the height of each lobe with a micrometer **(see illustration)** and compare the results to the minimum lobe height listed in this Chapter's Specifications. If damage is noted or wear is excessive, the camshaft must be replaced. Also, be sure to check the condition of the rocker arms, as described later in this Section.
11 Next, check the camshaft bearing oil clearances. Clean the camshafts, the bearing surfaces in the cylinder head and the bearing caps with a clean, lint-free cloth, then lay the

cams in place in the cylinder head, with the IN and EX marks on the gears facing away from each other and level with the valve cover gasket surface of the cylinder head **(see illustration)**. Engage the cam chain with the cam gears, so the camshafts don't turn as the bearing caps are tightened.
12 Cut ten strips of Plastigauge (type HPG-1) and lay one piece on each bearing journal, parallel with the camshaft centreline **(see illustration)**.
13 Install the bearing caps in their proper positions (the arrows on the caps must face toward the front of the engine and the

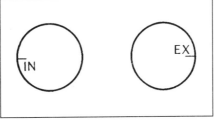

9.11 When installing the camshafts, the IN and EX marks on the camshaft sprockets should be positioned like this, with the marks aligned with the machined surface of the cylinder head

numbers on the caps must correspond with the numbers on the cylinder head) and install the bolts. Tighten the bolts in three steps, following the recommended sequence **(see illustration)**, to the torque listed in this Chapter's Specifications. While doing this, do not let the camshafts rotate!
14 Now unscrew the bolts, a little at a time, and carefully lift off the bearing caps.
15 To determine the oil clearance, compare the crushed Plastigauge (at its widest point) on each journal to the scale printed on the

9.12 Position a strip of Plastigauge on each cam bearing journal, parallel with the centerline of the camshaft

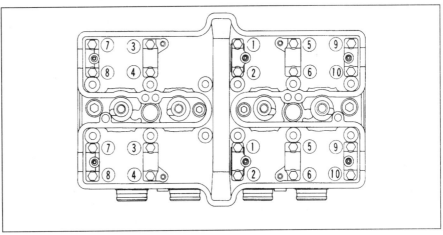

9.13 Camshaft bearing cap tightening sequence

Plastigauge container **(see illustration)**. Compare the results to this Chapter's Specifications. If the oil clearance is greater than specified, measure the diameter of the cam bearing journal with a micrometer **(see illustration)**. If the journal diameter is less than the specified limit, replace the camshaft with a new one and recheck the clearance. If the clearance is still too great, replace the cylinder head and bearing caps with new parts (see the Note that precedes Step 9).

16 Except in cases of oil starvation, the camshaft chain wears very little. If the chain has stretched excessively, which makes it difficult to maintain proper tension, replace it with a new one (see Section 28).

17 Check the sprockets for wear, cracks and other damage, replacing them if necessary. If the sprockets are worn, the chain is also worn, and also the sprocket on the crankshaft (which can only be remedied by replacing the crankshaft). If wear this severe is apparent, the entire engine should be disassembled for inspection.

18 Check the chain guides for wear or damage. If they are worn or damaged, the chain is worn out or improperly adjusted. Replacement of the guides requires removal of the cylinder head and cylinder block.

Installation

19 Make sure the bearing surfaces in the cylinder head and the bearing caps are clean, then apply a light coat of engine assembly lube or moly-based grease to each of them.

20 Apply a coat of moly-based grease to the camshaft lobes. Make sure the camshaft bearing journals are clean, then lay the camshafts in the cylinder head (do not mix them up), ensuring the marks on the cam sprockets are aligned properly **(see illustration 9.11)**.

21 Make sure the timing marks are aligned as described in Step 11, then mesh the chain with the camshaft sprockets. Count the number of chain link pins between the EX mark and the IN mark **(see illustration)**. There should be no slack in the chain between the two sprockets.

22 Carefully set the bearing caps in place (arrows pointing toward the front of the engine

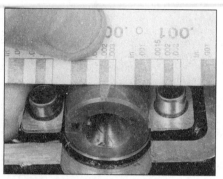

9.15a **Compare the width of the crushed Plastigauge to the scale on the Plastigauge container to obtain the clearance**

9.15b **Measure the cam bearing journal with a micrometer**

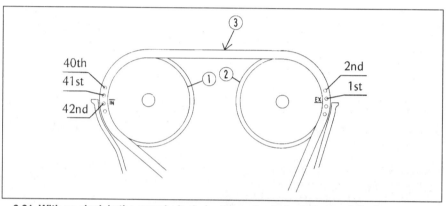

9.21 **With no slack in the camshaft chain, there should be forty-one link pins present between the EX and IN marks on the cam sprockets**

1 Intake sprocket
2 Exhaust sprocket

3 There should be no slack in this area when counting the link pins

and in their proper positions) **(see illustration 9.6)** and install the bolts. Tighten them in three steps, in the recommended sequence **(see illustration 9.13)**, to the torque listed in this Chapter's Specifications.

23 Insert your finger or a wood dowel into the cam chain tensioner hole and apply pressure to the cam chain. Check the timing marks to make sure they are aligned (see Step 11) and there are still the correct number of link pins between the EX and IN marks on the cam

sprockets. If necessary, change the position of the sprocket(s) on the chain to bring all of the marks into alignment.

Caution: If the marks are not aligned exactly as described, the valve timing will be incorrect and the valves may contact the pistons, causing extensive damage to the engine.

24 Remove the cap, spring and pushrod stop from the cam chain tensioner, then install the tensioner as described in Section 8.

25 Adjust the valve clearances (see Chapter 1).

26 The remainder of installation is the reverse of removal.

Rocker arm shafts and rocker arms

Removal

27 Remove the camshafts following the procedure given above. Be sure to keep tension on the camshaft chain.

28 Remove the rocker arm shaft locating pins **(see illustration)**.

29 Remove the rocker shaft plugs **(see illustration)**.

30 Thread a bolt (M8 x 1.8 x 30 mm) into the end of one of the rocker arm shafts and use it

9.28 **Remove the rocker arm shaft locating pin**

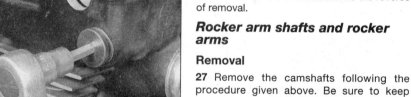

9.29 **Unscrew the rocker arm shaft plug – be sure to check the condition of the O-ring under the head of the plug**

9.30 Thread a bolt of the proper size and thread pitch into the rocker arm shaft and use it as a handle to pull the shaft out

9.31 After the shaft has been removed, remove the rocker arm and retaining spring – note how the springs are nearer to the center of the cylinder head

9.33 Inspect the rocker arms, especially the faces that contact the cam lobes, for wear

as a handle to pull the shaft out **(see illustration)**.
31 Remove the rocker arms and springs **(see illustration)**.
32 Repeat the above Steps to remove the other rocker arm shafts and rocker arms. Keep all of the parts in order so they can be reinstalled in their original locations.

Inspection

33 Clean all of the components with solvent and dry them off. Blow through the oil passages in the rocker arms with compressed air, if available. Inspect the rocker arm faces

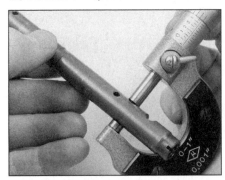

9.34a Measure the rocker arm shafts with a micrometer, in the area where the rocker arms ride

for pits, spalling, score marks and rough spots **(see illustration)**. Check the rocker arm-to-shaft contact areas and the adjusting screws, as well. Look for cracks in each rocker arm. If the faces of the rocker arms are damaged, the rocker arms and the camshafts should be replaced as a set.
34 Measure the diameter of the rocker arm shafts, in the area where the rocker arms ride, and compare the results with this Chapter's Specifications **(see illustration)**. Also measure the inside diameter of the rocker arms **(see illustrations)** and compare the results with this Chapter's Specifications. If either the shaft or the rocker arms are worn beyond the specified limits, replace them as a set.

Installation

35 Position a rocker arm and spring in the cylinder head, with the spring toward the centre of the cylinder head (see illustration 9.31).
36 Lubricate the rocker arm shaft with engine oil and slide it into the cylinder head and through the rocker arm and spring. Make sure the notch is aligned with the locating pin hole **(see illustration)**. Install the locating pin.
37 Check the O-ring under the head of the rocker arm shaft plug, replacing it if

necessary. Install the plug, tightening it to the torque listed in this Chapter's Specifications.
38 Repeat Steps 35, 36 and 37 to install the remaining rocker arms and shafts.
39 Install the camshafts following the procedure described earlier in this Section.

10 Cylinder head -
removal and installation

Caution: The engine must be completely cool before beginning this procedure, or the cylinder head may become warped.

Note: *This procedure can be performed with the engine in the frame. If the engine has been removed, ignore the steps which don't apply.*

Removal

1 Set the bike on its centrestand.
2 Remove the valve cover following the procedure given in Section 7.
3 Remove the radiator (see Chapter 3).
4 Remove the exhaust system (see Chapter 4).
5 Remove the camshafts (see Section 9).

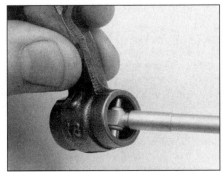

9.34b Measure the inside diameter of the rocker arm – in this case, a telescoping gauge is expanded against the bore of the rocker arm, then locked . . .

9.34c . . . and a micrometer is used to measure the gauge

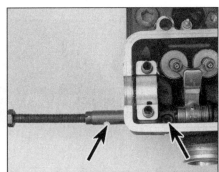

9.36 When installing the rocker arm shafts, align the notch in the shaft with the locating pin hole in the cylinder head (arrows)

10.6 Remove the cylinder block-to-cylinder head bolts (arrows) (rear bolts shown)

10.8 Remove the oil pipe banjo bolts (arrows)

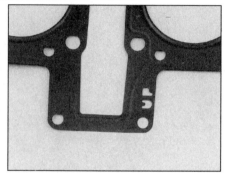

10.14 When installing the cylinder head gasket, make sure the UP mark is situated on the right-hand side of the engine (by the no. 3 cylinder)

6 Remove the cylinder block-to-cylinder head bolts **(see illustration)**. There are two on the rearside of the cylinder head and two on the front side of the cylinder head.

7 Loosen the cylinder head nuts, a little at a time, using the reverse order of the tightening sequence **(see illustration 10.17)**.

8 Remove the oil pipe banjo bolts and washers from the rear of the cylinder head **(see illustration)**.

9 Pull the cylinder head off the cylinder block. If the head is stuck, tap around the side of the head with a rubber mallet to jar it loose, or use two wooden dowels inserted into the intake or exhaust ports to lever the head off. Don't attempt to pry the head off by inserting a screwdriver between the head and the cylinder block - you'll damage the sealing surfaces.

10 Stuff a clean rag into the cam chain tunnel to prevent the entry of debris. Remove all of the washers from their seats, using a pair of needle-nose pliers.

11 Remove the two dowel pins from the cylinder block.

12 Check the cylinder head gasket and the mating surfaces on the cylinder head and block for leakage, which could indicate

warpage. Refer to Section 12 and check the flatness of the cylinder head.

13 Clean all traces of old gasket material from the cylinder head and block. Be careful not to let any of the gasket material fall into the crankcase, the cylinder bores or the water passages.

Installation

14 Install the two dowel pins over their studs, then lay the new gasket in place on the cylinder block. Make sure the UP mark on the gasket is positioned on the right-hand side of the engine **(see illustration)**. Never reuse the old gasket and don't use any type of gasket sealant.

15 Carefully lower the cylinder head over the studs. It is helpful to have an assistant support the camshaft chain with a piece of wire so it doesn't fall and become kinked or detached from the crankshaft. When the head is resting against the cylinder block, wire the cam chain to another component to keep tension on it.

16 Lubricate both sides of the head nut washers with engine oil and place them over the head studs.

17 Install the head nuts. Using the proper sequence **(see illustration)**, tighten the nuts to approximately half the torque listed in this Chapter's Specifications.

18 Using the same sequence, tighten the nuts to the torque listed in this Chapter's Specifications.

19 Install the four cylinder block-to-head bolts, tightening them to the initial torque, then to the final torque listed in this Chapter's Specifications.

20 Install the camshafts and the valve cover (see Sections 9 and 7).

21 Change the engine oil (see Chapter 1).

11 Valves/valve seats/valve guides - servicing

1 Because of the complex nature of this job and the special tools and equipment required, servicing of the valves, the valve seats and the valve guides (commonly known as a valve job) is best left to a professional.

2 The home mechanic can, however, remove and disassemble the head, do the initial cleaning and inspection, then reassemble and deliver the head to a dealer service department or properly equipped motorcycle repair shop for the actual valve servicing. Refer to Section 12 for those procedures.

3 The dealer service department will remove the valves and springs, recondition or replace the valves and valve seats, replace the valve guides, check and replace the valve springs, spring retainers and keepers/collets (as necessary), replace the valve seals with new ones and reassemble the valve components.

4 After the valve job has been performed, the head will be in like-new condition. When the head is returned, be sure to clean it again very thoroughly before installation on the engine to remove any metal particles or abrasive grit that may still be present from the valve service operations. Use compressed air, if available, to blow out all the holes and passages.

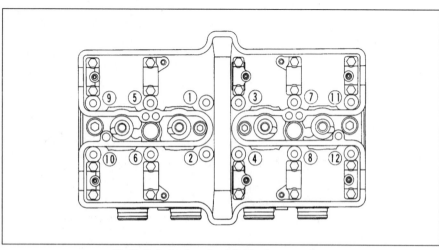

10.17 Cylinder head nut tightening sequence

12 Cylinder head and valves -
disassembly, inspection and
reassembly

1 As mentioned in the previous Section, valve servicing and valve guide replacement should be left to a dealer service department or motorcycle repair shop. However, disassembly, cleaning and inspection of the valves and related components can be done (if the necessary special tools are available) by the home mechanic. This way no expense is incurred if the inspection reveals that service work is not required at this time.

2 To properly disassemble the valve components without the risk of damaging them, a valve spring compressor is absolutely necessary. If the special tool is not available, have a dealer service department or motorcycle repair shop handle the entire process of disassembly, inspection, service or repair (if required) and reassembly of the valves.

Disassembly

3 Remove the rocker arm shafts and rocker arms (see Section 9). Store the components in

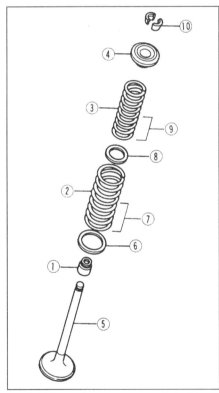

12.7c Valve components – exploded view

1 Oil seal
2 Outer spring
3 Inner spring
4 Valve spring
    retainer
5 Valve
6 Outer spring seat
7 Tightly wound
    coils
8 Inner spring seat
9 Tightly wound
    coils
10 Keepers

12.7a Compressing the valve springs with a valve spring compressor

such a way that they can be returned to their original locations without getting mixed up (labeled plastic bags work well).

4 Before the valves are removed, scrape away any traces of gasket material from the head gasket sealing surface. Work slowly and do not nick or gouge the soft aluminum of the head. Gasket removing solvents, which work very well, are available at most motorcycle shops and auto parts stores.

5 Carefully scrape all carbon deposits out of the combustion chamber area. A hand held wire brush or a piece of fine emery cloth can be used once the majority of deposits have been scraped away. Do not use a wire brush mounted in a drill motor, or one with extremely stiff bristles, as the head material is soft and may be eroded away or scratched by the wire brush.

6 Before proceeding, arrange to label and store the valves along with their related components so they can be kept separate and reinstalled in the same valve guides they are removed from (again, plastic bags work well for this).

7 Compress the valve spring on the first valve with a spring compressor, then remove the keepers/collets (see illustrations) from the valve groove. Do not compress the springs any more than is absolutely necessary. Carefully release the valve spring compressor and remove the retainer, springs and the valve from the head (see illustration). If the valve binds in the guide (won't pull through), push it

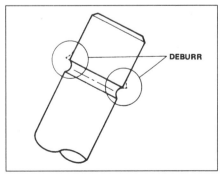

12.7d If the valve binds in the guide, deburr the area above the keeper groove

DEBURR

12.7b Remove the keepers with needle-nose pliers or tweezers

back into the head and deburr the area around the keeper groove with a very fine file or whetstone (see illustration).

8 Repeat the procedure for the remaining valves. Remember to keep the parts for each valve together so they can be reinstalled in the same location.

9 Once the valves have been removed and labeled, pull off the valve stem seals with pliers and discard them (the old seals should never be reused), then remove the spring seats.

10 Next, clean the cylinder head with solvent and dry it thoroughly. Compressed air will speed the drying process and ensure that all holes and recessed areas are clean.

11 Clean all of the valve springs, keepers/collets, retainers and spring seats with solvent and dry them thoroughly. Do the parts from one valve at a time so that no mixing of parts between valves occurs.

12 Scrape off any deposits that may have formed on the valve, then use a motorised wire brush to remove deposits from the valve heads and stems. Again, make sure the valves do not get mixed up.

Inspection

13 Inspect the head very carefully for cracks and other damage. If cracks are found, a new head will be required. Check the cam bearing surfaces for wear and evidence of seizure. Check the camshafts and rocker arms for wear as well (see Section 9).

14 Using a precision straightedge and a feeler gauge, check the head gasket mating surface for warpage. Lay the straightedge lengthwise, across the head and diagonally (corner-to-corner). intersecting the head bolt holes, and try to slip a 0.05 mm (0.002 in) feeler gauge under it, on either side of each combustion chamber (see illustration). If the feeler gauge can be inserted between the head and the straightedge, the head is warped and must either be machined or, if warpage is excessive, replaced with a new one.

15 Examine the valve seats in each of the combustion chambers. If they are pitted, cracked or burned, the head will require valve service that is beyond the scope of the home

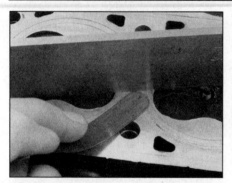

12.14 Lay a precision straightedge across the cylinder head and try to slide a feeler gauge of the specified thickness (equal to the maximum allowable warpage) under it

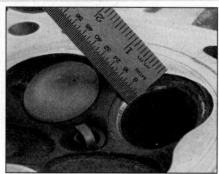

12.15 Measuring the valve seat width

12.16a Insert a small hole gauge into the valve guide and expand it so there is a slight drag when it's pulled out

mechanic. Measure the valve seat width **(see illustration)** and compare it to this Chapter's Specifications. If it is not within the specified range, or if it varies around its circumference, valve service work is required.

16 Clean the valve guides to remove any carbon buildup, then measure the inside diameters of the guides (at both ends and the centre of the guide) with a small hole gauge and a 0-to-25-mm micrometer **(see illustrations)**. Record the measurements for future reference. These measurements, along with the valve stem diameter measurements, will enable you to compute the valve stem-to-guide clearance. This clearance, when compared to the Specifications, will be one factor that will determine the extent of the

valve service work required. The guides are measured at the ends and at the centre to determine if they are worn in a bell-mouth pattern (more wear at the ends). If they are, guide replacement is an absolute must.

17 Carefully inspect each valve face for cracks, pits and burned spots. Check the valve stem and the keeper groove area for cracks **(see illustration)**. Rotate the valve and check for any obvious indication that it is bent. Check the end of the stem for pitting and excessive wear and make sure the bevel is the specified width. The presence of any of the above conditions indicates the need for valve servicing.

18 Measure the valve stem diameter **(see illustration)**. By subtracting the stem diameter from the valve guide diameter, the

valve stem-to-guide clearance is obtained. If the stem-to-guide clearance is greater than listed in this Chapter's Specifications, the guides and valves will have to be replaced with new ones. Also check the valve stem for bending. Set the valve in a V-block with a dial indicator touching the middle of the stem **(see illustration)**. Rotate the valve and note the reading on the gauge. If the stem runout exceeds the value listed in this Chapter's Specifications, replace the valve.

19 Check the end of each valve spring for wear and pitting. Measure the free length **(see illustration)** and compare it to this Chapter's Specifications. Any springs that are shorter than specified have sagged and should not be reused. Stand the spring on a flat surface and check it for squareness **(see illustration)**.

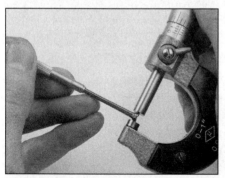

12.16b Measure the small hole gauge with a micrometer

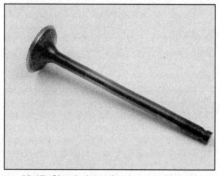

12.17 Check the valve face, stem and keeper groove for signs of wear and damage

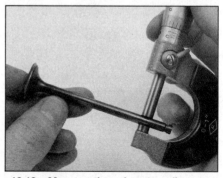

12.18a Measure the valve stem diameter with a micrometer

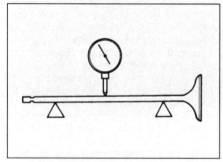

12.18b Check the valve stem for bends with a V-block (or blocks, as shown here) and a dial indicator

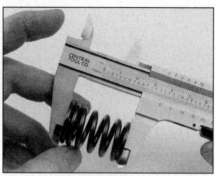

12.19a Measure the free length of the valve springs

12.19b Check the valve springs for squareness

12.23 Apply the lapping compound very sparingly, in small dabs, to the valve face only

12.24a A hose, pushed over the end of the valve, can be used to turn the valve back and forth

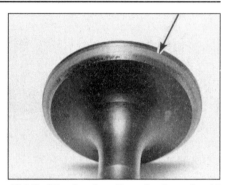

12.24b After lapping, the valve face should exhibit a uniform, unbroken contact pattern (arrow) . . .

20 Check the spring retainers and keepers/collets for obvious wear and cracks. Any questionable parts should not be reused, as extensive damage will occur in the event of failure during engine operation.

21 If the inspection indicates that no service work is required, the valve components can be reinstalled in the head.

Reassembly

22 Before installing the valves in the head, they should be lapped to ensure a positive seal between the valves and seats. This procedure requires fine valve lapping compound (available at auto parts stores) and a valve lapping tool. If a lapping tool is not available, a piece of rubber or plastic hose can be slipped over the valve stem (after the valve has been installed in the guide) and used to turn the valve.

23 Apply a small amount of fine lapping compound to the valve face **(see illustration)**, then slip the valve into the guide. **Note.** *Make sure the valve is installed in the correct guide and be careful not to get any lapping compound on the valve stem.*

24 Attach the lapping tool (or hose) to the valve and rotate the tool between the palms of your hands. Use a back-and-forth motion rather than a circular motion **(see illustration)**. Lift the valve off the seat and turn it at regular intervals to distribute the lapping compound properly. Continue the lapping procedure until the valve face and seat contact area is of uniform width and unbroken around the entire circumference of the valve face and seat **(see illustrations)**.

25 Carefully remove the valve from the guide and wipe off all traces of lapping compound. Use solvent to clean the valve and wipe the seat area thoroughly with a solvent soaked cloth. Repeat the procedure for the remaining valves.

26 Lay the spring seats in place in the cylinder head, then install new valve stem seals on each of the guides. Use an appropriate size deep socket to push the seals into place until they are properly seated. Don't twist or cock them, or they will not seal properly against the valve stems. Also, don't remove them again or they will be damaged.

27 Coat the valve stems with assembly lube or moly-based grease, then install one of them into its guide. Next, install the springs and retainers, compress the springs and install the keepers/collets. **Note:** *Install the springs with the tightly wound coils at the bottom (next to the spring seat). When compressing the springs with the valve spring compressor, depress them only as far as is absolutely necessary to slip the keepers/collets into place. Apply a small amount of grease to the keepers/collets* **(see illustration)** *to help hold them in place as the pressure is released from the springs. Make*

certain that the keepers/collets are securely locked in their retaining grooves.

28 Support the cylinder head on blocks so the valves can't contact the workbench top, then very gently tap each of the valve stems with a soft-faced hammer. This will help seat the keepers/collets in their grooves.

29 Once all of the valves have been installed in the head, check for proper valve sealing by pouring a small amount of solvent into each of the valve ports. If the solvent leaks past the valve(s) into the combustion chamber area, disassemble the valve(s) and repeat the lapping procedure, then reinstall the valve(s) and repeat the check. Repeat the procedure until a satisfactory seal is obtained.

13 Cylinder block - removal, inspection and installation

Removal

1 Following the procedure given in Section 10, remove the cylinder head. Make sure the crankshaft is positioned at Top Dead Centre (TDC) for cylinders 1 and 4.

2 Remove the water pipe from the rear of the cylinder head (see Chapter 3).

3 Remove the oil pipe mounting bolt and the cylinder block-to-crankcase nuts **(see illustration)**.

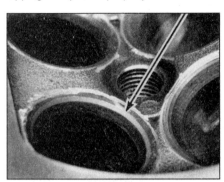

12.24c . . . and the seat should be the specified width (arrow), with a smooth, unbroken appearance

12.27 A small dab of grease will help hold the keepers in place on the valve while the spring is released

13.3 Remove the oil pipe bolt and the cylinder block-to-crankcase nuts (arrows)

13.4 Lift the cam chain front guide out

13.6 Remove the dowel pins from the crankcase

13.8 Measure the cylinder bore with a telescoping gauge (then measure the gauge with a micrometer)

4 Lift out the camshaft chain front guide **(see illustration)**.

5 Lift the cylinder block straight up to remove it. If it's stuck, tap around its perimeter with a soft-faced hammer. Don't attempt to pry between the block and the crankcase, as you will ruin the sealing surfaces.

6 Remove the dowel pins from the mating surface of the crankcase **(see illustration)**. Be careful not to let these drop into the engine. Stuff rags around the pistons and remove the gasket and all traces of old gasket material from the surfaces of the cylinder block and the cylinder head.

Inspection

Caution: Don't attempt to separate the liners from the cylinder block.

7 Check the cylinder walls carefully for scratches and score marks.

8 Using the appropriate precision measuring tools, check each cylinder's diameter near the top, centre and bottom of the cylinder bore, parallel to the crankshaft axis **(see illustration)**. Next, measure each cylinder's diameter at the same three locations across the crankshaft axis. Compare the results to this Chapter's Specifications. If the cylinder walls are tapered, out-of-round, worn beyond

the specified limits, or badly scuffed or scored, have them rebored and honed by a dealer service department or a motorcycle repair shop. If a rebore is done, oversize pistons and rings will be required as well. **Note:** *Kawasaki supplies pistons in one oversize only, +0.5 mm (+0.020 in).*

9 As an alternative, if the precision measuring tools are not available, a dealer service department or motorcycle repair shop will make the measurements and offer advice concerning servicing of the cylinders.

10 If they are in reasonably good condition and not worn to the outside of the limits, and if the piston-to-cylinder clearances can be maintained properly (see Section 14), then the cylinders do not have to be rebored; honing is all that is necessary.

11 To perform the honing operation you will need the proper size flexible hone with fine stones, or a "bottle brush" type hone, plenty of light oil or honing oil, some shop towels and an electric drill motor. Hold the cylinder block in a vise (cushioned with soft jaws or wood blocks) when performing the honing operation. Mount the hone in the drill motor, compress the stones and slip the hone into the cylinder. Lubricate the cylinder thoroughly, turn on the drill and move the

hone up and down in the cylinder at a pace which will produce a fine crosshatch pattern on the cylinder wall with the crosshatch lines intersecting at approximately a 60° angle. Be sure to use plenty of lubricant and do not take off any more material than is absolutely necessary to produce the desired effect. Do not withdraw the hone from the cylinder while it is running. Instead, shut off the drill and continue moving the hone up and down in the cylinder until it comes to a complete stop, then compress the stones and withdraw the hone. Wipe the oil out of the cylinder and repeat the procedure on the remaining cylinder. Remember, do not remove too much material from the cylinder wall. If you do not have the tools, or do not desire to perform the honing operation, a dealer service department or motorcycle repair shop will generally do it for a reasonable fee.

12 Next, the cylinders must be thoroughly washed with warm soapy water to remove all traces of the abrasive grit produced during the honing operation. Be sure to run a brush through the bolt holes and flush them with running water. After rinsing, dry the cylinders thoroughly and apply a coat of light, rust-preventative oil to all machined surfaces.

Installation

13 Lubricate the cylinder bores with plenty of clean engine oil. Apply a thin film of moly-based grease to the piston skirts.

14 Install the dowel pins, then lower a new cylinder base gasket over the studs, with the UP mark on the right-hand side of the engine. Some gaskets also have an arrow, which must point to the front of the engine **(see illustration)**.

15 Slowly rotate the crankshaft until all of the pistons are at the same level. Slide lengths of welding rod or pieces of a straightened-out coat hanger under the pistons, on both sides of the connecting rods **(see illustration)**. This will help keep the pistons level as the cylinder block is lowered onto them.

13.14 The cylinder base gasket must be installed with the UP mark on the right and the arrow pointing to the front of the engine

13.15 Slide rods under the pistons to support them as the cylinder block is lowered down

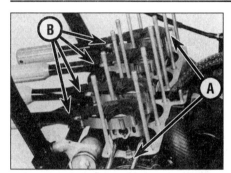

13.16 Compress the piston rings with piston ring compressors (shown here) or large hose clamps

A *Piston base (57001 – 147)*
B *Piston ring compressor assembly (57001 – 1094)*

16 Attach four piston ring compressors to the pistons and compress the piston rings **(see illustration)**. Large hose clamps can be used instead - just make sure they don't scratch the pistons, and don't tighten them too much.
17 Install the cylinder block over the studs and carefully lower it down until the piston crowns fit into the cylinder liners. While doing this, pull the camshaft chain up, using a hooked tool or a piece of coat hanger. Push down on the cylinder block, making sure the pistons don't get cocked sideways, until the bottom of the cylinder liners slide down past

14.3b Use needle-nose pliers and remove the piston pin circlips (note the rag used to keep foreign objects out of the crankcase)

14.3c If the piston pins won't come out, a pin removal tool will have to be used – if you don't have access to one of these, one can be fabricated from a threaded rod, nuts, washers and a piece of pipe

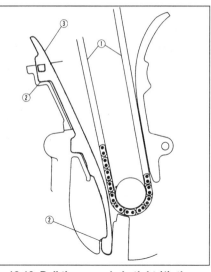

13.19 Pull the cam chain tight (1), then install the chain guide (3), making sure it fits correctly at each end (2)

the piston rings. A wood or plastic hammer handle can be used to gently tap the block down, but don't use too much force or the pistons will be damaged.
18 Remove the piston ring compressors or hose clamps, being careful not to scratch the pistons. Remove the rods from under the pistons.
19 Install the cam chain front guide **(see illustration)**.
20 Install the cylinder head and tighten the nuts (see Section 10).
21 Tighten the cylinder block-to-crankcase nuts to the torque listed in this Chapter's Specifications.
22 The remainder of installation is the reverse of removal.

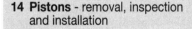

14 Pistons - removal, inspection and installation

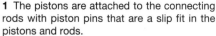

1 The pistons are attached to the connecting rods with piston pins that are a slip fit in the pistons and rods.
2 Before removing the pistons from the rods,

14.4 Use large rubber bands to keep the connecting rods from flopping around after the pistons are removed

14.3a Using a sharp scribe, scratch the cylinder numbers into the piston crowns – also note the arrow, which must point to the front

stuff a clean shop towel into each crankcase hole, around the connecting rods. This will prevent the circlips from falling into the crankcase if they are inadvertently dropped.

Removal

3 Using a sharp scribe, scratch the number of each piston into its crown **(see illustration)**. Each piston should also have an arrow pointing toward the front of the engine. If not, scribe an arrow into the piston crown before removal. Support the first piston, grasp the circlip with needle-nose pliers and remove it from the groove **(see illustration)**. If the pin won't come out, use a special piston pin removal tool (Kawasaki tool no. 57001-910) **(see illustration)**.
4 Push the piston pin out from the opposite end to free the piston from the rod. You may have to deburr the area around the groove to enable the pin to slide out (use a triangular file for this procedure). Repeat the procedure for the remaining pistons. Use large rubber bands to support the connecting rods **(see illustration)**.

Inspection

5 Before the inspection process can be carried out, the pistons must be cleaned and the old piston rings removed.
6 Using a piston ring installation tool, carefully remove the rings from the pistons **(see illustration)**. Do not nick or gouge the pistons in the process.

14.6 Remove the piston rings with a ring removal and installation tool

14.11 Check the piston pin bore and the piston skirt for wear, and make sure the oil holes are clear (arrows)

14.13 Measure the piston ring-to-groove clearance with a feeler gauge

14.14 Measure the piston diameter with a micrometer

7 Scrape all traces of carbon from the tops of the pistons. A hand-held wire brush or a piece of fine emery cloth can be used once the majority of the deposits have been scraped away. Do not, under any circumstances, use a wire brush mounted in a drill motor to remove deposits from the pistons; the piston material is soft and will be eroded away by the wire brush.

8 Use a piston ring groove cleaning tool to remove any carbon deposits from the ring grooves. If a tool is not available, a piece broken off the old ring will do the job. Be very careful to remove only the carbon deposits. Do not remove any metal and do not nick or gouge the sides of the ring grooves.

9 Once the deposits have been removed, clean the pistons with solvent and dry them thoroughly. Make sure the oil return holes below the oil ring grooves are clear.

10 If the pistons are not damaged or worn excessively and if the cylinders are not rebored, new pistons will not be necessary. Normal piston wear appears as even, vertical wear on the thrust surfaces of the piston and slight looseness of the top ring in its groove. New piston rings, on the other hand, should always be used when an engine is rebuilt.

11 Carefully inspect each piston for cracks around the skirt, at the pin bosses and at the ring lands **(see illustration)**.

12 Look for scoring and scuffing on the thrust faces of the skirt, holes in the piston crown and burned areas at the edge of the crown. If the skirt is scored or scuffed, the engine may have been suffering from overheating and/or abnormal combustion, which caused excessively high operating temperatures. The oil pump and cooling system should be checked thoroughly. A hole in the piston crown, an extreme to be sure, is an indication that abnormal combustion (pre-ignition) was occurring. Burned areas at the edge of the piston crown are usually evidence of spark knock (detonation). If any of the above problems exist, the causes must be corrected or the damage will occur again.

13 Measure the piston ring-to-groove clearance by laying a new piston ring in the ring groove and slipping a feeler gauge in beside it **(see illustration)**. Check the clearance at three or four locations around the

groove. Be sure to use the correct ring for each groove; they are different. If the clearance is greater then specified, new pistons will have to be used when the engine is reassembled.

14 Check the piston-to-bore clearance by measuring the bore (see Section 13) and the piston diameter. Make sure that the pistons and cylinders are correctly matched. Measure the piston across the skirt on the thrust faces at a 90° angle to the piston pin, about 1/2-inch (13 mm) up from the bottom of the skirt **(see illustration)**. Subtract the piston diameter from the bore diameter to obtain the clearance. If it is greater than specified, the cylinders will have to be rebored and new oversized pistons and rings installed.

15 If the appropriate precision measuring tools are not available, the piston-to-cylinder clearances can be obtained, though not quite as accurately, using feeler gauge stock. Feeler gauge stock comes in 12-inch lengths and various thicknesses and is generally available at auto parts stores. To check the clearance, select a 0.07 mm (0.002 in) feeler gauge and slip it into the cylinder along with the appropriate piston. The cylinder should be upside down and the piston must be positioned exactly as it normally would be. Place the feeler gauge between the piston and cylinder on one of the thrust faces (90° to the piston pin bore). The piston should slip through the cylinder (with the feeler gauge in place) with moderate pressure. If it falls through, or slides through easily, the clearance is excessive and a new piston will be required. If the piston binds at the lower end of the cylinder and is loose toward the top, the cylinder is tapered, and if tight spots are encountered as the piston/feeler gauge is rotated in the cylinder, the cylinder is out-of-round. Repeat the procedure for the remaining pistons and cylinders. Be sure to have the cylinders and pistons checked by a dealer service department or a motorcycle repair shop to confirm your findings before purchasing new parts.

16 Apply clean engine oil to the pin, insert it into the piston and check for freeplay by rocking the pin back-and-forth. If the pin is loose, new pistons and possibly new pins must be installed.

17 Refer to Section 15 and install the rings on the pistons.

Installation

Note: *When installing the pistons, install the pistons for cylinders 2 and 3 first.*

18 Install the pistons in their original locations with the arrows pointing to the front of the engine. Lubricate the pins and the rod bores with clean engine oil. Install new circlips in the grooves in the inner sides of the pistons (don't reuse the old circlips). Push the pins into position from the opposite side and install new circlips. Compress the circlips only enough for them to fit in the piston. Make sure the clips are properly seated in the grooves.

15 Piston rings - installation

1 Before installing the new piston rings, the ring end gaps must be checked.

2 Lay out the pistons and the new ring sets so the rings will be matched with the same piston and cylinder during the end gap measurement procedure and engine assembly.

3 Insert the top (no. 1) ring into the bottom of the first cylinder and square it up with the cylinder walls by pushing it in with the top of the piston. The ring should be about one inch above the bottom edge of the cylinder. To measure the end gap, slip a feeler gauge between the ends of the ring **(see illustration)**

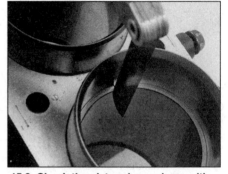

15.3 Check the piston ring end gap with a feeler gauge

15.5 If the end gap is too small, clamp a file in a vise and file the ring ends (from the outside in only) to enlarge the gap slightly

15.9a Installing the oil ring expander – make sure the ends don't overlap

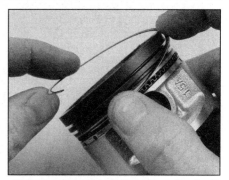

15.9b Installing an oil ring side rail – don't use a ring installation tool to do this

and compare the measurement to the Specifications.

4 If the gap is larger or smaller than specified, double check to make sure that you have the correct rings before proceeding.

5 If the gap is too small, it must be enlarged or the ring ends may come in contact with each other during engine operation, which can cause serious damage. The end gap can be increased by filing the ring ends very carefully with a fine file **(see illustration)**. When performing this operation, file only from the outside in.

6 Excess end gap is not critical unless it is greater than 1 mm (0.040 in). Again, double check to make sure you have the correct rings for your engine.

7 Repeat the procedure for each ring that will be installed in the first cylinder and for each ring in the remaining cylinder. Remember to keep the rings, pistons and cylinders matched up.

8 Once the ring end gaps have been checked/corrected, the rings can be installed on the pistons.

9 The oil control ring (lowest on the piston) is installed first. It is composed of three separate components. Slip the expander into the groove, then install the upper side rail **(see illustrations)**. Do not use a piston ring installation tool on the oil ring side rails as

they may be damaged. Instead, place one end of the side rail into the groove between the spacer expander and the ring land. Hold it firmly in place and slide a finger around the piston while pushing the rail into the groove. Next, install the lower side rail in the same manner.

10 After the three oil ring components have been installed, check to make sure that both the upper and lower side rails can be turned smoothly in the ring groove.

11 Install the no. 2 (middle) ring next. It can be readily distinguished from the top ring by its cross-section shape and the RN mark on it **(see illustration)**. Do not mix the top and middle rings.

12 To avoid breaking the ring, use a piston ring installation tool and make sure that the identification mark is facing up **(see illustration)**. Fit the ring into the middle groove on the piston. Do not expand the ring any more than is necessary to slide it into place.

13 Finally, install the no. 1 (top) ring in the same manner. Make sure the identifying mark is facing up.

14 Repeat the procedure for the remaining piston and rings. Be very careful not to confuse the no. 1 and no. 2 rings.

15 Once the rings have been properly installed, stagger the end gaps, including those of the oil ring side rails **(see illustration)**.

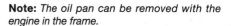

16 Oil pan -
removal and installation

Note: *The oil pan can be removed with the engine in the frame.*

Removal

1 Set the bike on its centrestand.
2 Drain the engine oil and remove the oil filter (see Chapter 1).
3 Remove the exhaust system (see Chapter 4).
4 Remove the banjo bolts that attach the oil cooler lines to the oil pan (see Chapter 3).

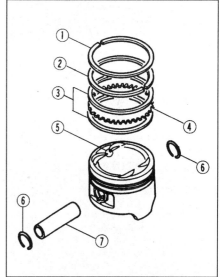

15.15 Piston and piston ring details – when installing the oil ring side rails, stagger them approximately 30 to 40-degrees on either side of the top compression ring

1 *Top compression ring*
2 *Second compression ring*
3 *Oil ring side rails*
4 *Oil ring expander*
5 *Arrow mark*
6 *Circlip*
7 *Piston pin*

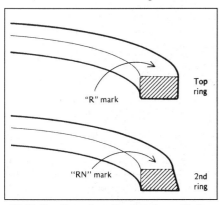

15.11 Don't confuse the top ring with the middle ring

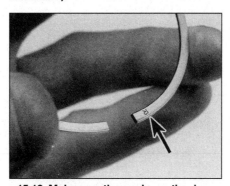

15.12 Make sure the marks on the rings (arrow) face up when the rings are installed on the pistons

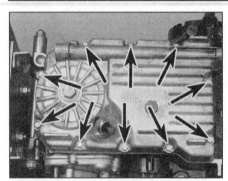

16.6 Loosen the oil pan bolts, a little at a time, in a criss-cross pattern

16.8 Check the seals (arrows) in the crankcase for deterioration

17.2 To check the oil pressure, remove the plug and connect an oil pressure gauge using the proper adapter

1 Oil pressure gauge   3 Plug
2 Adapter

5 Remove the small screw and disconnect the wire from the oil pressure switch (see Chapter 9).

6 Remove the oil pan bolts and detach the pan from the crankcase (see illustration).

7 Remove all traces of old gasket material from the mating surfaces of the oil pan and crankcase.

Installation

8 Check the small O-rings in the oil passages in the crankcase and the large O-ring around the oil filter hole (in the pan) for cracking and general deterioration (see illustration). Replace them if necessary. The flat side of the O-rings must face the crankcase.

9 Position a new gasket on the oil pan. A thin film of RTV sealant can be used to hold the gasket in place. Install the oil pan and bolts, tightening the bolts to the torque listed in this Chapter's Specifications, using a criss-cross pattern.

10 The remainder of installation is the reverse of removal. Install a new filter and fill the crankcase with oil (see Chapter 1 and Daily (pre-ride) checks), then run the engine and check for leaks.

17.9 Remove the bearing stop screws and detach the stop, then remove the oil pump bolt

1 Oil pump
2 Bolt
3 Bearing stop
4 Secondary shaft gear
5 Bearing stop screws

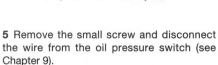

17 Oil pump - pressure check, removal, inspection and installation

Note: *The oil pump can be removed with the engine in the frame.*

Check

⚠️ **Warning: If the oil passage plug is removed when the engine is hot, hot oil will drain out - wait until the engine is cold before beginning this check (it must be cold to perform the relief valve opening pressure check, anyway).**

1 If you're working on an A or B model, remove the lower fairing (see Chapter 8).

2 Remove the plug at the bottom of the crankcase on the right-hand side and install an oil pressure gauge (see illustration).

3 Start the engine and watch the gauge while varying the engine rpm. The pressure should stay within the relief valve opening pressure listed in this Chapter's Specifications. If the pressure is too high, the relief valve is stuck closed. To check it, see Section 18.

4 If the pressure is lower than the standard, either the relief valve is stuck open, the oil pump is faulty, or there is other engine damage. Begin diagnosis by checking the relief valve (see Section 18), then the oil pump. If those items check out okay, chances

are the bearing oil clearances are excessive and the engine needs to be overhauled.

5 If the pressure reading is in the desired range, allow the engine to warm up to normal operating temperature and check the pressure again, at the specified engine rpm. Compare your findings with this Chapter's Specifications.

6 If the pressure is significantly lower than specified, check the relief valve and the oil pump.

Removal

7 Remove the oil pan (see Section 16).

8 Remove the clutch assembly (see Section 19).

9 Remove the mounting bolt and screws, then detach the bearing stop (see illustration).

10 Slide the oil pump toward the crankcase and remove it.

Inspection

11 Remove the snap-ring from the pump shaft (see illustration) and lift off the washer.

12 Push the shaft into the pump gear just far enough to remove the pin, then detach the gear from the shaft (see illustration).

13 Remove the oil pump cover screws and lift off the cover. Clean all traces of gasket material from the mating surfaces.

17.11 Remove the snap-ring with snap-ring pliers

17.12 Push the shaft in and remove the pin

17.17 Stake the screws to prevent them from loosening

1 Gear
2 Shaft
3 Pins
4 Screen
5 Oil pump body
6 Dowel pin
7 Outer rotor
8 Inner rotor
9 Gasket
10 Cover
11 Dowel pin
12 Washer
13 Snap-ring

17.14 Exploded view of the oil pump

14 Remove the oil pump shaft, pin, inner rotor and outer rotor from the pump **(see illustration)**. Mark the rotors so they can be installed in the same relative positions.
15 Wash all the components in solvent, then dry them off. Check the pump body, the rotors and the cover for scoring and wear. Make sure the pick-up screen isn't clogged. Kawasaki doesn't publish clearance specifications, so if any damage or uneven or excessive wear is evident, replace the pump. If you are rebuilding the engine, it's a good idea to install a new oil pump.
16 Reassemble the pump by reversing the removal steps, but before installing the cover, pack the cavities between the rotors with

petroleum jelly - this will ensure the pump develops suction quickly and begins oil circulation as soon as the engine is started. Be sure to use a new gasket.

Installation

17 Installation is the reverse of removal. Make sure the two dowel pins are in place before installing the bearing stop and screws. After tightening the mounting screws, stake them with a hammer and punch **(see illustration)**.

18 Oil pressure relief valve - removal, inspection and installation

Removal

1 Remove the oil pan (see Section 16).
2 Unscrew the relief valve from the crankcase **(see illustration)**.

Inspection

3 Clean the valve with solvent and dry it, using compressed air if available.
4 Using a wood or plastic tool, depress the steel ball inside the valve and see if it moves smoothly. Make sure it returns to its seat completely. If it doesn't, replace it with a new one (don't attempt to disassemble and repair it).

Installation

5 Apply a non-hardening thread locking compound to the threads of the valve and install it into the case, tightening it to the torque listed in this Chapter's Specifications.
6 The remainder of installation is the reverse of removal.

19 Clutch - removal, inspection and installation

Note: *The clutch can be removed with the engine in the frame.*

Removal

1 Set the bike on its centrestand. Remove the lower fairing (see Chapter 8).
2 Drain the engine oil (see Chapter 1).
3 Completely loosen the rear adjustment nut on the clutch cable at its bracket on the clutch cover. Pull the cable out of the bracket, then detach the cable end from the lever.
4 Remove the clutch cover bolts and take the cover off **(see illustration)**. If the cover is stuck, tap around its perimeter with a soft-face hammer.
5 Remove the clutch spring bolts **(see illustration)**. To prevent the assembly from

18.2 Location of the oil pressure relief valve (arrow)

19.4 Remove the clutch cover bolts (arrows)

19.5a Loosen the clutch spring bolts a little at a time, in a criss-cross pattern – a screwdriver wedged between a bolt and the clutch housing will prevent it from turning

19.5b Remove the spring plate and pushrod

19.6 Remove the clutch friction and steel plates

19.7a The clutch hub must be prevented from rotating so the hub nut can be removed – in this photo, the special Kawasaki tool is shown

1 Footpeg (the tool rests against it)
2 Clutch hub holding tool (no. 57001-305, or equivalent)
3 Hub nut
4 Clutch hub

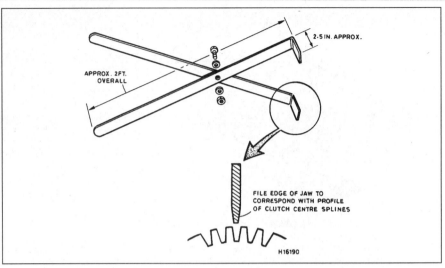

19.7b You can make your own clutch holding tool out of steel strap

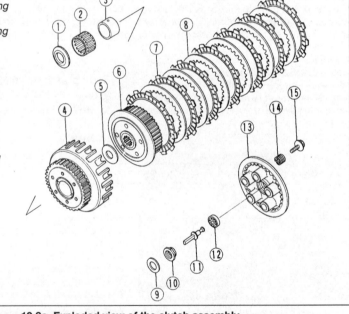

1 Spacer
2 Needle bearing
3 Collar
4 Clutch housing
5 Washer
6 Clutch hub
7 Friction plate
8 Steel plate
9 Washer
10 Hub nut
11 Pushrod
12 Bearing
13 Spring plate
14 Clutch spring
15 Bolt

19.8a Exploded view of the clutch assembly

turning, thread one of the cover mounting bolts into the case and wedge a screwdriver between the bolt and the clutch housing. Remove the clutch spring plate, bearing and pushrod **(see illustration)**.

6 Remove the clutch friction and steel plates from the clutch housing **(see illustration)**.

7 Remove the clutch hub nut, using a special holding tool (Kawasaki tool no. 57001-305 or 1243) to prevent the clutch housing from turning **(see illustration)**. An alternative to this tool can be fabricated from some steel strap, bent at the ends and bolted together in the middle **(see illustration)**.

8 Remove the thrust washer, clutch hub, washer, clutch housing, needle bearing, collar and spacer **(see illustrations)**.

Inspection

9 Examine the splines on both the inside and the outside of the clutch hub **(see illustration)**. If any wear is evident, replace the hub with a new one.

19.8b Remove the collar, bearing and spacer from the mainshaft

19.9 Check the clutch hub splines (arrows) for wear and distortion

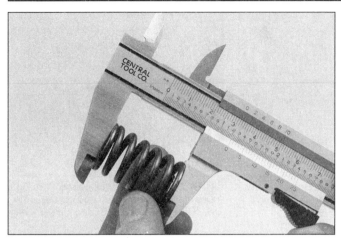

19.10 Measure the clutch spring free length

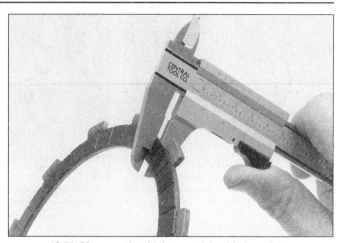

19.11 Measure the thickness of the friction plates

10 Measure the free length of the clutch springs **(see illustration)** and compare the results to this Chapter's Specifications. If the springs have sagged, or if cracks are noted, replace them with new ones as a set.

11 If the lining material of the friction plates smells burnt or if it is glazed, new parts are required. If the metal clutch plates are scored or discolored, they must be replaced with new ones. Measure the thickness of each friction plate **(see illustration)** and compare the results to this Chapter's Specifications. Replace with new parts any friction plates that are near the wear limit.

12 Lay the metal plates, one at a time, on a perfectly flat surface (such as a piece of plate glass) and check for warpage by trying to slip a 0.3 mm (0.019 in) feeler gauge between the flat surface and the plate **(see illustration)**. Do this at several places around the plate's circumference. If the feeler gauge can be slipped under the plate, it is warped and should be replaced with a new one.

13 Check the tabs on the friction plates for excessive wear and mushroomed edges.

They can be cleaned up with a file if the deformation is not severe.

14 Check the edges of the slots in the clutch housing for indentations made by the friction plate tabs **(see illustration)**. If the indentations are deep they can prevent clutch release, so the housing should be replaced with a new one. If the indentations can be removed easily with a file, the life of the housing can be prolonged to an extent. Also, check the primary gear teeth for cracks, chips and excessive wear. If the gear is worn or damaged, the clutch housing must be replaced with a new one. Check the bearing for score marks, scratches and excessive wear.

15 Check the bearing journal on the transmission mainshaft for score marks, heat discoloration and evidence of excessive wear. Check the clutch spring plate for wear and damage and make sure the pushrod is not bent (roll it on a perfectly flat surface or use V-blocks and a dial indicator).

16 Clean all traces of old gasket material from the clutch cover. If the release shaft seal

has been leaking, it can be replaced by removing the positioning bolt on the outside of the housing and pulling out the shaft. The seal can then be pried out and a new one driven in, using a hammer and a socket with an outside diameter slightly smaller than that of the seal.

Installation

17 Install the spacer over the transmission mainshaft, with the chamfered side facing in.

18 Lubricate the collar and the needle bearing with engine oil and slide them over the mainshaft.

19 Install the clutch housing, washer and the clutch hub. Install a new hub nut and tighten it to the torque listed in this Chapter's Specifications. Use the technique described in Step 7 to prevent the hub from turning.

20 Coat the clutch friction plates with engine oil. Install the clutch plates, starting with a friction plate and alternating them. There are seven friction plates and six steel plates.

21 Lubricate the pushrod and install it through the spring plate. Mount the spring

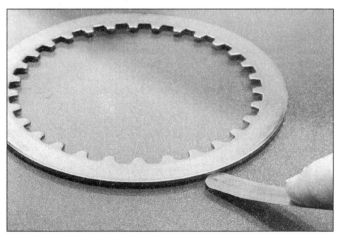

19.12 Check the metal plates for warpage

19.14 Check the slots in the clutch housing for indentations (minor damage can be removed with a file) and check the bearing for wear (arrows)

21.3 Remove the shift mechanism screws (arrows)

21.4a Spread the shift mechanism arm and the overshift limiter apart, then remove the mechanism

21.4b Be sure not to pull on the shift rod (arrow) – if you do, you'll have to separate the crankcase halves to reinstall the shift forks!

plate to the clutch assembly and install the springs and bolts, tightening them to the torque listed in this Chapter's Specifications in a criss-cross pattern.

22 Install the clutch cover and bolts, using a new gasket. Tighten the bolts, in a criss-cross pattern, to the torque listed in this Chapter's Specifications.

23 Connect the clutch cable to the release lever and adjust the freeplay (see Chapter 1).

24 Fill the crankcase with the recommended type and amount of engine oil (see Daily (pre-ride) checks).

20 Clutch cable - replacement

1 Disconnect the upper end of the clutch cable from the lever (**see Chapter 1, illustration 12.3a**).

2 Fully loosen the rear adjusting nut at the lower end of the clutch cable (see Chapter 1, Section 8 if necessary). Pull the cable through the bracket on the clutch cover, then detach the cable end from the release lever.

3 Before removing the cable from the bike, tape the lower end of the new cable to the upper end of the old cable. Slowly pull the

lower end of the old cable out, guiding the new cable down into position. Using this method will ensure the cable is routed correctly.

4 Lubricate the cable (see Chapter 1, Section 12). Reconnect the ends of the cable by reversing the removal procedure, then adjust the cable following the procedure given in Chapter 1, Section 8.

21 External shift mechanism - removal, inspection and installation

Removal

1 Set the bike on its centrestand.

2 Remove the shift lever, engine sprocket cover and the engine sprocket (see Chapter 6).

3 Position a drain pan under the shift mechanism cover. Remove the screws (**see illustration**) and detach the cover from the crankcase.

4 Spread the shift mechanism arm and overshift limiter to clear the shift drum, then pull the mechanism and shaft off (**see illustration**). Note: *Don't pull the shift rod out of the crankcase - the shift forks will fall into the oil pan, and the crankcase will have to be separated to reinstall them (**see illustration**).*

Inspection

5 Check the shift shaft for bends and damage to the splines (**see illustration**). If the shaft is bent, you can attempt to straighten it, but if the splines are damaged it will have to be replaced.

6 Check the condition of the return spring and the pawl spring. Replace them if they are cracked or distorted.

7 Check the shift mechanism arm and the overshift limiter for cracks, distortion and wear. If any of these conditions are found, replace the shift mechanism.

8 Make sure the return spring pin isn't loose. If it is, unscrew it, apply a non-hardening locking compound to the threads, then reinstall it and tighten it securely.

9 Check the condition of the seals in the cover. If they have been leaking, drive them out with a hammer and punch. New seals can be installed by driving them in with a socket (**see illustration**).

Installation

10 Slide the external shift mechanism into place, spreading the shift arm and the overshift limiter to clear the shift drum. Make sure the springs are positioned correctly.

11 Apply high-temperature grease to the lips of the seals. Wrap the splines of the shift shaft with electrical tape, so the splines won't damage the seal as the cover is installed.

12 Apply a thin coat of RTV sealant to the cover mating areas on the crankcase, where the halves of the crankcase join (**see illustration**).

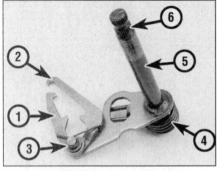

21.5 Shift mechanism details

1 Shift mechanism arm
2 Overshift limiter
3 Pawl spring
4 Return spring
5 Shift shaft
6 Splines

21.9 New seals can be driven into the cover with a socket having an outside diameter slightly smaller than the seal

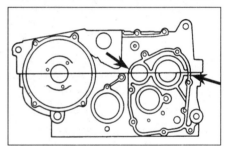

21.12 Apply RTV sealant to the shaded areas (arrows)

22.3 Hold the rotor from turning and loosen the secondary shaft nut

22.8 Loosen the upper crankcase half bolts (arrows), a little at a time, then remove them

22.12 Remove the primary chain tensioner bolt, spring and pin

13 Carefully guide the cover into place and install the screws, tightening them securely.
14 Install the engine sprocket and chain, engine sprocket cover and the shift lever (see Chapter 6).
15 Check the engine oil level and add some, if necessary (see Daily (pre-ride) checks).

22 Crankcase - disassembly and reassembly

1 To examine and repair or replace the crankshaft, connecting rods, bearings, transmission components and secondary shaft/starter motor clutch, the crankcase must be split into two parts.
2 Before this can be done, the water pump and coolant pipe (on the cylinder block) (see Chapter 3), the external shift mechanism (see Section 21), the starter motor, the alternator cover, the timing rotor and the pickup coil (see Chapter 9) must be removed.

Disassembly

3 Using a special alternator rotor holder or a pin-type spanner, hold the rotor stationary and loosen the secondary shaft nut **(see illustration)**.
4 Remove the alternator rotor and the stator (see Chapter 9).

5 If the crankcase is being separated to remove the crankshaft, remove the cylinder head, cylinder block and pistons (see Sections 10, 13 and 14).
6 Remove the clutch if you are separating the crankcase halves to disassemble the transmission main drive shaft (see Section 19).
7 If the engine is bolted to a mounting stand that fastens to the rear mounting bolt holes (upper and lower), remove it from the stand and fabricate a holder that will support the crankcase upside down, with the seam level.
8 Remove the twelve upper crankcase half bolts **(see illustration)**.
9 Turn the engine upside-down and remove the oil filter (see Chapter 1, if necessary).
10 Remove the oil pan (see Section 16) and retrieve the O-rings from the oil passages.
11 Remove the oil pump (see Section 17).
12 Remove the primary chain tensioner bolt **(see illustration)**.
13 Remove the secondary shaft nut.
14 Remove the secondary shaft from the case (see Section 27).
15 Remove the lower crankcase half bolts **(see illustration)**. Carefully pry the crankcase apart. Pry ONLY on the corners, in the areas indicated **(see illustration)**.
16 Lift the starter motor clutch/secondary sprocket assembly **(see illustration)**.

17 Refer to Sections 23 through 31 for information on the internal components of the crankcase.

Reassembly

18 Remove all traces of sealant from the crankcase mating surfaces. Be careful not to let any fall into the case as this is done.
19 Check to make sure the two dowel pins are in place in their holes in the mating surface of the upper crankcase half.
20 Insert the starter motor clutch gear into the secondary sprocket. Lift the primary chain up and guide the secondary sprocket/starter motor clutch assembly into place, making sure the chain meshes properly with the gear (see Section 27).
21 Pour some engine oil over the transmission gears, the crankshaft main bearings and the shift drum. Don't get any oil on the crankcase mating surface.
22 Apply a thin, even bead of Kawasaki Bond sealant (part no. 92104-1003) to the indicated areas of the crankcase mating surfaces **(see illustration)**. Also apply RTV sealant to the areas near the ends of the crankshaft seal areas (lay it over the Kawasaki Bond).
Caution: Don't apply an excessive amount of either type of sealant, as it will ooze out when the case halves are assembled and may obstruct oil passages.

22.15a Loosen the lower crankcase half bolts (arrows), a little at a time, then remove them

22.15b Pry the crankcase apart only on the corners, in the areas provided for this purpose

22.16 Lift up on the primary chain and remove the secondary sprocket, shaft and starter motor clutch assembly

23 Check the position of the shift drum - make sure it's in the neutral position **(see illustration)**.

24 Carefully place the lower crankcase half onto the upper crankcase half. While doing this, make sure the shift forks fit into their gear grooves, and guide the breather tube into its hole in the lower crankcase half **(see illustration)**.

25 Install the lower crankcase half bolts and tighten them so they are just snug.

26 In two steps, tighten the larger bolts (8 mm), in the indicated sequence, to the torque listed in this Chapter's Specifications **(see illustration)**.

27 Turn the case over and install the upper crankcase half bolts, tightening them to the torque listed in this Chapter's Specifications.

28 Turn the case over again and install the smaller (6 mm) bolts in the lower crankcase half, tightening them to the torque listed in this Chapter's Specifications.

29 Install the washer and sleeve on the secondary shaft, if they were removed. Lubricate the secondary shaft and install it through the case and the secondary sprocket/starter motor clutch assembly.

30 Turn the main drive shaft and the output shaft to make sure they turn freely. Install the shift lever on the shift shaft and, while turning the output shaft, shift the transmission through the gears, first through sixth, then back to first. If the transmission doesn't shift properly, the case will have to be separated again to correct the problem. Also make sure the crankshaft turns freely.

31 The remainder of installation is the reverse of removal.

23 Crankcase components - inspection and servicing

1 After the crankcases have been separated and the crankshaft, shift drum and forks and transmission components removed, the crankcases should be cleaned thoroughly with new solvent and dried with compressed air. All oil passages should be blown out with compressed air and all traces of old gasket sealant should be removed from the mating surfaces.

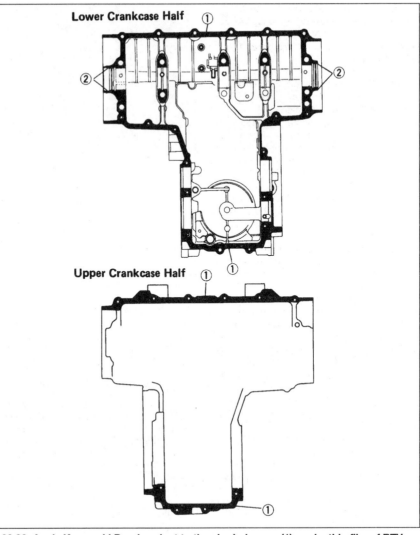

22.22 Apply Kawasaki Bond sealant to the shaded areas (1), and a thin film of RTV sealant to the areas on the sides of the crankshaft seal (2)

22.23 Make sure the ball is positioned in the neutral detent on the operating plate

22.24 When assembling the case halves, make sure the shift forks fit into their gear grooves (arrows) and the breather tube fits into its hole in the lower case half

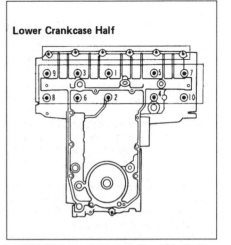

22.26 Tighten the larger (8 mm) bolts in the indicated sequence, in two passes, to the torque listed in this Chapter's Specifications

Caution: Be very careful not to nick or gouge the crankcase mating sunfaces or leaks will result. Check both crankcase sections very carefully for cracks and other damage.

2 Check the primary chain guides for wear - one is in the upper case half and the other is in the lower case half. If they appear to be worn excessively, replace them.

3 Check the ball and needle bearings in the case. If they don't turn smoothly, drive them out with a bearing driver or a socket having an outside diameter slightly smaller than that of the bearing. Before installing them, allow them to sit in the freezer overnight, and about fifteen-minutes before installation, place the case half in an oven, set to about 94° C (200° F), and allow it to heat up. The bearings are an interference fit, and this will ease installation. If you are installing the secondary shaft ball bearing on the left side of the lower case half, drive it in so it's recessed approximately 14 to 15 mm (9/16 to 19/32-inch) **(see illustration).**

 Warning: Before heating the case, wash it thoroughly with soap and water so no explosive fumes are present. Also, don't use a flame to heat the case.

4 Check all the studs in the case for damaged threads or bending. To replace a stud, thread two nuts onto it and tighten them against each other, then unscrew the stud, using a wrench on the lower nut. Before installing a stud, apply a non-hardening thread locking compound to the threads.

5 If any damage is found that can't be repaired, replace the crankcase halves as a set.

24 Main and connecting rod bearings - general note

1 Even though main and connecting rod bearings are generally replaced with new ones during the engine overhaul, the old bearings should be retained for close examination as they may reveal valuable information about the condition of the engine.

2 Bearing failure occurs mainly because of lack of lubrication, the presence of dirt or other foreign particles, overloading the engine and/or corrosion. Regardless of the cause of bearing failure, it must be corrected before the engine is reassembled to prevent it from happening again.

3 When examining the bearings, remove the main bearings from the case halves and the rod bearings from the connecting rods and caps and lay them out on a clean surface in the same general position as their location on the crankshaft journals. This will enable you to match any noted bearing problems with the corresponding side of the crankshaft journal.

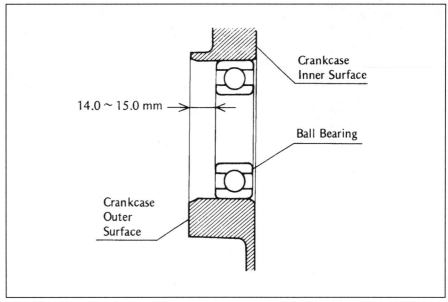

23.3 When installing the secondary shaft bearing in the left side of the case, drive it in to the depth shown

4 Dirt and other foreign particles get into the engine in a variety of ways. It may be left in the engine during assembly or it may pass through filters or breathers. It may get into the oil and from there into the bearings. Metal chips from machining operations and normal engine wear are often present. Abrasives are sometimes left in engine components after reconditioning operations such as cylinder honing, especially when parts are not thoroughly cleaned using the proper cleaning methods. Whatever the source, these foreign objects often end up imbedded in the soft bearing material and are easily recognized. Large particles will not imbed in the bearing and will score or gouge the bearing and journal. The best prevention for this cause of bearing failure is to clean all parts thoroughly and keep everything spotlessly clean during engine reassembly. Frequent and regular oil and filter changes are also recommended.

5 Lack of lubrication or lubrication breakdown has a number of interrelated causes. Excessive heat (which thins the oil), overloading (which squeezes the oil from the bearing face) and oil leakage or throw off (from excessive bearing clearances, worn oil pump or high engine speeds) all contribute to lubrication breakdown. Blocked oil passages will also starve a bearing and destroy it. When lack of lubrication is the cause of bearing failure, the bearing material is wiped or extruded from the steel backing of the bearing. Temperatures may increase to the point where the steel backing and the journal turn blue from overheating.

6 Riding habits can have a definite effect on bearing life. Full throttle low speed operation, or lugging the engine, puts very high loads on bearings, which tend to squeeze out the oil film. These loads cause the bearings to flex, which produces fine cracks in the bearing face (fatigue failure). Eventually the bearing material will loosen in pieces and tear away from the steel backing. Short trip driving leads to corrosion of bearings, as insufficient engine heat is produced to drive off the condensed water and corrosive gases produced. These products collect in the engine oil, forming acid and sludge. As the oil is carried to the engine bearings, the acid attacks and corrodes the bearing material.

7 Incorrect bearing installation during engine assembly will lead to bearing failure as well. Tight fitting bearings which leave insufficient bearing oil clearances result in oil starvation. Dirt or foreign particles trapped behind a bearing insert result in high spots on the bearing which lead to failure.

8 To avoid bearing problems, clean all parts thoroughly before reassembly, double check all bearing clearance measurements and lubricate the new bearings with engine assembly lube or moly-based grease during installation.

25 Crankshaft and main bearings - removal, inspection and installation

Removal

1 Crankshaft removal is a simple matter of lifting it out of place once the crankcase has been separated and the starter motor clutch/secondary sprocket assembly has been removed. Before removing the crankshaft check the endplay. This can be done with a dial indicator mounted in-line with the crankshaft, or feeler gauges inserted between the no. 2 crankcase main journal

25.1 Measure the endplay with a feeler gauge inserted between the no. 2 crank journal and the case web – if the endplay isn't as listed in this Chapter's Specifications, the case halves must be replaced

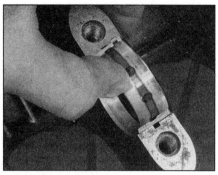

25.2 To remove a main bearing insert, push it sideways and lift it out

25.8 Lay the Plastigauge strips (arrow) on the journals, parallel to the crankshaft centerline

(see illustration). Compare your findings with this Chapter's Specifications. If the endplay is excessive, the case halves must be replaced.

2 The main bearing inserts can be removed from their saddles by pushing their centres to the side, then lifting them out **(see illustration)**. Keep the bearing inserts in order. The main bearing oil clearance should be checked, however, before removing the inserts (see Step 8).

Inspection

3 Mark and remove the connecting rods from the crankshaft (see Section 26).
4 Clean the crankshaft with solvent, using a rifle-cleaning brush to scrub out the oil passages. If available, blow the crank dry with compressed air. Check the main and connecting rod journals for uneven wear, scoring and pits.
5 Check the camshaft chain gear and the primary chain gear on the crankshaft for chipped teeth and other wear. If any undesirable conditions are found, replace the crankshaft. Check the chains as described in Section 28.

6 Check the rest of the crankshaft for cracks and other damage. It should be magnafluxed to reveal hidden cracks - a dealer service department or motorcycle machine shop will handle the procedure.
7 Set the crankshaft on V-blocks and check the runout with a dial indicator touching the centre main journal, comparing your findings with this Chapter's Specifications. If the runout exceeds the limit, replace the crank.

Main bearing selection

8 To check the main bearing oil clearance, clean off the bearing inserts (and reinstall them, if they've been removed from the case) and lower the crankshaft into the upper half of the case. Cut five pieces of Plastigauge (type HPG-1) and lay them on the crankshaft main journals, parallel with journal axis **(see illustration)**.
9 Very carefully, guide the lower case half down onto the upper case half. Install the large (8 mm) bolts and tighten them, using the recommended sequence, to the torque listed in this Chapter's Specifications (see Section 22). Don't rotate the crankshaft!
10 Now, remove the bolts and carefully lift the lower case half off. Compare the width of the crushed Plastigauge on each journal to the scale

printed on the Plastigauge envelope to obtain the main bearing oil clearance **(see illustration)**. Write down your findings, then remove all traces of Plastigauge from the journals, using your fingernail or the edge of a credit card.
11 If the oil clearance falls into the specified range, no bearing replacement is required (provided they are in good shape). If the clearance is within 0.038 mm (0.0015-in) and the service limit of 0.08 mm (0.003-in), replace the bearing inserts with inserts that have blue paint marks, then check the oil clearance once again. Always replace all of the inserts at the same time.
12 The clearance might be slightly greater than the standard clearance, but that doesn't matter, as long as it isn't greater than the maximum clearance or less than the minimum clearance.
13 If the clearance is greater than the service limit listed in this Chapter's Specifications, measure the diameter of the crankshaft journals with a micrometer **(see illustration)** and compare your findings with this Chapter's Specifications. Also, by measuring the diameter at a number of points around each journal's circumference, you'll be able to determine whether or not the journal is out-of-round. Take the measurement at each end of

25.10 Measuring the width of the crushed Plastigauge (be sure to use the correct scale – standard and metric are included)

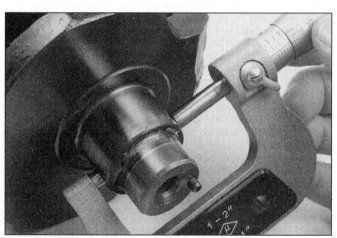

25.13 Measure the diameter of each crankshaft journal at several points to detect taper and out-of-round conditions

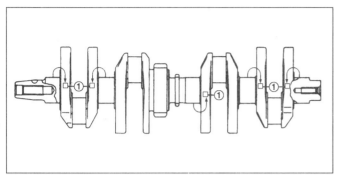

25.15 Use the marks ("1" or none) on the crankshaft (arrows) . . .

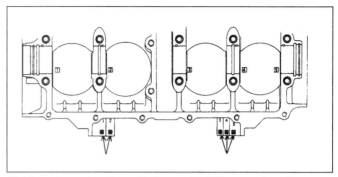

25.16 . . . in conjunction with the marks ("0" or none) on the upper case half (arrows) . . .

Crankcase Main Bearing Bore Diameter Marking	Crankshaft Main Journal Diameter Marking	Bearing Insert*		
		Size Color	Part Number	Journal Nos.
○	1	Brown	13034-1016	2, 4
			13034-1066	1, 3, 5
○	None	Black	13034-1017	2, 4
None	1		13034-1065	1, 3, 5
None	None	Blue	13034-1018	2, 4
			13034-1064	1, 3, 5

25.17 . . . to determine the proper main bearing sizes

the journal, near the crank throws, to determine if the journal is tapered.

14 If any crank journal has worn down past the service limit, replace the crankshaft.

15 If the diameters of the journals aren't less than the service limit but differ from the original markings on the crankshaft (see illustration), apply new marks with a hammer and punch.

If the journal measures between 31.984 to 31.992 mm (1.2601 to 1.2604-in) don't make any marks on the crank (there shouldn't be any marks there, anyway).

If the journal measures between 31.993 to 32.000 mm (1.2605 to 1.2608-in), make a "1" mark on the crank in the area indicated (if it's not already there).

16 Remove the main bearing inserts and assemble the case halves (see Section 22). Using a telescoping gauge and a micrometer, measure the diameters of the main bearing bores, then compare the measurements with the marks on the upper case half (see illustration).

If the bores measure between 36.000 to 36.008 mm (1.4184 to 1.4187-in), there should be a "0" mark in the indicated areas.

If the bores measure between 36.009 to 36.016 mm (1.4187 to 1.4190-in), there shouldn't be any marks in the indicated areas.

17 Using the marks on the crank and the marks on the case, determine the bearing sizes required by referring to the accompanying bearing selection chart (see illustration).

Installation

18 Separate the case halves once again. Clean the bearing saddles in the case halves, then install the bearing inserts in their webs in the case (see illustration). The bearing inserts for journals 2 and 4 have oil grooves.

When installing the bearings, use your hands only - don't tap them into place with a hammer.

19 Lubricate the bearing inserts with engine assembly lube or moly-based grease.

20 Install the connecting rods, if they were removed (see Section 26).

21 Install new oil seals to the ends of the crankshaft (the lips of the seal must face the crankshaft). Be sure to lubricate the lips of the seals with high-temperature grease before sliding them into place.

22 Loop the camshaft chain and the primary chain over the crankshaft and lay them onto their gears.

23 Carefully lower the crankshaft into place, making sure the ribs on the seal outer diameters seat in the grooves in the case (see illustration).

24 Assemble the case halves (see Section 22) and check to make sure the crankshaft and the transmission shaft turns freely.

26 Connecting rods and bearings - removal, inspection and installation

Removal

1 Before removing the connecting rods from the crankshaft, measure the side clearance of each rod with a feeler gauge (see illustration). If the clearance on any rod is greater than that listed in this Chapter's

25.18 Make sure the tabs on the bearing inserts fit into the notches in the web

25.23 When installing the crankshaft, make sure the ribs on each seal seat in the grooves in the case (arrows)

26.1 Check the connecting rod side clearance with a feeler gauge

26.2 Using a hammer and a punch, make matching cylinder number marks on the connecting rod and its cap

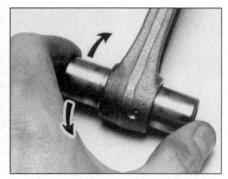

26.5 Checking the piston pin and connecting rod bore for wear

Specifications, that rod will have to be replaced with a new one.

2 Using a centre punch, mark the position of each rod and cap, relative to its position on the crankshaft **(see illustration)**.

3 Unscrew the bearing cap nuts, separate the cap from the rod, then detach the rod from the crankshaft. If the cap is stuck, tap on the ends of the rod bolts with a soft face hammer to free them.

4 Separate the bearing inserts from the rods and caps, keeping them in order so they can be reinstalled in their original locations. Wash the parts in solvent and dry them with compressed air, if available.

Inspection

5 Check the connecting rods for cracks and other obvious damage. Lubricate the piston pin for each rod, install it in the proper rod and check for play **(see illustration)**. If it is loose, replace the connecting rod and/or the pin.

6 Refer to Section 24 and examine the connecting rod bearing inserts. If they are scored, badly scuffed or appear to have been seized, new bearings must be installed. Always replace the bearings in the connecting rods as a set. If they are badly damaged, check the corresponding crankshaft journal. Evidence of extreme heat, such as discoloration, indicates that lubrication failure has occurred. Be sure to thoroughly check the oil pump and pressure relief valve as well as all oil holes and passages before reassembling the engine.

7 Have the rods checked for twist and bending at a dealer service department or other motorcycle repair shop.

Bearing selection

8 If the bearings and journals appear to be in good condition, check the oil clearances as follows:

9 Start with the rod for the number one cylinder. Wipe the bearing inserts and the connecting rod and cap clean, using a lint-free cloth.

10 Install the bearing inserts in the connecting rod and cap. Make sure the tab on the bearing engages with the notch in the rod or cap.

11 Wipe off the connecting rod journal with a lint-free cloth. Lay a strip of Plastigauge (type HPG-1) across the top of the journal, parallel with the journal axis **(see illustration 25.8)**.

12 Position the connecting rod on the bottom of the journal, then install the rod cap and nuts. Tighten the nuts to the torque listed in this Chapter's Specifications, but don't allow the connecting rod to rotate at all.

13 Unscrew the nuts and remove the connecting rod and cap from the journal, being very careful not to disturb the Plastigauge. Compare the width of the crushed Plastigauge to the scale printed in the Plastigauge envelope **(see illustration 25.10)** to determine the bearing oil clearance.

14 If the clearance is within the range listed in this Chapter's Specifications and the bearings are in perfect condition, they can be reused. If the clearance is within 0.059 mm (0.0023-in) and the service limit of 0.10 mm (0.0039-in), replace the bearing inserts with inserts that have blue paint marks, then check the oil clearance once again. Always replace all of the inserts at the same time.

15 The clearance might be slightly greater than the standard clearance, but that doesn't matter, as long as it isn't greater than the maximum clearance or less than the minimum clearance.

16 If the clearance is greater than the service limit listed in this Chapter's Specifications, measure the diameter of the connecting rod journal with a micrometer and compare your findings with this Chapter's Specifications. Also, by measuring the diameter at a number of points around the journal's circumference, you'll be able to determine whether or not the journal is out-of-round. Take the measurement at each end of the journal to determine if the journal is tapered.

17 If any journal has worn down past the service limit, replace the crankshaft.

18 If the diameter of the journal isn't less than the service limit but differs from the original markings on the crankshaft **(see illustration)**, apply new marks with a hammer and punch.

If the journal measures between 32.984 to 32.992 mm (1.2995 to 1.2998-in) don't make any marks on the crank (there shouldn't be one there anyway).

If the journal measures between 32.993 to 33.000 mm (1.2999 to 1.3002-in), make a "0" mark on the crank in the area indicated (if not already there).

19 Remove the bearing inserts from the connecting rod and cap, then assemble the cap to the rod. Tighten the nuts to the torque listed in this Chapter's Specifications.

20 Using a telescoping gauge and a micrometer, measure the inside diameter of the connecting rod **(see illustration)**. The mark on the connecting rod (if any) should coincide with the measurement, but if

26.20a Assemble the connecting rod and measure the diameter of the bore with a telescoping gauge – then measure the gauge with a micrometer

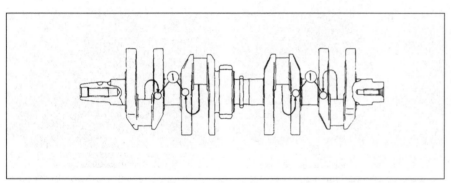

26.18 The marks on the crank throws (1) should coincide with the diameters of the connecting rod journals

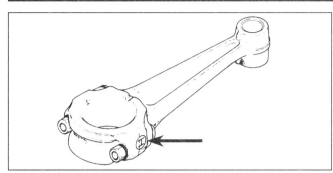

26.20b The mark (or lack of a mark) on the connecting rod (arrow), in conjunction with the mark on the crank throw . . .

Con-rod Big End Bore Diameter Marking	Crankpin Diameter Marking	Bearing Insert	
		Size Color	Part Number
O	None	Blue	13034-1067
O	O	Black	13034-1068
None	None		
None	O	Brown	13034-1069

26.21 . . . can be used, along with this chart, to determine the correct connecting rod bearing inserts to install

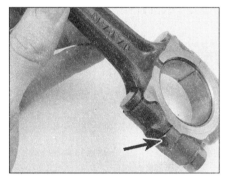

26.24 When installing the connecting rods, make sure your matchmarks are on the same side and the JAPAN casting points toward the tapered end of the crankshaft – the letter (arrow) is a weight grade mark

it doesn't, make a new mark **(see illustration)**.

If the inside diameter measures between 36.000 to 36.008 mm (1.4184 to 1.4187-in), don't make any mark on the rod (there shouldn't be any there anyway).

If the inside diameter measures between 36.009 to 36.016 mm (1.4187 to 1.4190-in), make a 0 mark on the rod (it should already be there).

21 By referring to the accompanying chart **(see illustration)**, select the correct connecting rod bearing inserts.

22 Repeat the bearing selection procedure for the remaining connecting rods.

Installation

23 Wipe off the bearing inserts and connecting rods and caps. Install the inserts into the rods and caps, using your hands only, making sure the tabs on the inserts engage with the notches in the rods and caps. When all the inserts are installed, lubricate them with engine assembly lube or moly-based grease. Don't get any lubricant on the mating surfaces of the rod or cap.

24 Assemble each connecting rod to its proper journal, making sure the previously applied matchmarks correspond to each other and the JAPAN casting points to the alternator rotor end of the crankshaft (the tapered end) **(see illustration)**. Also, the letter present at the rod/cap seam on one side of the connecting rod is a weight mark. If new rods are being installed and they don't all have the same letter on them, two rods with the same letter should be installed on one side of the crank, and the letters on the other two rods should match each other. This will minimize vibration.

25 When you're sure the rods are positioned correctly, tighten the nuts to the torque listed in this Chapter's Specifications.

26 Turn the rods on the crankshaft. If any of them feel tight, tap on the bottom of the connecting rod caps with a hammer - this should relieve stress and free them up. If it doesn't, recheck the bearing clearance.

27 As a final step, recheck the connecting

rod side clearances (see Step 1). If the clearances aren't correct, find out why before proceeding with engine assembly.

27 Secondary sprocket, shaft and starter motor clutch - removal, inspection and installation

Removal

1 Drain the cooling system and the engine oil (see Chapter 1). Remove the water pump (see Chapter 3). Also remove the water pump base - it's fastened to the crankcase with two bolts.

2 Remove the alternator cover. Using a special alternator rotor holder or a pin-type spanner, hold the rotor stationary and loosen the secondary shaft nut **(see illustration 22.3)**.

3 Remove the engine (see Section 5).

4 Separate the crankcase halves (see Section 22).

5 Remove the nut from the secondary shaft. Using a soft-face hammer, tap the secondary shaft out, toward the right side of the case, until the bearing on the right side of the case is free **(see illustration)**.

6 Hold the starter motor clutch assembly and pull the secondary shaft out of the case **(see illustration)**. There is a sleeve on the shaft that fits into the ball bearing on the left side of the case - be careful not to lose it. There's also a washer on the shaft.

7 Detach the primary chain from the starter motor clutch/secondary sprocket remove the clutch assembly **(see illustration)**.

27.5 Use a plastic hammer to drive the secondary shaft from the case

27.6 Hold the secondary sprocket/starter motor clutch assembly and pull the secondary shaft out – be careful not to lose the sleeve that fits into the bearing in the left side of the case, or the washer on the shaft

27.7 Lift the primary chain and remove the secondary sprocket/starter motor clutch assembly

27.8 Try to turn the starter motor clutch gear back and forth – it should only turn in one direction

Inspection

Starter motor clutch

8 Hold the starter motor clutch and attempt to turn the starter motor clutch gear back and forth **(see illustration)**. It should only turn in one direction.

9 If the starter motor clutch turns freely in both directions, or if it's locked up, disassemble it and inspect the components **(see illustration)**. To do this, remove the snap-ring and pull off the secondary sprocket, remove the Allen-head bolts and detach the inner coupling from the clutch.

10 Check the springs for wear, the rollers for scoring and pitting, and the rubber dampers for deterioration. Check the teeth on the sprockets for cracks and chips. Check the bushing in the starter motor clutch gear for scoring or heat discoloration. Replace parts as necessary.

Starter motor idler gear

11 Inspect the teeth on the starter motor idler gear for cracks and chips **(see illustration)**. Turn the idler gear to make sure it spins freely. If the idler gear exhibits any undesirable conditions, replace it. To remove the idler, pry the clip from the idler shaft, slide the shaft out and remove the gear. Coat the shaft with engine assembly lube or moly-based grease before installing it.

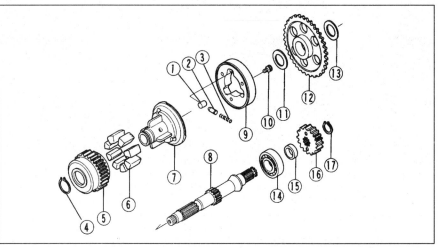

27.9 Exploded view of the secondary sprocket/starter motor clutch and shaft assembly

1 Roller	6 Rubber dampers	10 Allen bolts	14 Ball bearing
2 Spring cap	7 Inner coupling	11 Thrust washer	assembly
3 Spring	8 Secondary shaft	12 Starter motor	15 Collar
4 Snap-ring	9 Starter motor	clutch gear	16 Secondary shaft
5 Secondary	clutch	13 Thrust washer	gear
sprocket			17 Snap-ring

Secondary shaft and bearings

12 Check the splines, gear and threads on the shaft for wear or damage.

13 Turn the bearing and feel for tight spots and roughness. If necessary, pull the bearing off the shaft, using a bearing puller **(see illustration)**. The new bearing can be tapped onto the shaft, using a piece of pipe with an inside diameter large enough to fit over the shaft and contact the inner race of the bearing.

14 If the gear needs to be replaced, remove the snap-ring from the end of the shaft **(see illustration)** and pull the gear off, using the same set-up described in Step 13 **(see illustration 27.13)**. Don't lose the collar on the shaft. When installing the gear, lubricate the splines on the shaft with engine oil and carefully tap the gear into place using a section of pipe.

15 Check the secondary shaft bearing in the right side of the lower case half. If it turns

roughly or feels tight, replace it by referring to Section 23.

Installation

16 Lubricate the rollers in the starter motor clutch with engine oil and assemble the clutch gear to it. Make sure the washer is present between the two components.

17 Lift the primary chain and guide the starter motor clutch assembly into place.

18 Lubricate the secondary shaft with engine oil and install the washer and sleeve. Guide the shaft into the case and through the secondary sprocket/starter clutch assembly.

19 Using a soft-face hammer, tap the shaft into the case until the bearing is completely seated. Install the secondary shaft nut, hold the alternator rotor stationary and tighten the nut to the torque listed in this Chapter's Specifications. You may want to wait until the engine is mounted in the frame before you tighten this nut, unless you have an assistant to help you.

27.11 If it's necessary to remove the idler gear from the case, remove the clip (arrow), slide the shaft out and lift the gear from the case

27.13 Use a bearing puller setup like this to remove the bearing from the shaft (it can also be used to pull the gear off the shaft, if it's turned around)

27.14 Remove the snap-ring from the end of the shaft with snap-ring pliers

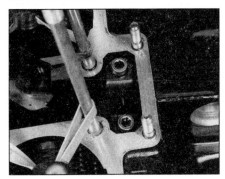

28.6 The camshaft chain rear guide is retained to the crankcase by two Allen-head bolts

28.8 The chain guide in the upper case half is secured to the case with two Allen-head bolts

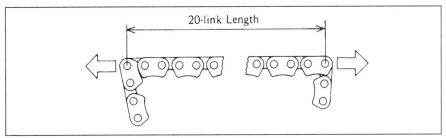

28.9 When checking the camshaft chain or the primary chain, measure the length of twenty links and compare the results to this Chapter's Specifications

20 The remainder of installation is the reverse of the removal procedure. Be sure to fill the cooling system with the proper coolant mixture and the crankcase with the recommended engine oil (see Chapter 1).

28 Primary chain, camshaft chain and guides - removal, inspection and installation

Removal

Primary chain and camshaft chain

1 Remove the engine (see Section 5).
2 Separate the crankcase halves (see Section 22).
3 Remove the crankshaft (see Section 25).
4 Remove the chains from the crankshaft.

Chain guides

5 The cam chain front guide can be lifted from the cylinder block after the head has been removed (see Section 13).
6 The cam chain rear guide is fastened to the crankcase with a bracket and two bolts **(see illustration)**. Remove the bolts and detach the guide and bracket from the case.
7 The primary chain guide in the lower case half is removed in a similar manner.
8 The primary chain guide in the upper case half is secured by two Allen-head bolts **(see illustration)**.

Inspection

Primary chain and camshaft chain

9 The primary chain and camshaft chains are checked in a similar manner. Pull the chain tight to eliminate all slack and measure the length of twenty links, pin-to-pin **(see illustration)**. Compare your findings to this Chapter's Specifications.
10 Also check the chains for binding and obvious damage.
11 If the twenty-link length is not as specified, or there is visible damage, replace the chain.

Chain guides

12 Check the guides for deep grooves, cracking and other obvious damage, replacing them if necessary.

Installation

13 Installation of these components is the reverse of the removal procedure. When installing the brackets for the cam chain rear

29.2 With the plunger of the dial indicator contacting a gear tooth, move the gear back-and-forth within its freeplay while holding its companion gear still

guide and the primary chain guides, apply a non-hardening thread locking compound to the threads of the bolts. Tighten the bolts to the torque listed in this Chapter's Specifications. Apply engine oil to the faces of the guides and to the chains.

29 Transmission shafts - removal and installation

Removal

1 Remove the engine and clutch, then separate the case halves (see Sections 5, 19 and 22).
2 Before removing either shaft, check the backlash of each set of gears. To do this, mount a dial indicator with the plunger of the indicator touching a tooth on one of the gears, then move the gear back and forth within its freeplay, holding its companion gear stationary **(see illustration)**. Check each set of gears, recording the measurements, and compare the results to this Chapter's Specifications. If the backlash between any pair of gears exceeds the limit, replace both gears (see Section 30).
3 The shafts can simply be lifted out of the upper half of the case. If they are stuck, use a soft-face hammer and gently tap on the bearings on the ends of the shafts to free them. The shaft nearest the rear of the case is the output shaft - the other shaft is the main drive shaft.
4 Refer to Section 30 for information pertaining to transmission shaft service and Section 31 for information pertaining to the shift drum and forks.

Installation

5 Check to make sure the set pins and rings are present in the upper case half, where the shaft bearings seat **(see illustration)**.
6 Carefully lower each shaft into place. The holes in the needle bearing outer races must engage with the set pins, and the grooves in

29.5 Make sure the set pins and rings (arrows) are installed in their proper positions

29.6 When lowering the shafts into place, the hole in the needle bearing outer race must fit over the set pin, and groove in the ball bearing outer race must fit over the set ring (arrows)

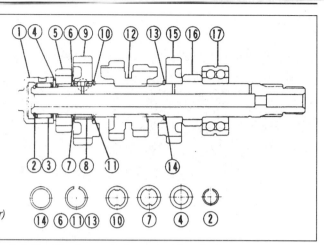

1 Bearing outer race
2 Snap-ring
3 Needle bearing
4 Thrust washer
5 Second gear
6 Snap-ring
7 Toothed washer
8 Bushing
9 Sixth gear
10 Toothed washer
11 Snap-ring
12 Third/fourth gear
13 Snap-ring
14 Washer
15 Fifth gear
16 Drive shaft (first gear)
17 Ball bearing

30.2a Details of the transmission main drive shaft

the ball bearing outer races must engage with the set rings (see illustration).

7 The remainder of installation is the reverse of removal.

30 Transmission shafts - disassembly, inspection and reassembly

> **HAYNES HiNT** When disassembling the transmission shafts, place the parts on a long rod or thread a wire through them to keep them in order and facing the proper direction.

1 Remove the shafts from the case (see Section 29).

Main drive shaft

Disassembly

2 Remove the needle bearing outer race, then remove the snap-ring from the end of the shaft and slide the needle bearing off (see illustrations).

3 Remove the thrust washer and slide second gear off the shaft (see illustration).

4 Remove the snap-ring (see illustration), toothed washer, sixth gear and bushing (see illustration).

5 Slide the next toothed washer off and remove the snap-ring (see illustration).

6 Remove the third/fourth gear cluster from the shaft (see illustration).

7 Remove the next snap-ring, then slide the washer and fifth gear off the shaft (see illustration).

Inspection

8 Wash all of the components in clean solvent and dry them off. Rotate the ball bearing on the shaft, feeling for tightness, rough spots, excessive looseness and

30.2b Remove the snap-ring from the end of the shaft and slide the needle bearing off

30.3 Slide the thrust washer and second gear from the shaft

30.4a Remove the snap-ring . . .

30.4b . . . followed by the washer, sixth gear and bushing

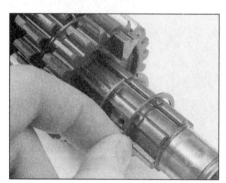

30.5 Remove the toothed washer and snap-ring from the shaft . . .

30.6 . . . then slide the third/fourth gear cluster off the shaft

30.7 Remove the snap-ring, washer and fifth gear

30.9 Measure the width of the shift fork grooves with a caliper

30.11 If the gear dogs and dog holes (arrows) show signs of excessive wear, replace the gears as a set

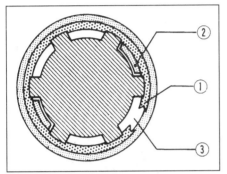

30.13a When installing snap-rings on the main drive shaft, align the opening In the snap-ring with a spline groove

1   Snap-ring
2   Tab on toothed washer
3   Spline groove

30.13b Align the oil hole in the bushing with the hole in the shaft (arrows)

listening for noises. If any of these conditions are found, replace the bearing. This will require the use of a hydraulic press or a bearing puller setup. If you don't have access to these tools, take the shaft and bearing to a Kawasaki dealer or other motorcycle repair

shop and have them press the old bearing off the shaft and install the new one.

9 Measure the shift fork groove between third and fourth gears **(see illustration)**. If the groove width exceeds the figure listed in this Chapter's Specifications, replace the third/fourth gear assembly, and also check the third/fourth gear shift fork (see Section 31).

10 Check the gear teeth for cracking and other obvious damage. Check the bushing

surface in the inner diameter of sixth gear for scoring or heat discoloration. If it's damaged, replace it (check with your Kawasaki dealer - they may be able to replace the bushing insert).

11 Inspect the dogs and the dog holes on the gears for excessive wear **(see illustration)**. Replace the paired gears as a set if necessary.

12 Check the needle bearing and race for wear or heat discoloration and replace them if necessary.

Reassembly

13 Reassembly is the basically the reverse of the disassembly procedure, but take note of the following points:

a) Always use new snap-rings and align the opening of the ring with a spline groove **(see illustration)**.

b) When installing the bushing for sixth gear to the shaft, align the oil hole on the shaft with the oil hole in the bushing **(see illustration)**.

c) Lubricate the components with engine oil before assembling them.

Output shaft

Disassembly

14 Remove the needle bearing outer race **(see illustration)**.

15 Remove the snap-ring from the end of the shaft and slide the needle bearing off **(see illustration)**.

30.15 Remove the snap-ring from the end of the shaft and slide the needle bearing off

1   Collar
2   Ball bearing
3   Output shaft
4   Second gear
5   Toothed washer
6   Snap-ring
7   Sixth gear
8   Snap-ring
9   Toothed washer
10 Fourth gear
11 Bushing
12 Third gear
13 Toothed washer
14 Snap-ring
15 Steel ball
16 Fifth gear
17 First gear
18 Thrust washer
19 Needle bearing
20 Bearing outer race
21 Snap-ring

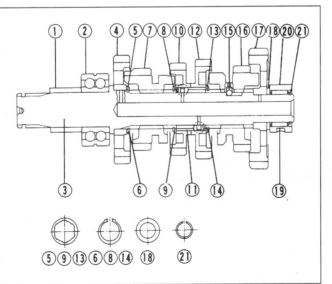

30.14 Details of the transmission output shaft

30.16 Hold the output shaft vertically and spin fifth gear (which will also turn the shaft) while pulling up

30.17 Remove the snap-ring, toothed washer, third gear, bushing and fourth gear

30.18 Remove the washer, snap-ring and sixth gear

16 Remove the washer, first gear and fifth gear from the shaft. Fifth gear has three steel balls in it for the positive neutral finder mechanism. To remove the gear, grasp third gear and hold the shaft in a vertical position with one hand, and with the other hand, spin the shaft back and forth, holding onto fifth gear and pulling up **(see illustration)**. *Caution: Don't pull the gear up too hard or fast - the balls will fly out of the gear.*

17 Remove the snap-ring, toothed washer, third gear, bushing and fourth gear from the shaft **(see illustration)**.

18 Remove the toothed washer, snap-ring and sixth gear **(see illustration)**.

19 Remove the next snap-ring, toothed washer and second gear.

Inspection

20 Refer to Steps 8 through 12 for the inspection procedures. They are the same, except when checking the shift fork groove width you'll be checking it on fifth gear and sixth gears.

Reassembly

21 Reassembly is basically the reverse of the disassembly procedure, but take note of the following points:

a) *Always use new snap-rings and align the opening of the ring with a spline groove (see illustration).*

30.19 Remove the snap-ring, washer and second gear

b) *When installing the bushing for third and fourth gear, align the oil hole in the bushing with the hole in the shaft.*

c) *When installing fifth gear, don't use grease to hold the balls in place - to do so would impair the positive neutral finder mechanism. Just set the balls in their holes (the holes that they can't pass through), keep the gear in a vertical position and carefully set it on the shaft (engine oil will help keep them in place). The spline grooves that contain the holes with the balls must be aligned with the slots in the shaft spline grooves.*

d) *Lubricate the components with engine oil before assembling them.*

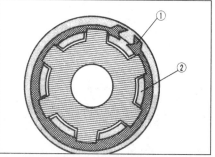

30.21 Install the snap-rings so the opening in the ring (1) is aligned with a spline groove (2)

31 Shift drum and forks - removal, inspection and installation

Removal

1 Remove the engine and separate the crankcase halves (see Sections 5 and 22).

2 Support the shift forks and pull the shift rod out to the left **(see illustration)**.

3 Remove the shift drum positioning bolt and remove the spring and pin **(see illustration)**.

4 Straighten the lock tab on the shift drum guide bolt, then remove the bolt and tab **(see illustration)**.

31.2 Hold the shift forks and pull the shift rod out

31.3 Remove the shift drum positioning bolt and lift out the spring and pin

31.4 Remove the guide bolt

31.5 Remove the cotter pin and guide pin

31.6 Remove the snap-ring and slide the operating plate off the shift drum

31.7 Pull the shift drum out slightly and remove the fifth/sixth gear shift fork

5 Remove the cotter pin (split pin) and pull out the guide pin **(see illustration)**.
6 Remove the operating plate snap-ring from the end of the shift drum, then slide the shift drum off **(see illustration)**.
7 Pull the shift drum out of the case far enough to remove the fifth/sixth shift fork **(see illustration)**, then slide the shift drum out of the case.

Inspection

8 Check the edges of the grooves in the drum for signs of excessive wear. Measure the widths of the grooves and compare your findings to this Chapter's Specifications.
9 Check the shift forks for distortion and wear, especially at the fork ears. Measure the thickness of the fork ears and compare your findings with this Chapter's Specifications **(see illustration)**. If they are discolored or severely worn they are probably bent. If damage or wear is evident, check the shift fork groove in the corresponding gear as well. Inspect the guide pins and the shaft bore for excessive wear and distortion and replace any defective parts with new ones.
10 Check the shift fork shafts for evidence of wear, galling and other damage. Make sure the shift forks move smoothly on the shafts. If the shafts are worn or bent, replace them with new ones.

Installation

11 Installation is the reverse of removal, noting the following points:

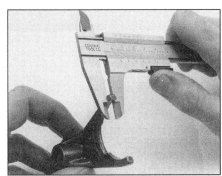

31.9 Measure the thickness of the shift fork ears

a) *Install the shift drum part-way into the case and install the fifth/sixth shift fork (the long end goes onto the drum first).*
b) *Be sure to use a new cotter pin (split pin) and install it correctly (see illustration).*
c) *Lubricate all parts with engine oil before installing them.*
d) *Tighten the guide bolt and the positioning bolt to the torque listed in this Chapter's Specifications.*

32 Initial start-up after overhaul

Note: *Make sure the cooling system is checked carefully (especially the coolant level) before starting and running the engine.*
1 Make sure the engine oil level is correct, then remove the spark plugs from the engine. Place the engine STOP switch in the Off position and unplug the primary (low tension) wires from the coil.
2 Turn on the key switch and crank the engine over with the starter until the oil pressure indicator light goes off (which indicates that oil pressure exists). Reinstall the spark plugs, connect the wires and turn the switch to On.
3 Make sure there is fuel in the tank, then turn the fuel tap to the Prime position and operate the choke.

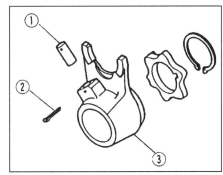

31.11 When installing the cotter pin, insert it from the side shown

1 Shift fork guide pin   3 Shift fork
2 Cotter pin

4 Start the engine and allow it to run at a moderately fast idle until it reaches operating temperature.
Caution: If the oil pressure indicator light doesn't go off, or it comes on while the engine is running, stop the engine immediately
5 Check for oil leaks and make sure the transmission and controls, especially the brakes, function properly before road testing the machine. Refer to Section 33 for the recommended break-in procedure.
6 Upon completion of the road test, and after the engine has cooled down completely, recheck the valve clearances (see Chapter 1).

33 Recommended break-in procedure

1 Any rebuilt engine needs time to break-in, even if parts have been installed in their original locations. For this reason, treat the machine gently for the first few miles to make sure oil has circulated throughout the engine and any new parts installed have started to seat.
2 Even greater care is necessary if the engine has been rebored or a new crankshaft has been installed. In the case of a rebore, the engine will have to be broken in as if the machine were new. This means greater use of the transmission and a restraining hand on the throttle until at least 500 miles (800 km) have been covered. There's no point in keeping to any set speed limit - the main idea is to keep from lugging the engine and to gradually increase performance until the 500 mile (800 km) mark is reached. These recommendations can be lessened to an extent when only a new crankshaft is installed. Experience is the best guide, since it's easy to tell when an engine is running freely.
3 If a lubrication failure is suspected, stop the engine immediately and try to find the cause. If an engine is run without oil, even for a short period of time, irreparable damage will occur.

Notes

Chapter 2 Part B
Engine, clutch and transmission (750 models)

Contents

Degrees of difficulty

| **Easy,** suitable for novice with little experience | | **Fairly easy,** suitable for beginner with some experience | | **Fairly difficult,** suitable for competent DIY mechanic | | **Difficult,** suitable for experienced DIY mechanic | | **Very difficult,** suitable for expert DIY or professional | |

Specifications

General

Bore	68.0 mm (2.68 in)
Stroke	51.5 mm (2.03 in)
Displacement	748 cc
Compression ratio	11.2 : 1

Camshaft and rocker arms

Lobe height (intake and exhaust)
    Standard .... 33.623 to 33.731 mm (1.324 to 1.328 in)
    Minimum .... 33.52 mm 1.319 in)
Bearing oil clearance
    Journals 1 and 4
        Standard .... 0.028 to 0.071 mm (0.0011 to 0.0027 in)
        Maximum .... 0.16 mm (0.0063 in)
    Journals 2 and 3
        Standard .... 0.078 to 0.121 mm (0.0030 to 0.0047 in)
        Maximum .... 0.21 mm (0.008 in)
Journal diameter
    Journals 1 and 4
        Standard .... 23.950 to 23.972 mm (0.9429 to 0.9437 in)
        Minimum .... 23.92 mm (0.9417 in)
    Journals 2 and 3
        Standard .... 23.900 to 23.922 mm (0.9409 to 0.9418 in)
        Minimum .... 23.87 mm (0.9397 in)
Bearing journal inside diameter
    Standard .... 24.000 to 24.021 mm (0.9448 to 0.9457 in)
    Maximum .... 24.08 mm (0.9480 in)
Camshaft runout
    Standard .... 0.02 mm (0.0007 in)
    Maximum .... 0.1 mm (0.0039 in)
Camshaft chain 20-link length
    Standard .... 127.0 to 127.4 mm (5.000 to 5.015 in)
    Maximum .... 128.9 mm (5.074 in)

Cylinder head, valves and valve springs

Cylinder head warpage limit .... 0.05 mm (0.002 in)
Valve stem runout limit .... 0.02 mm (0.0008 in)
Valve stem diameter
    Intake
        Standard .... 4.975 to 4.990 mm (0.1958 to 0.1964 in)
        Minimum .... 4.96 mm (0.1952 in)
    Exhaust
        Standard .... 4.955 to 4.970 mm (0.1950 to 0.1956 in)
        Minimum .... 4.94 mm (0.1944 in)
Valve guide inside diameter (intake and exhaust)
    Standard .... 5.000 to 5.012 mm (0.1968 to 0.1974 in)
    Maximum .... 5.08 mm (0.2 in)
Valve seat width (intake and exhaust) .... 0.5 to 1.0 mm (0.020 to 0.040 in)
Valve spring free length (inner)
    Standard .... 36.3 mm (1.429 in)
    Minimum .... 35.0 mm (1.378 in)
Valve spring free length (outer)
    Standard .... 40.4 mm (1.590 in)
    Minimum .... 39 mm (1.535 in)

Cylinder block

Bore diameter
    Standard .... 68.000 to 68.012 mm (2.677 to 2.6776 in)
    Maximum .... 68.1 mm (2.681 in)
Taper limit .... 0.05 mm (0.002 in)
Out-of-round limit .... 0.05 mm (0.002 in)

Pistons

Piston diameter
    Standard .... 67.942 to 67.958 mm (2.6748 to 2.6755 in)
    Minimum .... 67.79 mm (2.6688 in)
Piston-to-cylinder clearance .... 0.042 to 0.070 mm (0.001 to 0.002 in)
Top ring side clearance
    Standard .... 0.03 to 0.07 mm (0.001 to 0.003 in)
    Maximum .... 0.17 mm (0.006 in)
Second ring side clearance
    Standard .... 0.02 to 0.06 mm (0.0007 to 0.0023 in)
    Maximum .... 0.16 mm (0.006 in)

Pistons (continued)

Top ring groove width
Standard . . . . 1.02 to 1.04 mm (0.0410 to 0.0409 in)
Maximum . . . . 1.12 mm (0.044 in)
Second ring groove width
Standard . . . . 1.01 to 1.093 mm (0.0397 to 0.0405 in)
Maximum . . . . 1.11 mm (0.0437 in)
Oil ring groove width
Standard . . . . 2.51 to 2.53 mm (0.0988 to 0.0996 in)
Maximum . . . . 2.60 mm (0.1023 in)
Top and second ring thickness
Standard . . . . 0.97 to 0.99 mm (0.0381 to 0.0389 in)
Minimum . . . . 0.90 mm (0.035 in)
Top and second ring end gap
Standard . . . . 0.20 to 0.35 mm (0.0078 to 0.0137 in)
Maximum . . . . 0.7 mm (0.0275 in)

Crankshaft and bearings

Main bearing oil clearance
Standard . . . . 0.020 to 0.044 mm (0.0008 to 0.0017 in)
Maximum . . . . 0.08 mm (0.0031 in)
Main bearing journal diameter
No mark on crank throw . . . . 33.984 to 33.992 mm (1.3379 to 1.3382 in)
"1" mark on crank throw . . . . 33.993 to 34.000 mm (1.3383 to 1.3385 in)
Connecting rod side clearance
Standard . . . . 0.013 to 0.300 mm (0.0005 to 0.0118 in)
Maximum . . . . 0.50 mm (0.0196 in)
Connecting rod bearing oil clearance
Standard . . . . 0.046 to 0.076 mm (0.0018 to 0.0029 in)
Maximum . . . . 0.11 mm (0.0043 in)
Connecting rod big-end bore diameter
No mark on side of rod . . . . 38.000 to 38.008 mm (1.496 to 1.4963 in)
"0" mark on side of rod . . . . 38.009 to 38.016 mm (1.4964 to 1.4966 in)
Connecting rod journal (crankpin) diameter
No mark on crank throw . . . . 34.984 to 34.992 mm (1.3773 to 1.3776 in)
"0" mark on crank throw . . . . 34.993 to 35.000 mm (1.3776 to 1.3779 in)

Oil pump and relief valve

Oil pressure (warm) . . . . 43 to 57 psi @ 4000 rpm
Relief valve opening pressure . . . . 63 to 85 psi

Clutch

Spring free length
Standard . . . . 36.4 mm (1.433 in)
Minimum . . . . 35.1 mm (1.381 in)
Friction plate thickness
Standard . . . . 2.9 to 3.1 mm (0.114 to 0.122 in)
Minimum . . . . 2.8 mm (0.110 in)
Friction and steel plate warpage
Standard . . . . 0.2 mm (0.008 in)
Maximum . . . . 0.3 mm (0.012 in)

Transmission

Shift fork groove width in gears
Standard . . . . 5.05 to 5.15 mm (0.1988 to 0.2027 in)
Maximum . . . . 5.3 mm (0.2086 in)
Shift fork ear thickness
Standard . . . . 4.9 to 5.0 mm (0.1929 to 0.1968 in)
Minimum . . . . 4.8 mm (0.1889 in)
Shift fork guide pin diameter
Standard . . . . 5.9 to 6.0 mm (0.2322 to 0.2362 in)
Maximum . . . . 5.8 mm (0.2283 in)
Shift drum groove width
Standard . . . . 6.05 to 6.20 mm (0.2381 to 0.2440 in)
Maximum . . . . 6.3 mm (0.2480 in)

Torque specifications

Valve cover bolts	10 Nm (87 in-lbs)
Camshaft bearing cap bolts	12 Nm (104 in-lbs)
Camshaft gear bolts	15 Nm (11 ft-lbs)*
Camshaft chain tensioner cap	5 Nm (43 in-lbs)
Camshaft chain tensioner mounting bolts	12 Nm (104 in-lbs)
Camshaft chain tensioner rear guide bracket bolts	12 Nm (104 in-lbs)
Cylinder head bolts	39 Nm (29 ft-lbs)**
Clutch damper bolts	10 Nm (87 in-lbs)
Clutch spring bolts	10 Nm (87 in-lbs)
Clutch hub nut	136 Nm (100 ft-lbs)
Clutch master cylinder clamp bolts	9 Nm (78 in-lbs)
Clutch slave cylinder bleed valve	8 Nm (69 in-lbs)
Clutch line banjo bolts	30 Nm (22 ft-lbs)
Oil pan bolts	12 Nm (104 in-lbs)
Oil pipe-to-cylinder head union bolts	12 Nm (104 in-lbs)
Oil pipe-to-oil cooler union bolts	24 Nm (18 ft-lbs)
Relief valve-to-oil pan	15 Nm (11 ft-lbs)
Long engine mounting bolts	45 Nm (33 ft-lbs)
Short engine mounting bolts	18 Nm (13.5 ft-lbs)
Crankcase bolts	
6 mm bolts marked "12" on head (F4 models)	22 Nm (16 ft-lbs)
6 mm bolts not marked "12" on head	12 Nm (104 in-lbs)
8 mm bolts	27 Nm (20 ft-lbs)
Connecting rod nuts	37 Nm (27 ft-lbs)
Chain guide bracket bolts	12 Nm (104 in-lbs)*
Shift drum mounting plate bolt	Not specified*

Apply a non-permanent thread locking agent to the threads.
**Apply clean engine oil to both sides of the head bolt washers. Tighten the bolts evenly in the proper sequence (see text).*

1 General information

The ZX750 engine/transmission unit is of the water-cooled, inline, four-cylinder design, installed transversely across the frame. The sixteen valves are operated by double overhead camshafts which are chain driven off the crankshaft. Separate rocker arms are used for each valve.

The engine/transmission assembly is constructed from aluminum alloy. The crankcase is divided horizontally.

The crankcase incorporates a wet sump, pressure-fed lubrication system which uses a chain-driven, dual-rotor oil pump, an oil filter and bypass valve assembly, a relief valve and an oil pressure switch. Also contained in the crankcase is the idler shaft and the starter motor clutch. The idler shaft turns the alternator pulley, which in turn drives the alternator through a ribbed belt.

Power from the crankshaft is routed to the transmission via the clutch, which is of the wet, multi-plate type and is gear-driven off the crankshaft. The transmission is a six-speed, constant-mesh unit.

2 Operations possible with the engine in the frame

The components and assemblies listed below can be removed without having to remove the engine from the frame. If, however, a number of areas require attention at the same time, removal of the engine is recommended.

Gear selector mechanism external
* components*
Water pump
Starter motor
Alternator
Clutch assembly and slave cylinder
Oil pan, oil pump and relief valve
Valve cover, camshafts and rocker arms
Cam chain tensioner
Cylinder head
Cylinder block and pistons

3 Operations requiring engine removal

It is necessary to remove the engine/ transmission assembly from the frame and separate the crankcase halves to gain access to the following components:

Crankshaft, connecting rods and bearings
Transmission shafts
Shift drum and forks
Idler shaft and starter motor clutch
Camshaft chain
Alternator/starter chain

4 Major engine repair - general note

1 Refer to Chapter 2A, Section 4.

5 Engine - removal and installation

Note: *Engine removal and installation should be done with the aid of an assistant to avoid damage or injury that could occur if the engine is dropped. A hydraulic floor jack should be used to support and lower the engine if possible (they can be rented at low cost).*

Removal

1 Set the bike on its centrestand.
2 Remove the seat, the fuel tank and the baffle plates (see Chapter 4 and 8).
3 Remove the upper and lower fairings (see Chapter 8).
4 Drain the coolant and the engine oil (see Chapter 1).
5 Remove the ignition coils (see Chapter 5).
6 Remove the carburettors (see Chapter 4) and plug the intake openings with rags.
7 Remove the radiator, radiator hoses, oil lines and oil cooler (see Chapter 3).
8 Remove the exhaust system (see Chapter 4).
9 Remove the clutch slave cylinder (see Section 20).
10 Remove the engine sprocket cover, remove the sprocket retaining nut and detach the sprocket and chain from the engine (see Chapter 6). **Note:** *The chain can't be slipped off the output shaft until the engine is removed from the frame.*
11 Mark and disconnect the wires from the oil pressure switch, neutral switch and the

5.12 A ground wire securing bolt goes in this hole at the right rear of the crankcase

starter motor. Unplug the alternator, sidestand and pickup coil electrical connectors (see Chapters 5 and 9).

12 Remove the bolt securing the ground (earth) wire to the right rear of the crankcase **(see illustration)**.

13 Support the engine with a floor jack and a wood block.

14 Remove the engine mounting bolts/nuts and bracket bolts **(see illustration)**.

15 With the engine still supported, make sure no wires or hoses are still attached to the engine assembly.

16 Raise the engine slightly then, with the help of an assistant, slide the engine out to the right so the output shaft clears the drive chain. It would be helpful to have another jack, or a small table or platform that is the same height as the bottom frame tube, which the engine can be slid onto as it is removed.

Installation

17 Installation is basically the reverse of removal. Note the following points:
  a) *Don't tighten any of the engine mounting bolts until they all have been installed.*
  b) *Use new gaskets at all exhaust pipe connections.*
  c) *Tighten the engine mounting bolts and bracket bolts to the torques listed in this Chapter's Specifications.*
  d) *Adjust the drive chain, throttle cables and choke cable following the procedures in Chapter 1.*

---

6   Engine disassembly and reassembly - general information

1 Before disassembling the engine, clean the exterior with a degreaser and rinse it with water. A clean engine will make the job easier and prevent the possibility of getting dirt into the internal areas of the engine.

2 In addition to the precision measuring tools mentioned earlier, you will need a torque wrench, a valve spring compressor, oil gallery brushes, a piston ring removal and installation tool, a piston ring compressor, a pintype

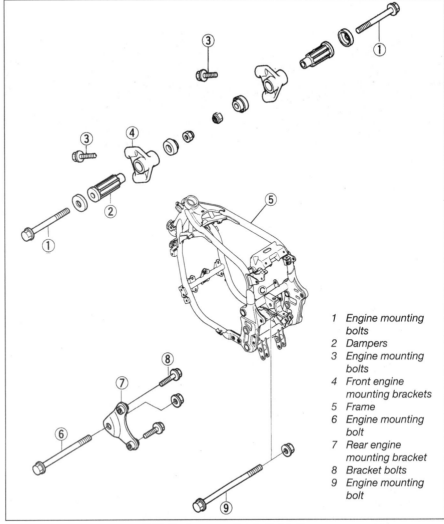

1 *Engine mounting bolts*
2 *Dampers*
3 *Engine mounting bolts*
4 *Front engine mounting brackets*
5 *Frame*
6 *Engine mounting bolt*
7 *Rear engine mounting bracket*
8 *Bracket bolts*
9 *Engine mounting bolt*

5.14 Engine mounting bolt details

spanner wrench and a clutch holder tool (which is described in Section 19). Some new, clean engine oil of the correct grade and type, some engine assembly lube (or moly-based grease), a tube of Kawasaki Bond liquid gasket (part no. 92104-1003) or equivalent, and a tube of RTV (silicone) sealant will also be required. Although it may not be considered a tool, some Plastigauge (type HPG-1) should also be obtained to use for checking bearing oil clearances **(see illustrations)**.

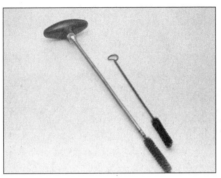

6.2a A selection of brushes is required for cleaning holes and passages in the engine components

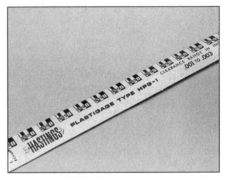

6.2b Type HPG-1 Plastigauge is needed to check the crankshaft, connecting rod and camshaft oil clearances

6.3 An engine stand can be made from short lengths of 2 x 4 lumber and lag bolts or nails

7.5 Remove the bolts (arrows) and lift the valve cover off

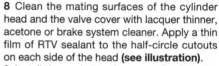

7.6 The upper chain guide is mounted inside the valve cover

3 An engine support stand made from short lengths of 2 x 4's bolted together will facilitate the disassembly and reassembly procedures **(see illustration)**. The perimeter of the mount should be just big enough to accommodate the engine oil pan. If you have an automotive-type engine stand, an adapter plate can be made from a piece of plate, some angle iron and some nuts and bolts.

4 When disassembling the engine, keep "mated" parts together (including gears, cylinders, pistons, etc. that have been in contact with each other during engine operation). These "mated" parts must be reused or replaced as an assembly.

5 Engine/transmission disassembly should be done in the following general order with reference to the appropriate Sections.

Remove the alternator (see Chapter 9)
Remove the cylinder head
Remove the cylinder block
Remove the pistons
Remove the clutch
Remove the oil pan
Remove the external shift mechanism
Separate the crankcase halves
Remove the idler pulley shaft and starter motor
Remove the crankshaft and connecting rods
Remove the transmission shafts/gears
Remove the shift drum/forks

6 Reassembly is accomplished by reversing the general disassembly sequence.

7 Valve cover - removal and installation

Note: *The valve cover can be removed with the engine in the frame. If the engine has been removed, ignore the steps which don't apply.*

Removal

1 Set the bike on its centrestand.
2 Remove the fuel tank (see Chapter 4).
3 If necessary for access, remove the upper and lower fairings (see Chapter 8). Remove the upper and side baffle plates.
4 Remove the air suction valve and the vacuum switching valve (see Chapter 1).
5 Remove the valve cover bolts **(see illustration)**.
6 Lift the cover off the cylinder head. If it's stuck, don't attempt to pry it off - tap around the sides of it with a plastic hammer to dislodge it. Check the chain guide in the centre of the cover - if it's excessively worn, pry it out and install a new one **(see illustration)**.

Installation

7 Peel the rubber gasket from the cover. If it's cracked, hardened, has soft spots or shows signs of general deterioration, replace it with a new one.

8 Clean the mating surfaces of the cylinder head and the valve cover with lacquer thinner, acetone or brake system cleaner. Apply a thin film of RTV sealant to the half-circle cutouts on each side of the head **(see illustration)**.
9 Install the gasket to the cover. Position the cover on the cylinder head, making sure the gasket doesn't slip out of place.
10 Check the rubber seals on the valve cover bolts, replacing them if necessary. Install the bolts, tightening them evenly, to the torque listed in this Chapter's Specifications.
11 The remainder of installation is the reverse of removal.

8 Camshaft chain tensioner - removal and installation

Removal

Caution: Once you start to remove the tensioner bolts, you must remove the tensioner all the way and reset it before tightening the bolts. The tensioner extends and locks in place, so if you loosen the bolts part way and then retighten them, the tensioner or cam chain will be damaged.

1 Loosen the tensioner cap bolt while the tensioner is still installed **(see illustration)**.
2 Remove the tensioner mounting bolts and take it off the engine **(see illustration)**.

7.8 Apply a thin film of sealant to the half-circle cutouts (arrows)

8.1 Loosen the tensioner cap bolt (centre arrow), then remove the tensioner mounting bolts (outer arrows) . . .

8.2 . . . and take the tensioner off

8.3 Cam chain tensioner details

1 O-rings
2 Tensioner body
3 Tensioner cap bolt

3 Remove the tensioner cap bolt and O-ring **(see illustration)**.

Installation

4 Check the O-ring on the tensioner body for cracks or hardening. It's a good idea to replace this O-ring whenever the tensioner cap is removed.

Original tensioner

5 Place the tensioner mounting bolts where you can reach them with one hand, while the other hand holds the tensioner in position as described in Step 7.
6 Press the end of the rod that contacts the chain into the tensioner body. At the same time, turn the other end of the rod clockwise with a screwdriver until the rod protrudes about 10 mm (3/8-inch) from the tensioner body.
Caution: Don't turn the rod counterclockwise (anti-clockwise) or it may separate from the tensioner. If this happens it can't be reassembled.
7 Place the tensioner in position on the engine. Push it firmly against the engine, remove the screwdriver, and install the mounting bolts finger-tight.
Caution: If the tensioner moves away from the engine before you tighten the bolts, the rod will extend too far. If this happens (or you think it might have happened), remove the tensioner and repeat Step 6, then continue with Step 7.
8 Tighten the mounting bolts to the torque listed in this Chapter's Specifications.

New tensioner

9 New tensioners come with a keeper that fits in the tensioner rod slot and holds the rod in the correct position for installation **(see illustration)**.
10 Place the tensioner on the engine. Install the mounting bolts and tighten them to the torque listed in this Chapter's Specifications.
11 Pull the keeper out with needle nosed pliers. **Note:** *Save the keeper and place it in*

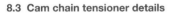

8.9 Install a keeper (A) in the tensioner rod slot to hold it in the correct position

your toolbox for future use. You can use it to hold the tensioner rod in position next time you install the tensioner, leaving both hands free.

Original or new tensioner

12 Install the tensioner cap and O-ring. Tighten the cap to the torque listed in this Chapter's Specifications.

---

9 Camshafts, rocker arm shafts and rocker arms - removal, inspection and installation

Note: *This procedure can be performed with the engine in the frame.*

9.5a Loosen the cam bearing cap bolts (arrows) evenly . . .

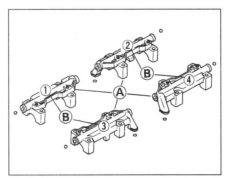

9.5c The camshaft bearing caps (A) are marked with numbers (B) that correspond to numbers on the cylinder head

Camshafts

Removal

1 Remove the valve cover following the procedure given in Section 7.
2 Remove the cam chain tensioner (see Section 8).
3 Remove the pickup coil cover (see Chapter 5).
4 Position the engine at Top Dead Centre (TDC) for cylinders 1 and 4 (see Chapter 1, Valve clearances - check and adjustment, for the TDC locating procedure).
5 Unscrew the bearing cap bolts for one of the camshafts, a little at a time, until they are all loose, then unscrew the bearing cap bolts for the other camshaft **(see illustration)**. Remove the bolts and lift off the bearing caps **(see illustration)**. Note the numbers on the bearing caps which correspond to the numbers on the cylinder head **(see illustration)**. When you reinstall the caps, be sure to install them in the correct positions.

Caution: If the bearing cap bolts aren't loosened evenly, the camshaft may bind.

6 Pull up on the camshaft chain and carefully guide the camshaft out **(see illustration)**. With the chain still held taut, remove the other camshaft. Look for marks on the camshafts. The intake camshaft should have an IN mark and the exhaust camshaft should have an EX

9.5b . . . and lift the caps off, noting the positions of the dowels (arrows)

9.6a Lift the camshaft chain, disengage it from the sprocket and guide the camshaft out

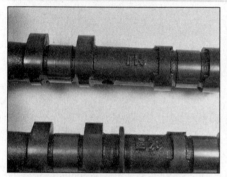

9.6b The intake and exhaust camshafts are identified by cast marks

9.7 Tie the camshaft chain up so it doesn't slip down off the crankshaft sprocket

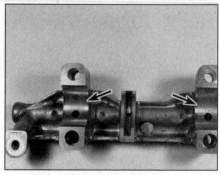

9.9 Check the bearing surfaces (arrows) for wear or damage

mark (see illustration). If you can't find these marks, label the camshafts to ensure they are installed in their original locations. Note: *Don't remove the sprockets from the camshafts unless absolutely necessary.*

7 While the camshafts are out, don't allow the chain to go slack - if you do, it will become detached from the gear on the crankshaft and may bind between the crankshaft and case, which could cause damage to these components. Wire the chain to another component to prevent it from dropping down (see illustration). Also, cover the top of the cylinder head with a rag to prevent foreign objects from falling into the engine.

Inspection

8 Camshaft and sprocket inspection are the same as for ZX600 models (see Chapter 2A).

Installation

9 Make sure the bearing surfaces in the cylinder head and the bearing caps are clean, then apply a light coat of engine assembly lube or moly-based grease to each of them (see illustration).

10 Apply a coat of moly-based grease to the camshaft lobes. Make sure the camshaft bearing journals are clean, then lay the camshafts in the cylinder head (do not mix them up), ensuring the marks on the cam sprockets are aligned properly (see illustrations).

11 Make sure the timing marks are aligned as described in Step 4, then mesh the chain with the camshaft sprockets. Count the number of chain link pins between the EX mark and the IN mark (see illustration). There should be no

9.10a Each camshaft has a mark (IN or EX) and a line next to it . . .

slack in the chain between the two sprockets.

12 Carefully set the bearing caps in place in their proper positions (see illustrations 9.5b and 9.5c) and install the bolts. Snug all of the bolts evenly, then tighten them in a criss-cross pattern to the torque listed in this Chapter's Specifications.

13 Insert your finger or a wood dowel into the cam chain tensioner hole and apply pressure to the cam chain. Check the timing marks to make sure they are aligned (see Step 4) and there are still the correct number of link pins between the EX and IN marks on the cam sprockets. If necessary, change the position of the sprocket(s) on the chain to bring all of the marks into alignment.

Caution: If the marks are not aligned exactly as described, the valve timing will be incorrect and the valves may contact the pistons, causing extensive damage to the engine.

9.10b . . . when installing the camshafts, the IN and EX marks on the camshaft sprockets should be positioned like this, with the marks aligned with the valve cover gasket surface on the cylinder head

14 Install the cam chain tensioner as described in Section 8.

15 Adjust the valve clearances (see Chapter 1).

16 The remainder of installation is the reverse of removal.

Rocker arm shafts and rocker arms

Removal

17 Remove the camshafts following the procedure given above. Be sure to keep tension on the camshaft chain.

18 Lift the rocker arms out of the shafts (see illustration). Place the rocker arms in order in

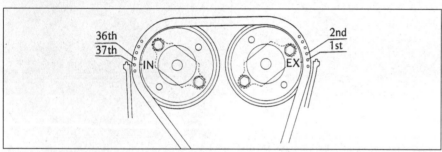

9.11 With no slack in the camshaft chain, there should be 37 pins of the chain between the sprocket marks

9.18a Lift the rocker arms out of the shafts . . .

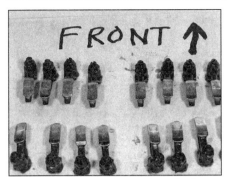

9.18b . . . and place them in order in a holder (a good method is to punch small holes in the side of a cardboard box, then press the ball pivots of the rocker arms into the holes)

9.19 Remove the rocker shaft retaining bolts

9.20a Tap the rocker shaft outward with a punch . . .

a labeled holder so they can be reinstalled in their original positions **(see illustration)**.

19 Remove the rocker shaft retaining bolts **(see illustration)**.

20 Tap the rocker shaft and its end plug out of the engine with a soft metal drift **(see illustrations)**.

21 Repeat the above Steps to remove the other rocker arm shafts and rocker arms. Keep all of the parts in order so they can be reinstalled in their original locations.

Inspection

22 Clean all of the components with solvent and dry them off. Blow through the oil passages in the rocker arms with compressed air, if available. Inspect the rocker arm faces for pits, spalling, score marks and rough spots **(see illustration)**. Check the rocker arm-to-shaft contact areas and the adjusting screws, as well. Look for cracks in each rocker arm. If the faces of the rocker arms are damaged, the rocker arms and the camshafts should be replaced as a set.

Installation

23 Lubricate the rocker arm shaft with engine

9.20b . . . remove the shaft seal as the rocker shaft pushes it out . . .

oil and slide it into the cylinder head. Make sure the retaining bolt hole in the shaft is aligned with the hole in the cylinder head **(see illustration)**. Install the retaining bolt.

24 Coat the circumference of the rocker shaft plug with a film of silicone sealant, then tap it into its bore in the cylinder head.

25 Lay the rocker arms in their notches in the shaft.

26 Repeat Steps 23, 24 and 25 to install the remaining rocker arms and shafts.

27 Install the camshafts following the procedure described earlier in this Section.

9.20c . . . then grasp the rocker shaft and pull it out the rest of the way

10 Cylinder head -
removal and installation

Caution: The engine must be completely cool before beginning this procedure, or the cylinder head may become warped.
Note: *This procedure can be performed with the engine in the frame. If the engine has been removed, ignore the steps which don't apply.*

Removal

1 Set the bike on its centrestand.

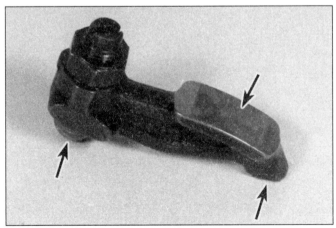

9.22 Check the camshaft contact surface, valve contact surface and pivot ball (arrows) for wear or damage

9.23 Align the retaining bolt hole in the cylinder head with the corresponding hole in the rocker shaft (arrows)

10.10a Gently pry the oil lines loose from the cylinder head; inspect the O-ring on the fitting and replace it if its condition is in doubt

10.10b Remove the banjo bolts that secure the external oil lines . . .

10.10c . . . there's a sealing washer on each side of the banjo fitting; replace these whenever they're removed

2 Remove the fairing and fuel tank (see Chapter 4 and 8).

3 Remove the carburettors (see Chapter 4).

4 Remove the horns (see Chapter 9).

5 Remove the valve cover following the procedure given in Section 7.

6 Remove the cam chain tensioner following the procedure given in Section 8.

7 Remove the radiator (see Chapter 3).

8 Remove the exhaust system (see Chapter 4).

9 Remove the camshafts (see Section 9).

10 Detach the oil lines from the top and front of the cylinder head **(see illustrations)**.

11 Loosen the cylinder head bolts, a little at a time, using the reverse order of the tightening sequence **(see illustration)**.

12 Lift out the cylinder head bolts and their steel washers once they have all been loosened **(see illustration)**. **Note:** *Four of the bolts are longer than the others; these are installed in the centre holes on the exhaust (front) side of the engine* **(holes 1, 3, 5 and 7 in illustration 10.11)**. *Store the bolts in a holder to keep the four long bolts in their correct locations* **(see illustration)**.

13 Pull the cylinder head off the cylinder block. If the head is stuck, tap around the side of the head with a rubber mallet to jar it loose, or use two wooden dowels inserted into the intake or exhaust ports to lever the head off. Don't attempt to pry the head off by inserting a screwdriver between the head and the cylinder block - you'll damage the sealing surfaces.

14 Stuff a clean rag into the cam chain tunnel to prevent the entry of debris. Remove all of the head bolt washers from their seats, using a pair of needle-nose pliers.

15 Remove the two dowel pins from the cylinder block **(see illustration)**.

16 Check the cylinder head gasket and the mating surfaces on the cylinder head and block for leakage, which could indicate warpage. Refer to Chapter 2A and check the flatness of the cylinder head.

17 Clean all traces of old gasket material from the cylinder head and block. Be careful not to let any of the gasket material fall into the crankcase, the cylinder bores or the water passages.

Installation

18 Install the two dowel pins over their studs, then lay the new gasket in place on the cylinder block. Make sure the UP mark on the gasket is positioned on the right-hand side of the engine. Never reuse the old gasket and don't use any type of gasket sealant.

19 Carefully lower the cylinder head onto the cylinder block. It is helpful to have an assistant support the camshaft chain with a piece of wire so it doesn't fall and become kinked or detached from the crankshaft. When the head is resting against the cylinder block, wire the cam chain to another component to keep tension on it.

20 Lubricate both sides of the head bolt washers with engine oil and place them on the bolts.

21 Install the head bolts, making sure the four longer bolts are installed in the correct holes **(1, 3, 5 and 7 in illustration 10.11)**. Using the proper sequence **(see illustration 10.11)**, tighten the bolts to approximately half the torque listed in this Chapter's Specifications

22 Using the same sequence, tighten the bolts to the torque listed in this Chapter's Specifications.

23 The remainder of installation is the reverse of the removal steps.

24 Change the engine oil (see Chapter 1).

10.11 Cylinder head bolt TIGHTENING sequence

10.12a Lift out the cylinder head bolts and washers . . .

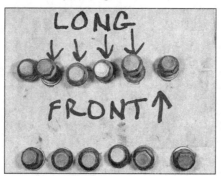

10.12b . . . and place them in a holder so the four longer bolts can be identified (holes punched in the side of a cardboard box will work)

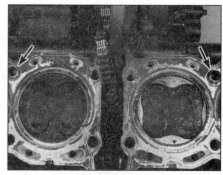

10.15 Note the locations of the dowels (arrows)

13.2a Remove the Allen bolt from the upper side of the water pipe . . .

13.2b . . . and from the lower side, then pull the pipe fittings out of the block

11 Valves/valve seats/valve guides - servicing

1 Because of the complex nature of this job and the special tools and equipment required, servicing of the valves, the valve seats and the valve guides (commonly known as a valve job) is best left to a professional.

2 The home mechanic can, however, remove and disassemble the head, do the initial cleaning and inspection, then reassemble and deliver the head to a dealer service department or properly equipped motorcycle repair shop for the actual valve servicing. Refer to Chapter 2A for those procedures.

3 The dealer service department will remove the valves and springs, recondition or replace the valves and valve seats, replace the valve guides, check and replace the valve springs, spring retainers and keepers/collets (as necessary), replace the valve seals with new ones and reassemble the valve components.

4 After the valve job has been performed, the head will be in like new condition. When the head is returned, be sure to clean it again very thoroughly before installation on the engine to remove any metal particles or abrasive grit that may still be present from the valve service operations. Use compressed air, if available, to blow out all the holes and passages.

12 Cylinder head and valves - disassembly, inspection and reassembly

1 As mentioned in the previous Section, valve servicing and valve guide replacement should be left to a dealer service department or motorcycle repair shop. However, disassembly, cleaning and inspection of the valves and related components can be done (if the necessary special tools are available) by the home mechanic. This way no expense is incurred if the inspection reveals that service work is not required at this time.

2 Procedures are the same as for the ZX600 cylinder head; refer to Chapter 2A for procedures and this Chapter's Specifications.

13 Cylinder block - removal, inspection and installation

Removal

1 Following the procedure given in Section 10, remove the cylinder head. Make sure the crankshaft is positioned at Top Dead Centre (TDC) for cylinders 1 and 4.

2 Remove the water pipe from the rear of the cylinder block (see illustrations).

3 Lift out the camshaft chain front guide (see illustration).

4 Lift the cylinder block straight up off the pistons to remove it. If it's stuck, tap around its perimeter with a soft-faced hammer. Don't attempt to pry between the block and the crankcase, as you will ruin the sealing surfaces.

5 Remove the dowel pins from the mating surface of the crankcase (see illustration). Be careful not to let these drop into the engine. Stuff rags around the pistons and remove the gasket and all traces of old gasket material from the surfaces of the cylinder block and the cylinder head.

Inspection

6 Cylinder block inspection and honing are the same as for ZX600 models. Refer to Chapter 2A for procedures and the Specifications at the beginning of this chapter.

Installation

7 Lubricate the cylinder bores with plenty of clean engine oil. Apply a thin film of moly-based grease to the piston skirts.

8 Install the dowel pins, then lower a new cylinder base gasket over the studs, with the UP mark on the right-hand side of the engine. Some gaskets also have an arrow, which must point to the front of the engine.

9 Slowly rotate the crankshaft until all of the pistons are at the same level. Slide lengths of welding rod or pieces of a straightened-out coat hanger under the pistons, on both sides of the connecting rods (see illustration). This

13.3 Lift the front (exhaust side) chain guide out

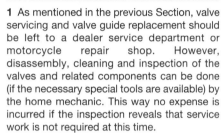

13.5 There's a dowel for the cylinder block at each front corner of the crankcase (arrow)

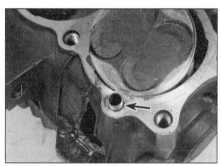

13.9 Slip a rod (A) beneath the front and rear of the pistons; be sure the rear (intake side) cam chain guide is in place (B)

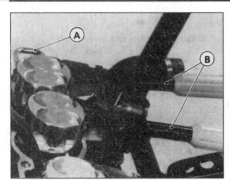

13.10 If you're experienced and very careful, you can guide the cylinder block over the pistons using only a screwdriver, but it's a good idea to use ring compressors

A *Piston support rods*
B *Ring compressors*

will help keep the pistons level as the cylinder block is lowered onto them.
10 Attach four piston ring compressors to the pistons and compress the piston rings **(see illustration)**. Large hose clamps can be used instead - just make sure they don't scratch the pistons, and don't tighten them too much.
11 Make sure the intake side cam chain guide is installed **(see illustration 13.9)**.
12 Position the cylinder block over the engine and carefully lower it down until the piston crowns fit into the cylinder liners. While

16.6 The oil lines are secured to the pan by a banjo bolt (arrow); be sure to use new sealing washers on installation

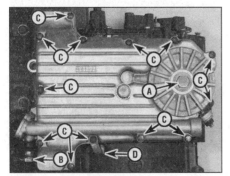

16.8 Remove the oil filter bolt (A), banjo bolt (B), and oil pan bolts (C); one of the bolts secures a wiring harness retainer (D)

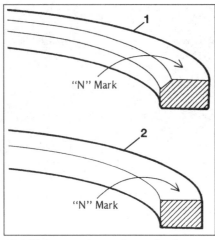

15.1 Make sure the marks on the rings are upward when the rings are installed

1 *Top ring*
2 *Second ring*

doing this, pull the camshaft chain up, using a hooked tool or a piece of coat hanger. Push down on the cylinder block, making sure the pistons don't get cocked sideways, until the bottom of the cylinder liners slide down past the piston rings. A wood or plastic hammer handle can be used to gently tap the block down, but don't use too much force or the pistons will be damaged.
13 Remove the piston ring compressors or hose clamps, being careful not to scratch the pistons. Remove the rods from under the pistons.
14 Install the cam chain front guide **(see illustration 13.3)**.
15 The remainder of installation is the reverse of removal.

14 Pistons - removal, inspection and installation

1 These procedures are the same as for ZX600 models. Refer to Chapter 2A for procedures and the Specifications at the beginning of this Chapter.

16.10 It's a good idea to replace the O-rings (arrows) whenever the oil pan is removed

15 Piston rings - installation

1 Piston ring installation is the same as for ZX600 models, except that ring profiles differ **(see illustration)**. Refer to Chapter 2A for details.

16 Oil pan - removal and installation

Note: *The oil pan can be removed with the engine in the frame.*

Removal
1 Set the bike on its centrestand.
2 Drain the engine oil and remove the oil filter (see Chapter 1).
3 Remove the upper and lower fairings (see Chapter 8).
4 Remove the radiator and oil cooler (see Chapter 3).
5 Remove the exhaust system (see Chapter 4).
6 Remove the banjo bolt that attaches the oil cooler lines to the oil pan **(see illustration)**.
7 Remove the small screw and disconnect the wire from the oil pressure switch (see Chapter 9).
8 Remove the oil pan bolts and detach the pan from the crankcase **(see illustration)**.
9 Remove all traces of old gasket material from the mating surfaces of the oil pan and crankcase.

Installation
10 Check the small O-rings in the oil passages in the crankcase and the large O-ring around the oil filter hole (in the pan) for cracking and general deterioration **(see illustration)**. Replace them if necessary. The flat side of the O-rings must face the crankcase.
11 Position a new gasket on the oil pan. A thin film of RTV sealant can be used to hold the gasket in place. Install the oil pan and bolts, making sure the bolts are in the correct holes and the wiring harness retainer is installed on the proper bolt **(see illustration)**.

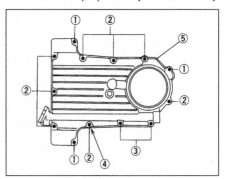

16.11 Oil pan bolt locations

1 *40 mm bolts*
2 *25 mm bolts*
3 *35 mm bolts*
4 *Wiring harness retainer*
5 *Oil pan*

Tighten the bolts to the torque listed in this Chapter's Specifications, using a criss-cross pattern.

12 The remainder of installation is the reverse of removal. Install a new filter and fill the crankcase with oil (see Chapter 1), then run the engine and check for leaks.

17 Oil pump - pressure check, removal, inspection and installation

Note: *The oil pump can be removed with the engine in the frame.*

Pressure check

Warning: If the oil gallery plug is removed when the engine is hot, hot oil will drain out - wait until the engine is cold before beginning this check (it must be cold to perform the relief valve opening pressure check, anyway).

1 If necessary for access, remove the lower fairing (see Chapter 8).
2 Remove the plug at the bottom of the crankcase on the right-hand side and install an oil pressure gauge **(see illustration)**.
3 Start the engine and watch the gauge while varying the engine rpm. The pressure should stay within the relief valve opening pressure listed in this Chapter's Specifications. If the

17.2 Remove the oil gallery plug (arrow) and install a pressure gauge

pressure is too high, the relief valve is stuck closed. To check it, see Section 18.
4 If the pressure is lower than the standard, either the relief valve is stuck open, the oil pump is faulty, or there is other engine damage. Begin diagnosis by checking the relief valve (see Section 18), then the oil pump. If those items check out okay, chances are the bearing oil clearances are excessive and the engine needs to be overhauled.
5 If the pressure reading is in the desired range, allow the engine to warm up to normal operating temperature and check the pressure again, at the specified engine rpm. Compare your findings with this Chapter's Specifications.

17.9 Work the pickup O-ring free of the crankcase and take the pickup out

6 If the pressure is significantly lower than specified, check the relief valve and the oil pump.

Removal

7 Remove the oil pan (see Section 16).
8 Remove the clutch assembly (see Section 19).
9 Remove the oil pickup **(see illustration)**.
10 Turn the crankshaft so the drive slot and tab in the oil pump shaft and water pump shaft are vertical **(see illustration)**.
11 Remove the oil pump sprocket bolt and chain guide bolt, then take out the sprocket **(see illustrations)**.
12 Remove the oil pump mounting bolts, then take the pump and holder out **(see illustration)**.

17.10 Place the oil pump drive tab (arrow) and the water pump drive slot in a vertical position, so they'll be aligned on installation of the water pump

17.11a Remove the oil pump sprocket bolt (arrow); it's located behind the clutch housing

17.11b Use a socket (arrow) to remove the chain guide bolt . . .

17.11c . . . lift out the chain guide . . .

17.11d . . . then disengage the oil pump sprocket from the chain and lift it out

17.12 Loosen the holder-to-oil pump bolt (A) if you plan to separate the pump and holder, then remove the mounting bolts (B) and lift the oil pump out

17.13 Remove the cover screws (arrows); use an impact driver if necessary

Inspection

13 Remove the oil pump cover screws and lift off the cover **(see illustration)**. Thoroughly clean the mating surfaces.

14 Remove the oil pump shaft, pin, inner rotor and outer rotor from the pump **(see illustration)**. Mark the rotors so they can be installed in the same relative positions.

15 Wash all the components in solvent, then dry them off. Check the pump body, the rotors and the cover for scoring and wear. Make sure the pick-up screen isn't clogged. Kawasaki doesn't publish clearance specifications, so if any damage or uneven or excessive wear is evident, replace the pump. If you are rebuilding the engine, it's a good idea to install a new oil pump.

16 Reassemble the pump by reversing the removal steps, but before installing the cover, pack the cavities between the rotors with petroleum jelly - this will ensure the pump develops suction quickly and begins oil circulation as soon as the engine is started. Be sure to use a new gasket.

Installation

17 If you removed the oil pump sprocket, engage it with the chain before installing the oil pump.

17.18 Lift the latch (A) clear of the grooves in the tensioner pushrod, then compress the pushrod (B) against the spring – when the small hole appears at point C, slip a thin piece of wire into it to hold the pushrod compressed

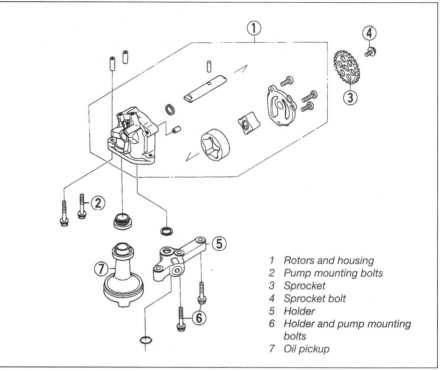

1	Rotors and housing
2	Pump mounting bolts
3	Sprocket
4	Sprocket bolt
5	Holder
6	Holder and pump mounting bolts
7	Oil pickup

17.14 Oil pump details

18 Hold the oil pump and unlatch the stopper from the alternator chain tensioner pushrod **(see illustration)**. Press the alternator chain tensioner pushrod into the oil pump, then slip a piece of thin wire into the pushrod hole to hold the pushrod in the compressed position (the wire will be removed after the oil pump is installed).

19 Position the oil pump on the engine. Coat the threads of the oil pump bolts with non-permanent thread locking agent, then install the bolts and tighten them to the torque listed in this Chapter's Specifications. Remove the wire from the tensioner to allow it to take up chain slack.

20 Install the oil pickup so the flat side is toward the oil pump holder **(see illustration)**.

21 The remainder of installation is the reverse of the removal steps.

18 Oil pressure relief valve - removal, inspection and installation

Removal

1 Remove the oil pan (see Section 16).

2 Unscrew the relief valve from the oil pan **(see illustration)**.

Inspection

3 Clean the valve with solvent and dry it, using compressed air if available.

4 Using a wood or plastic tool, depress the steel ball inside the valve and see if it moves smoothly. Make sure it returns to its seat completely. If it doesn't, replace it with a new one (don't attempt to disassemble and repair it).

17.20 Install the pickup with its flat side (arrow) toward the oil pump

18.2 The oil pressure relief valve is mounted in the oil pan (arrow)

19.3a Loosen the clutch cover bolts (arrows) evenly in a criss-cross pattern . . .

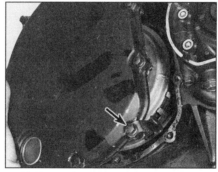

19.3b . . . then remove the cover; one lower bolt secures a wiring harness clip (arrow)

19.4a Remove the clutch spring bolts . . .

Installation

5 Apply a non-hardening thread locking compound to the threads of the valve and install it into the oil pan, tightening it to the torque listed in this Chapter's Specifications.
6 The remainder of installation is the reverse of removal.

19 Clutch - removal, inspection and installation

Note: *The clutch can be removed with the engine in the frame.*

Removal

1 Set the bike on its centrestand and remove the lower fairing (see Chapter 8).
2 Drain the engine oil (see Chapter 1).

19.4b . . . take out the springs . . .

3 Remove the clutch cover bolts and take the cover off **(see illustrations)**. If the cover is stuck, tap around its perimeter with a soft-face hammer.

19.4c . . . and lift off the spring plate

4 Remove the clutch spring bolts **(see illustration)**. Remove the clutch springs, spring plate, bearing and push piece **(see illustrations)**.

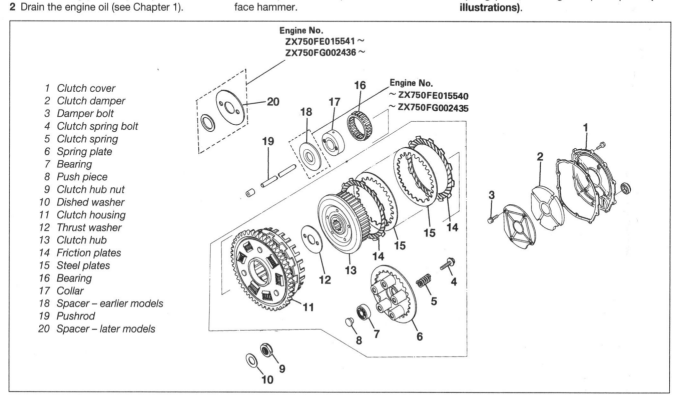

1 Clutch cover
2 Clutch damper
3 Damper bolt
4 Clutch spring bolt
5 Clutch spring
6 Spring plate
7 Bearing
8 Push piece
9 Clutch hub nut
10 Dished washer
11 Clutch housing
12 Thrust washer
13 Clutch hub
14 Friction plates
15 Steel plates
16 Bearing
17 Collar
18 Spacer – earlier models
19 Pushrod
20 Spacer – later models

19.4d Clutch details

19.5a Take out the first friction plate . . .

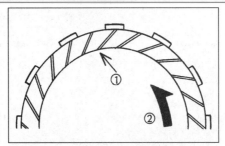

19.5b . . . noting the direction of the oil
grooves; some are directional

1 Oil grooves
2 Direction of rotation

19.5c Take out the first steel plate, then
continue removing friction and steel plates

HAYNES
HiNT

An old steel plate and friction plate can
be drilled and bolted together to make a
holding tool . .

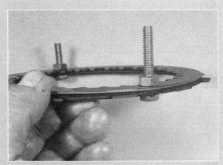

. . . the bolts should be long enough to
serve as removal handles . . .

. . . slip the bolted plates between the
clutch hub and housing in their normal
installed position, with the bolt shafts
facing outward

*If the clutch plates will be replaced, you can drill holes in a friction plate and a steel plate and bolt them together as they would be
when installed. Slip the bolted plates into their installed positions; the clutch hub will be locked to the clutch housing.*

5 Remove the outermost clutch friction plate
from the clutch housing **(see illustration)**.
Note the direction of the oil grooves in the
friction plates **(see illustration)**; on some
models they are directional and must face the
same way during installation. Remove the
steel plate **(see illustration)**, then continue
removing the friction and steel plates until all
are removed.

6 Bend back the staked portion of the clutch
hub nut with a hammer and sharp punch
(see illustration). Remove the clutch hub
nut, using a special holding tool (Kawasaki
tool no. 57001-305) to prevent the clutch
housing from turning **(see illustration 19.7a
in Chapter 2A)**. An alternative to this tool
can be fabricated from some steel strap,
bent at the ends and bolted together in the

middle **(see illustration 19.7b in Chapter
2A)**. Shift the transmission into a low gear
and have an assistant hold the rear brake on.
Unscrew the nut and remove it **(see
illustration)**.

7 Remove the dished washer, clutch hub and
thrust washer **(see illustrations)**. Thread a 6
mm bolt or tap into the collar and pull it out
(see illustration).

19.6a If the clutch hub is secured by a
staked nut, bend back the staked portion
(arrow) with a hammer and sharp punch

19.6b Remove the clutch hub nut

19.7a Remove the dished washer . . .

19.7b . . . the clutch hub . . .

19.7c . . . and thrust washer

19.7d Thread a 6 mm bolt or tap into one of the holes in the collar . . .

19.7e . . . and pull the collar out

19.8a Slide the clutch housing sideways to disengage its teeth from the crankshaft . . .

19.8b . . . and lift out the bearing . . .

8 Remove the clutch housing and bearing **(see illustration)**. Remove the spacer behind the clutch housing **(see illustrations)**.

Inspection

9 This is the same as for ZX600 models (see Chapter 2A). Wear tolerances are listed in this Chapter's Specifications.

Installation

10 Install the spacer over the transmission mainshaft, with the chamfered side facing in **(see illustration)**.
11 Lubricate the collar and the needle bearing with engine oil and slide them over the mainshaft.
12 Install the clutch housing, thrust washer and the clutch hub. The dished side of the washer faces in **(see illustration 19.10)**.

13 If the hub nut is a self-locking type, install a new nut. If the hub nut has a separate lockwasher, install a new lockwasher. Tighten the hub nut to the torque listed in this Chapter's Specifications. Use the technique described in Step 6 to prevent the hub from turning. After tightening, stake the nut or bend the lockwasher against one of the flats so the nut can't loosen.
14 Coat the clutch friction plates with engine oil. Install the clutch plates, starting with a friction plate and alternating them. There are eight friction plates and seven steel plates. Make sure the oil grooves in the friction plates are facing the proper direction **(see illustration 19.5b)**.
15 Lubricate the pushrod and install it through the spring plate. Mount the spring

plate to the clutch assembly and install the springs and bolts, tightening them to the torque listed in this Chapter's Specifications in a criss-cross pattern.
16 Install the clutch cover and bolts, using a new gasket. Tighten the bolts, in a criss-cross pattern, to the torque listed in this Chapter's Specifications.
17 Fill the crankcase with the recommended type and amount of engine oil (see Chapter 1).

19.8c . . . the clutch housing . . .

19.8d . . . and the spacer

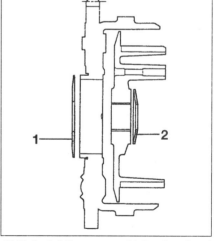

19.10 Install the spacer with its chamfered side toward the engine; install the washer with its dished side toward the engine

1 Spacer   2 Washer

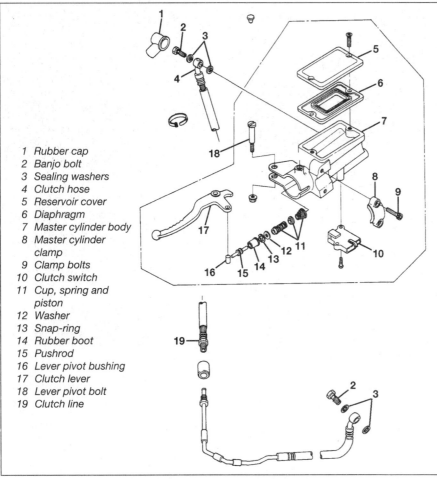

1 Rubber cap
2 Banjo bolt
3 Sealing washers
4 Clutch hose
5 Reservoir cover
6 Diaphragm
7 Master cylinder body
8 Master cylinder clamp
9 Clamp bolts
10 Clutch switch
11 Cup, spring and piston
12 Washer
13 Snap-ring
14 Rubber boot
15 Pushrod
16 Lever pivot bushing
17 Clutch lever
18 Lever pivot bolt
19 Clutch line

20.2 Clutch master cylinder details

20.18 The slave cylinder (A) is secured by two mounting bolts (B)

14 Install the rubber boot and pushrod.
15 When you install the clutch lever, align the hole in the lever bushing with the pushrod.

Master cylinder installation

16 Installation is the reverse of the removal steps, with the following additions:
a) Make sure the UP mark on the master cylinder clamp is upright and the arrow points upward.
b) Tighten the clamp bolts to the torque listed in this Chapter's Specifications. Tighten the upper bolt first, then the lower bolt. There will be a small gap between the master cylinder and the clamp at the bottom.
c) Use a new sealing washer on each side of the banjo bolt fitting and tighten the banjo bolt to the torque listed in this Chapter's Specifications.
d) Fill and bleed the master cylinder as described below. Operate the clutch and check for fluid leaks.

Slave cylinder removal

17 If you're removing the slave cylinder for overhaul, loosen the banjo fitting bolt while the slave cylinder is still mounted on the engine. If you're just removing it for access to other components, leave the hydraulic line connected.
18 Remove the mounting bolts and take the slave cylinder off (see illustration).
19 If you're removing the slave cylinder for overhaul, remove the banjo fitting bolt and detach the hydraulic line. Place the end of the line in a container to catch dripping fluid.
20 If the hydraulic line is still connected, push the piston as far into the bore as it will go. Hold the piston in, slowly squeeze the clutch lever to the handlebar and tie the clutch lever in that position. Otherwise, the slave cylinder piston will fall out.

Slave cylinder overhaul

21 Let the pressure of the slave cylinder spring push the piston out of the cylinder, then remove the spring.
22 Separate the spring from the piston.
23 Thoroughly clean all parts in clean brake fluid (don't use any type of petroleum-based solvent).

---

20 Clutch hydraulic system - component removal, overhaul and installation

Caution: Brake and clutch fluid will damage paint. Wipe up any spills immediately and wash the area with soap and water.

Master cylinder removal

1 Disconnect the electrical connector for the clutch switch beneath the master cylinder.
2 Place a towel under the master cylinder to catch any spilled fluid, then remove the union bolt from the master cylinder fluid line (see illustration).
3 Remove the master cylinder clamp bolts and take the cylinder body off the handlebar.

Master cylinder overhaul

4 Remove the lever pivot bolt and nut and take off the lever (see illustration 20.2).
5 Remove the cover and rubber diaphragm from the reservoir.
6 Remove the rubber boot and pushrod from the master cylinder.
7 Remove the snap-ring, then dump out the

piston and secondary cup, primary cup and spring. If they won't come out, blow compressed air into the fluid line hole.

⚠️ Warning: The piston may shoot out forcefully enough to cause injury. Point the piston at a block of wood or a pile of rags inside a box and apply air pressure gradually. Never point the end of the cylinder at yourself, including your fingers.

8 Thoroughly clean all of the components in clean brake fluid (don't use any type of petroleum-based solvent).
9 Check the piston and cylinder bore for wear, scratches and rust. If the piston shows these conditions, replace it and both rubber cups as a set. If the cylinder bore has any defects, replace the entire master cylinder.
10 Install the spring in the cylinder bore, wide end first.
11 Coat a new cup with brake fluid and install it in the cylinder, wide end first.
12 Coat the piston with brake fluid and install it in the cylinder.
13 Install the washer. Press the piston into the bore and install the snap ring to hold it in place.

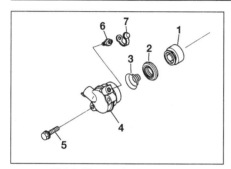

20.24 Slave cylinder details

1 Piston	5 Mounting bolt
2 Seal	6 Bleed valve
3 Spring	7 Bleed valve cap
4 Cylinder body	

24 Check the cylinder bore and piston for wear, scratches and rust. If the piston shows these conditions, replace it and the seal as a set. If the cylinder bore has any defects, replace the entire slave cylinder. If the piston and bore are good, carefully remove the seal from the piston and install a new one with its lip facing into the bore **(see illustration)**.

Slave cylinder Installation

25 Installation is the reverse of the removal procedure with the following additions:
   a) *Use new sealing washers on the clutch fluid line.*
   b) *Tighten the mounting bolts and fluid line banjo bolt to the torques listed in this Chapter's Specifications.*
   c) *Bleed the clutch (see below).*
   d) *Operate the clutch and check for fluid leaks.*

Bleeding the clutch

26 Place the bike on its centrestand and point the front wheel straight ahead.
27 Remove the master cylinder cover and diaphragm. Top up the master cylinder with fluid to the upper edge of the fluid level window.
28 Remove the cap from the bleed valve on the slave cylinder. Place a box wrench (ring spanner) over the bleed valve. Attach a rubber tube to the valve fitting and put the other end

of the tube in a container. Pour enough clean brake fluid into the container to cover the end of the tube.
29 Slowly squeeze the clutch lever several times. At the same time, tap on the clutch fluid line, starting at the bottom and working your way to the top. Stop when no more air bubbles can be seen rising from the bottom of the reservoir.
30 Squeeze the clutch lever several times until you feel an increase in the effort required to pull the lever, then hold it in.
31 With the lever held in, quickly open the bleed valve to let air and fluid escape, then close it.
32 Repeat Steps 30 and 31 until there aren't any more bubbles in the fluid flowing into the container. **Note:** *Keep an eye on the fluid level in the reservoir. If it drops too low, air will be sucked into the line and the procedure will have to be repeated.*
33 Replenish the master cylinder with fluid, then reinstall the diaphragm and cover and tighten the screws securely.

---

21 External shift mechanism - removal, inspection and installation

Removal

1 Set the bike on its centrestand.

21.7 Remove the cover screws (arrows) with an impact driver

2 Drain the engine oil and coolant (see Chapter 1).
3 Remove the lower fairing (see Chapter 8).
4 Remove the clutch slave cylinder (see Section 20).
5 Remove the shift lever, engine sprocket cover and the engine sprocket (see Chapter 6). Disconnect the wire from the neutral switch.
6 Remove the alternator (see Chapter 9).
7 Position a drain pan under the shift mechanism cover. Remove the screws **(see illustration)** and detach the cover from the crankcase. **Note:** *The two flathead screws at the rear of the cover can be very difficult to remove, even with an impact driver. If you need to get the bike back in service quickly, it would be a good idea to start early on a day when your local Kawasaki dealership is open so you can buy new screws if the old ones are ruined during removal.*
8 Pull the shift mechanism arm (against the pressure of the pawl spring) toward the shift shaft until the pointed tips clear the shift drum, then pull the mechanism and shaft off **(see illustrations)**.
Caution: Don't pull the shift rod out of the crankcase - the shift forks will fall into the oil pan, and the crankcase will have to be separated to reinstall them.
9 Note the positions of the gear and neutral positioning levers and their springs **(see illustration)**. Remove the nuts that secure the levers and lift them off **(see illustrations)**. The

21.8a Compress the shift mechanism arm until the tips are clear . . .

21.8b . . . then take the shift mechanism out

21.9a Remove the nuts (arrows) that secure the gear and neutral positioning levers . . .

21.9b . . . remove the collar . . .

21.9c . . . the lever . . .

21.9d . . . the spring . . .

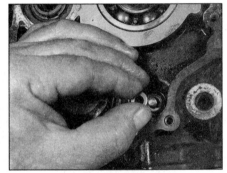

21.9e . . . and the washer

levers are interchangeable, but it's a good idea to label them for return to their original locations, since they will take a wear pattern during use.

Inspection

10 Check the shift shaft for bends and damage to the splines. If the shaft is bent, you can attempt to straighten it, but if the splines are damaged it will have to be replaced.
11 Check the condition of the return spring and the pawl spring. Replace them if they are cracked or distorted.
12 Check the shift mechanism arm and the overshift limiter for cracks, distortion and wear. If any of these conditions are found, replace the shift mechanism.

21.14a Check the condition of the pushrod seal (shown with cover installed) . . .

21.14b . . . and the shift shaft seal; remove the cover and replace them if they've been leaking

13 Make sure the return spring pin isn't loose in the crankcase. If it is, unscrew it, apply a non-hardening locking compound to the threads, then reinstall it and tighten it securely.
14 Check the condition of the seals in the cover **(see illustrations)**. If they have been leaking, drive them out with a hammer and punch. New seals can be installed by driving them in with a socket.

Installation

15 Slide the external shift mechanism into place, lifting the shift arm and the overshift limiter to clear the shift drum **(see illustration)**.
16 Install the gear and neutral positioning levers and make sure the springs are positioned correctly **(see illustration 21.9a)**.
17 Apply high-temperature grease to the lips of the seals. Wrap the splines of the shift shaft with electrical tape, so the splines won't damage the seal as the cover is installed.
18 Apply a thin coat of RTV sealant to the cover mating areas on the crankcase, where the halves of the crankcase join **(see illustration)**.
19 Carefully guide the cover into place. Apply non-permanent thread locking agent to the threads of the two flathead screws along the rear of the cover, then install all of the screws, tightening them securely. Reconnect the neutral switch wire.

21.15 The shift mechanism should look like this when it's installed

20 Install the engine sprocket and chain (see Chapter 6) and engine sprocket cover.
21 Install the shift lever (see Chapter 6).
22 Check the engine oil level and add some, if necessary (see Daily (pre-ride) checks).

22 Crankcase - disassembly and reassembly

Disassembly

1 To examine and repair or replace the crankshaft, connecting rods, bearings, transmission components and idler shaft/starter motor clutch, the crankcase must be split into two parts.
2 Remove the alternator and starter motor (see Chapter 9).
3 On the right side of the engine, remove the clutch (see Section 19), pickup coil cover and pickup coils (see Chapter 5).
4 On the left side of the engine, remove the water pump (see Chapter 3) and external shift mechanism (see Section 21).
5 On the bottom of the engine, remove the oil filter, oil pan, pump and pickup (see Chapter 1 and Sections 16 and 17).
6 If you're planning to remove the crankshaft, remove the cylinder head, cylinder block and pistons (see Sections 10, 13 and 14).

21.18 Apply a small amount of sealant to the crankcase parting line where it meets the cover mating surface (arrows)

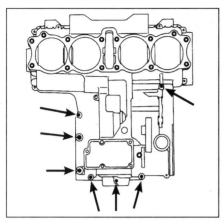

22.7 Upper crankcase bolts

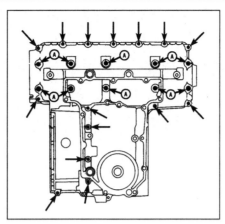

22.9a The lower crankcase 8 mm bolts (A) and 6 mm bolts (arrows)

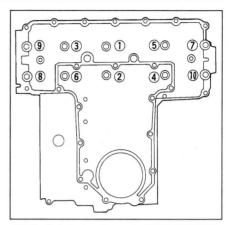

22.9b Crankcase 8 mm bolt TIGHTENING sequence

22.10a There's a pry point at each end of the crankcase . . .

22.10b . . . pry only at these points to prevent damaging the gasket surfaces

22.11 Lift the lower case half off the upper half

7 Remove the upper crankcase bolts (see illustration).

8 Unbolt the oil pump sprocket, then slide the sprocket and chain off together. Remove the chain guide and separate the sprocket from the chain (see Section 17).

9 Turn the crankcase over and remove the lower crankcase bolts (see illustration). Loosen the 8 mm bolts that secure the crankshaft in the reverse order of the tightening sequence (see illustration).

10 Pry the crankcase halves apart at the pry points only (see illustrations). *Caution: Don't pry between the crankcase halves or the mating surfaces will be gouged, resulting in an oil leak.*

11 Lift the lower crankcase half off the upper half (see illustration).

12 Refer to Sections 23 through 31 for information on the internal components of the crankcase.

Reassembly

13 Remove all traces of sealant from the crankcase mating surfaces. Be careful not to let any fall into the case as this is done.

14 Check to make sure the two dowel pins are in place in their holes in the mating surface of the upper crankcase half.

15 Pour some engine oil over the transmission gears, the crankshaft main bearings and the shift drum. Don't get any oil on the crankcase mating surface.

16 Apply a thin, even bead of Kawasaki Bond sealant (part no. 92104-1003) to the indicated areas of the crankcase mating surfaces (see

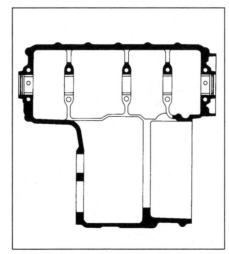

22.16 Apply sealant to the shaded areas

illustration). Also apply RTV sealant to the areas near the ends of the crankshaft seal areas (lay it over the Kawasaki Bond).

Caution: Don't apply an excessive amount of either type of sealant, as it will ooze out when the case halves are assembled and may obstruct oil passages.

17 Check the position of the shift drum - make sure it's in the neutral position (see illustration).

18 Carefully place the lower crankcase half onto the upper crankcase half. While doing

22.17 Position the shift drum and forks in the neutral position

this, make sure the shift forks fit into their gear grooves **(see illustration)**.

19 Install the lower crankcase half bolts and tighten them so they are just snug.

20 In two steps, tighten the larger bolts (8 mm), in the indicated sequence, to the torque listed in this Chapter's Specifications **(see illustration 22.9b)**.

21 Turn the case over and install the upper crankcase half bolts, tightening them to the torque listed in this Chapter's Specifications.

22 Turn the case over again and install the smaller (6 mm) bolts in the lower crankcase half, tightening them to the torque listed in this Chapter's Specifications.

23 Turn the main drive shaft and the output shaft to make sure they turn freely. Install the shift lever on the shift shaft and, while turning the output shaft, shift the transmission through the gears, from first to sixth, then back to first. The positive neutral finder prevents the transmission from being shifted past neutral into second gear unless the output shaft is turning at a fairly high rate of speed, which can be difficult. If the transmission doesn't shift properly, the case will have to be separated again to correct the problem. Also make sure the crankshaft turns freely.

24 The remainder of installation is the reverse of removal.

23 Crankcase components - inspection and servicing

1 After the crankcases have been separated and the crankshaft, shift drum and forks and transmission components removed, the crankcases should be cleaned thoroughly with new solvent and dried with compressed air. All oil passages should be blown out with compressed air and all traces of old gasket sealant should be removed from the mating surfaces.

Caution: Be very careful not to nick or gouge the crankcase mating surfaces or leaks will result. Check both crankcase sections very carefully for cracks and other damage.

22.18 Make sure the forks engage the gear grooves (arrows)

2 Check the cam chain and alternator/starter chain guides for wear - one is in the upper case half and the other is in the lower case half. If they appear to be worn excessively, replace them.

3 Check the ball bearings in the case. If they don't turn smoothly, drive them out with a bearing driver or a socket having an outside diameter slightly smaller than that of the bearing. Before installing them, allow them to sit in the freezer overnight, and about fifteen-minutes before installation, place the case half in an oven, set to about 94∞ C (200∞ F), and allow it to heat up. The bearings are an interference fit, and this will ease installation.

> ⚠️ *Warning: Before heating the case, wash it thoroughly with soap and water so no explosive fumes are present. Also, don't use a flame to heat the case.*

4 Remove the breather cover and lift out the separator screen **(see illustrations)**. Clean the screen with solvent. Replace it if it's clogged or damaged.

5 If any damage is found that can't be repaired, replace the crankcase halves as a set.

24 Main and connecting rod bearings - general note

1 Even though main and connecting rod bearings are generally replaced with new ones during the engine overhaul, the old bearings

should be retained for close examination as they may reveal valuable information about the condition of the engine.

2 Bearing failure occurs mainly because of lack of lubrication, the presence of dirt or other foreign particles, overloading the engine and/or corrosion. Regardless of the cause of bearing failure, it must be corrected before the engine is reassembled to prevent it from happening again.

3 When examining the bearings, remove the main bearings from the case halves and the rod bearings from the connecting rods and caps and lay them out on a clean surface in the same general position as their location on the crankshaft journals. This will enable you to match any noted bearing problems with the corresponding side of the crankshaft journal.

4 Dirt and other foreign particles get into the engine in a variety of ways. It may be left in the engine during assembly or it may pass through filters or breathers. It may get into the oil and from there into the bearings. Metal chips from machining operations and normal engine wear are often present. Abrasives are sometimes left in engine components after reconditioning operations such as cylinder honing, especially when parts are not thoroughly cleaned using the proper cleaning methods. Whatever the source, these foreign objects often end up imbedded in the soft bearing material and are easily recognized. Large particles will not imbed in the bearing and will score or gouge the bearing and journal. The best prevention for this cause of bearing failure is to clean all parts thoroughly and keep everything spotlessly clean during engine reassembly. Frequent and regular oil and filter changes are also recommended.

5 Lack of lubrication or lubrication breakdown has a number of interrelated causes. Excessive heat (which thins the oil), overloading (which squeezes the oil from the bearing face) and oil leakage or throw off (from excessive bearing clearances, worn oil pump or high engine speeds) all contribute to lubrication breakdown. Blocked oil passages will also starve a bearing and destroy it. When lack of lubrication is the cause of bearing failure, the bearing material is wiped or extruded from the steel backing of the

23.4a Remove the breather cover bolts (arrows) . . .

23.4b . . . and lift off the cover, noting the locations of the dowels (arrows) . . .

23.4c . . . then lift out the screen

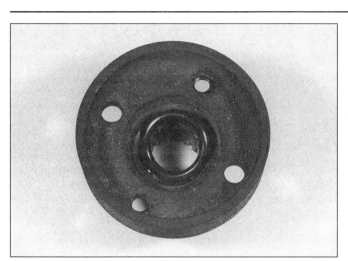

27.3 There's an O-ring behind the pulley bolt

27.4 Thread a pair of 6 mm bolts into the holes in the pulley, then with the pulley bolt removed, tighten the 6 mm bolts against the case to push the pulley off

bearing. Temperatures may increase to the point where the steel backing and the journal turn blue from overheating.

6 Riding habits can have a definite effect on bearing life. Full throttle low speed operation, or lugging (laboring) the engine, puts very high loads on bearings, which tend to squeeze out the oil film. These loads cause the bearings to flex, which produces fine cracks in the bearing face (fatigue failure). Eventually the bearing material will loosen in pieces and tear away from the steel backing. Short trip driving leads to corrosion of bearings, as insufficient engine heat is produced to drive off the condensed water and corrosive gases produced. These products collect in the engine oil, forming acid and sludge. As the oil is carried to the engine bearings, the acid attacks and corrodes the bearing material.

7 Incorrect bearing installation during engine assembly will lead to bearing failure as well. Tight fitting bearings which leave insufficient bearing oil clearances result in oil starvation. Dirt or foreign particles trapped behind a bearing insert result in high spots on the bearing which lead to failure.

8 To avoid bearing problems, clean all parts thoroughly before reassembly, double check all bearing clearance measurements and

lubricate the new bearings with engine assembly lube or moly-based grease during installation.

25 Crankshaft and main bearings - removal, inspection and installation

1 Refer to Chapter 2A for service procedures and this Chapter's Specifications.

2 Note that the main journal diameter marks are located on the first, second, fourth, seventh and eighth crank throws (counting from the left end of the crankshaft). Part numbers are available from Kawasaki dealers.

26 Connecting rods and bearings - removal, inspection and installation

1 Refer to Chapter 2A for service procedures and this Chapter's Specifications.

2 The connecting rod journal diameter marks are located on the second, fourth, fifth and seventh crank throws (counting from the left end of the crankshaft). Part numbers are available from Kawasaki dealers

27 Idler pulley, shaft and starter motor clutch - removal, inspection and installation

Pulley removal

1 Remove the clutch slave cylinder, engine sprocket cover and alternator (see Section 20, Chapter 6 and Chapter 9).

2 Keep the pulley from turning. If the engine is in the bike, shift the transmission into a low gear and have an assistant hold the rear brake on. If the engine has been removed, hold the pulley from turning with a strap wrench.

3 Unscrew the pulley bolt and remove the O-ring **(see illustration)**.

4 Thread a pair of 6 mm bolts into the threaded holes in the pulley and tighten them against the engine to push the pulley off **(see illustration)**.

Idler shaft, gear and starter clutch removal

5 Remove the engine, separate the crankcase halves and remove the transmission shafts (see Sections 5, 22 and 29).

6 Remove the oil pump drive chain guide (see Section 17).

7 Remove the idler shaft oil nozzle and bearing retainer **(see illustrations)**.

27.7a Remove the oil nozzle bolts (arrows) . . .

27.7b . . . and work the O-ring free of the case . . .

27.7c . . . then unbolt the bearing retainer (arrow)

27.9 Support the starter clutch and remove the idler shaft

27.10a Lift the starter clutch out

27.10b Disengage the idler shaft sprocket from the chain and remove the sprocket

27.11 Starter idle gear details

A Starter idle gear   *C Short collar*
B Long collar   *D Snap-ring*

8 Remove the idler shaft pulley (see Steps 2 through 4 above).
9 Support the starter clutch chain with one hand and pull the idler shaft out of the crankcase **(see illustration)**.
10 Lift the starter clutch and idler shaft sprocket out of the crankcase **(see illustrations)**.
11 Remove the snap-ring from the idle gear shaft **(see illustration)**. Pull the shaft out of the crankcase and lift out the starter idle gear.

27.12 Hold the starter clutch and try to rotate the gear; it should turn easily in one direction, but not at all in the other direction

Inspection

Starter motor clutch

12 Hold the starter motor clutch and attempt to turn the starter motor clutch gear back and forth **(see illustration)**. It should turn in one direction only.
13 If the starter motor clutch turns freely in both directions, or if it's locked up, replace it.
14 Remove the snap-ring and washer and slide off the gear. Check the rollers for scoring and pitting and the retainers for damage.

Check the gear teeth for cracks and chips. Check the needle roller bearing in the gear for wear or damage. Replace parts as necessary.

Starter motor idler gear

15 Inspect the teeth on the starter motor idler gear for cracks and chips. Turn the idler gear to make sure it spins freely. If the idler gear exhibits any undesirable conditions, replace it. Remove the idler gear as described in Step 11. Coat the shaft with engine assembly lube or moly-based grease before installing it.

Idler shaft and bearings

16 Check the splines, sprocket, and threads on the shaft for wear or damage **(see illustration)**.
17 Turn the bearing and feel for tight spots and roughness. If necessary, pull the bearing off the shaft, using a bearing puller having removed the snap-ring and washer from the end of the shaft. The new bearing can be tapped onto the shaft, using a piece of pipe with an inside diameter large enough to fit over the shaft and contact the inner race of the bearing.
18 Check the idler shaft bearing in the crankcase **(see illustration)**. Replace it if it's worn or damaged.

27.16 Inspect the idler shaft splines and sprocket; if the bearing is worn or damaged, have it pressed off

27.18 Replace the idler shaft bearing if it's worn or damaged

28.4 Inspect the alternator/starter and camshaft chains

28.6 The cam chain rear (exhaust side) guide is secured by two Allen bolts (arrows)

28.7 The alternator/starter chain guide is secured by a single Allen bolt

Installation

19 Installation is the reverse of the removal steps, with the following additions:
a) Coat all parts with clean engine oil.
b) Be sure to place the long and short idle gear collars on the correct sides of the gear (see illustration 27.11).

28 Alternator/starter chain, camshaft chain and guides - removal and installation

Removal

Alternator/starter chain and camshaft chain

1 Remove the engine (see Section 5).
2 Separate the crankcase halves (see Section 22).
3 Remove the crankshaft (see Chapter 2).
4 Remove the chains from the crankshaft (see illustration).

Chain guides

5 The cam chain front guide can be lifted from the cylinder block after the head has been removed (see Section 13).
6 The cam chain rear guide is fastened to the crankcase with a bracket and two bolts (see illustration). Remove the bolts and detach the guide and bracket from the case.

7 The alternator/starter chain guide in the upper case half is secured by a single Allen bolt (see illustration).

Inspection

Camshaft chain

8 Pull the chain tight to eliminate all slack and measure the length of twenty links, pin-to-pin (see illustration 28.9 in Chapter 2A). Compare your findings to this Chapter's Specifications.
9 Also check the chain for binding and obvious damage.
10 If the twenty-link length is not as specified, or there is visible damage, replace the chain.

Chain guides

11 Check the guides for deep grooves, cracking and other obvious damage, replacing them if necessary.

Installation

12 Installation of these components is the reverse of the removal procedure. When installing the bracket for the cam chain rear guide, apply a non-hardening thread locking compound to the threads of the bolts. Tighten the bolts to the torque listed in this Chapter's Specifications. Apply engine oil to the faces of the guides and to the chains.

29 Transmission shafts - removal and installation

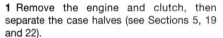

Removal

1 Remove the engine and clutch, then separate the case halves (see Sections 5, 19 and 22).
2 The shafts can simply be lifted out of the upper half of the case (see illustrations). If they are stuck, use a soft-face hammer and gently tap on the bearings on the ends of the shafts to free them. The shaft nearest the rear of the case is the output shaft - the other shaft is the main drive shaft.
3 Refer to Section 30 for information pertaining to transmission shaft service and 31 for information pertaining to the shift drum and forks.

Installation

4 Check to make sure the set pins and rings are present in the upper case half, where the shaft bearings seat (see illustration).
5 Carefully lower each shaft into place. The holes in the needle bearing outer races must engage with the set pins, and the grooves in the ball bearing outer races must engage with the set rings.
6 The remainder of installation is the reverse of removal.

29.2a Lift out the main drive shaft . . .

29.2b . . . and the output shaft

29.4 Be sure the set pins and rings (arrows) are in position

30.2a Slide off the needle bearing outer race . . .

30.2b . . . then remove the snap-ring and bearing

30.3 Remove the thrust washer and second gear

30 Transmission shafts -
disassembly, inspection and reassembly

1 Remove the shafts from the case (see Section 29).

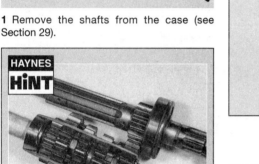

HAYNES
HiNT

When disassembling the transmission shafts, place the parts on a long rod or thread a wire through them to keep them in order and facing the proper direction.

30.4a Slide off sixth gear . . .

30.4b . . . and its bushing

Main drive shaft

Disassembly

2 Remove the needle bearing outer race, then remove the snap-ring from the end of the shaft and slide the needle bearing off **(see illustrations)**.
3 Remove the thrust washer and slide second gear off the shaft **(see illustration)**.
4 Remove sixth gear and bushing **(see illustrations)**.
5 Slide the toothed washer off and remove the snap-ring **(see illustration)**. To keep the snap-ring from bending as it's expanded, hold the back of it with pliers **(see illustration)**.
6 Remove the third/fourth gear cluster from the shaft **(see illustration)**.
7 Remove the next snap-ring, then slide the washer, fifth gear and its bushing off the shaft **(see illustrations)**.

30.5a Remove the snap-ring and toothed washer . . .

30.5b . . . holding the back of the snap-ring with pliers will prevent it from twisting

30.6 Slide off third-fourth gear

30.7a Remove the snap-ring . . .

30.7b . . . the toothed washer and fifth gear . . .

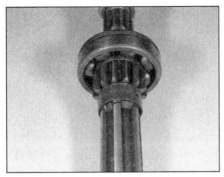

30.7c . . . and the bushing

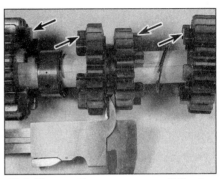

30.9 Measure the gear grooves; if they're too wide, replace the gear – also replace the gear if the dogs (arrows) are worn

Inspection

8 Wash all of the components in clean solvent and dry them off. Rotate the ball bearing on the shaft, feeling for tightness, rough spots, excessive looseness and listening for noises. If any of these conditions are found, replace the bearing. This will require the use of an hydraulic press or a bearing puller setup. If you don't have access to these tools, take the shaft and bearing to a Kawasaki dealer or other motorcycle repair shop and have them press the old bearing off the shaft and install the new one.

9 Measure the shift fork groove between third and fourth gears **(see illustration)**. If the groove width exceeds the figure listed in this Chapter's Specifications, replace the third/fourth gear assembly, and also check the third/fourth gear shift fork (see Section 31).

10 Check the gear teeth for cracking and other obvious damage. Check the bushing and surface in the inner diameter of the fifth and sixth gears for scoring or heat discoloration. If either one is damaged, replace it.

11 Inspect the dogs and the dog holes in the gears for excessive wear **(see illustration)**. Replace the paired gears as a set if necessary.

12 Check the needle bearing and race for wear or heat discoloration and replace them if necessary.

Reassembly

13 Reassembly is the basically the reverse of the disassembly procedure, but take note of the following points:
a) Always use new snap-rings and align the opening of the ring with a spline groove **(see illustration 30.21** in Chapter 2A). Face the sharp side of the snap-ring toward the gear being secured; the rounded side faces away from the gear.
b) When installing the gear bushings on the shaft, align the oil hole in the shaft with the oil hole in the bushing **(see illustration)**.

30.11 Inspect the bushing (left arrow) in gears so equipped; replace the gear if it's worn – if the edges of the slots (right arrow) are rounded, replace the gear

c) Lubricate the components with engine oil before assembling them.

Output shaft

Disassembly

14 Remove the needle bearing outer race and slide the needle bearing off **(see illustrations)**.

15 Remove the thrust washer and first gear from the shaft **(see illustration)**.

16 Remove fifth gear from the shaft. Fifth gear has three steel balls in it for the positive neutral finder mechanism. These lock fifth gear to the shaft unless it is spun rapidly

30.13 Be sure to align the bushing oil hole with the shaft oil hole (arrows)

30.14a Slide off the bearing outer race . . .

30.14b . . . and the bearing

30.15 Remove the thrust washer (arrow) and first gear

30.16a Hold third gear (A) with one hand and spin the transmission shaft while lifting up on fifth gear (B); it may take several tries to disengage fifth gear from the shaft, but it will slide off easily once it is disengaged

30.16b These three balls ride in slots in the transmission shaft; they must be flung outward by centrifugal force before fifth gear can be removed

enough to fling the balls outward. To remove fifth gear, grasp third gear and hold the shaft in a vertical position with one hand, and with the other hand, spin the shaft back and forth, holding onto fifth gear and pulling up (see illustration). After fifth gear is removed, collect the three steel balls (see illustration). *Caution: Don't pull the gear up too hard or fast - the balls will fly out of the gear.*

17 Remove the snap-ring, toothed washer, third gear, bushing and fourth gear from the shaft (see illustrations).

18 Remove the toothed washer, snap-ring and sixth gear (see illustrations).

19 Remove the next snap-ring, toothed washer, second gear and its bushing (see illustrations).

30.17a Remove the snap-ring . . .

30.17b . . . the toothed washer . . .

30.17c . . . third gear . . .

30.17d . . . its bushing and fourth gear

30.18a Remove the toothed washer . . .

30.18b . . . the snap-ring . . .

30.18c . . . and sixth gear

30.19a Remove the snap-ring . . .

30.19b . . . the toothed washer and second gear . . .

30.19c . . . and the bushing

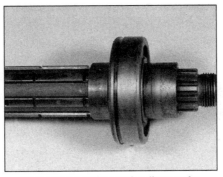

30.20 The bearing and collar can be left on the shaft unless they're worn or damaged

20 The ball bearing and collar can remain on the shaft unless they need to be replaced **(see illustration)**.

Inspection

21 Refer to Steps 8 through 12 for the inspection procedures. They are the same, except when checking the shift fork groove width you'll be checking it on fifth gear and sixth gears.

Reassembly

22 Reassembly is basically the reverse of the disassembly procedure, but take note of the following points:

   a) Always use new snap-rings and align the

opening of the ring with a spline groove **(see illustration 30.21 in Chapter 2A)**. Face the sharp side of each snap-ring toward the gear being secured; face the rounded side of snap-ring away from the gear.

   b) When installing the bushing for third and fourth gear and second gear, align the oil hole in the bushing with the hole in the shaft.

   c) When installing fifth gear, don't use grease to hold the balls in place - to do so would impair the positive neutral finder mechanism. Just set the balls in their holes (the holes that they can't pass through), keep the gear in a vertical

position and carefully set it on the shaft (engine oil will help keep them in place). The spline grooves that contain the holes with the balls must be aligned with the slots in the shaft spline grooves.

   d) Lubricate the components with engine oil before assembling them.

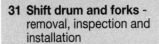

31 Shift drum and forks - removal, inspection and installation

Removal

1 Remove the engine, remove the external shift mechanism and separate the crankcase halves (see Sections 5, 21 and 22).
2 Remove the retaining plates for the shift drum and shift rod **(see illustrations)**.
3 Support the shift forks and pull the shift rods out **(see illustrations)**. The output shaft forks and the shift rods are interchangeable, but it's a good idea to assemble them as they were in the engine so they can be returned to their original positions **(see illustration)**.
4 Slide the shift drum out of the crankcase **(see illustration)**.

Inspection

5 Check the edges of the grooves in the drum for signs of excessive wear **(see**

31.2a Remove the shift drum retainer bolts (arrows) . . .

31.2b . . . and the shift rod retainer bolt on the other side of the case

31.3a Slide out the shift rod with the single fork . . .

31.3b . . . and the shift rod with two forks . . .

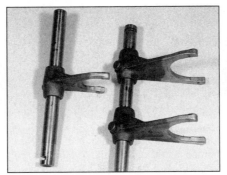

31.3c . . . and reassemble them so they can be returned to their original positions

31.4 Slide the shift drum out of the case

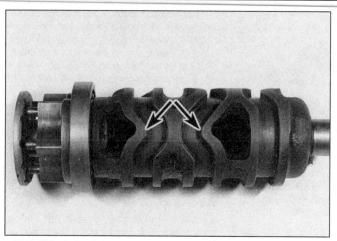

31.5 Check the fork grooves for wear, especially at their points (arrows)

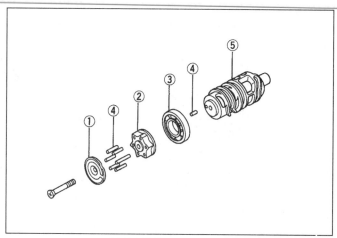

31.6a Shift drum details

1 Pin plate
2 Bearing holder
3 Bearing

4 Pins
5 Shift drum

illustration). Measure the widths of the grooves and compare your findings to this Chapter's Specifications.

6 Remove the Phillips screw from the end of the shift drum and disassemble the drum (see illustrations). Check the pin plate and pins for wear or damage and replace them as necessary. Spin the bearing and check for roughness, noise or looseness. Replace the bearing if defects are found. Reassemble the shift drum, making sure the short dowel pin aligns with the hole in the bearing holder and the one longer pin fits in the recess in the pin plate (see illustration).

7 Check the shift forks for distortion and wear, especially at the fork ears. Measure the thickness of the fork ears and compare your findings with this Chapter's Specifications (see illustration). If they are discolored or severely worn they are probably bent. If

damage or wear is evident, check the shift fork groove in the corresponding gear as well. Inspect the guide pins and the shaft bore for excessive wear and distortion and replace any defective parts with new ones.

8 Check the shift fork shafts for evidence of wear, galling and other damage. Make sure the shift forks move smoothly on the shafts. If the shafts are worn or bent, replace them with new ones.

Installation

9 Installation is the reverse of removal, noting the following points:
a) Lubricate all parts with engine oil before installing them.
b) Use non-permanent thread locking agent on the threads of the shift drum and shift rod retaining plates. Tighten the bolts securely.

32 Initial start-up after overhaul

Note: Make sure the cooling system is checked carefully (especially the coolant level) before starting and running the engine.

1 Make sure the engine oil level is correct, then remove the spark plugs from the engine. Place the engine STOP switch in the Off position and unplug the primary (low tension) wires from the coil.

2 Turn on the key switch and crank the engine over with the starter until the oil pressure indicator light goes off (which indicates that oil pressure exists). Reinstall the spark plugs, connect the wires and turn the STOP switch to On.

3 Make sure there is fuel in the tank, then

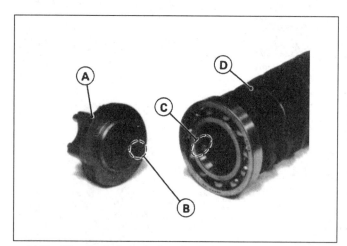

31.6b Align the hole in the bearing holder with the shift drum dowel pin

A Bearing holder
B Hole
C Dowel pin
D Shift drum

31.7 Measure the thickness of the shift fork ears and replace the shift forks if they're worn

push the button on the fuel tap several times to prime the carburettors and operate the choke.

4 Start the engine and allow it to run at a moderately fast idle until it reaches operating temperature.

 Warning: If the oil pressure indicator light doesn't go off, or it comes on while the engine is running, stop the engine immediately.

5 Check carefully for oil leaks and make sure the transmission and controls, especially the brakes, function properly before road testing the machine. Refer to Section 33 for the recommended break-in procedure.

6 Upon completion of the road test, and after the engine has cooled down completely, recheck the valve clearances (see Chapter 1).

33 Recommended break-in procedure

1 Any rebuilt engine needs time to break-in, even if parts have been installed in their original locations. For this reason, treat the machine gently for the first few miles to make sure oil has circulated throughout the engine and any new parts installed have started to seat.

2 Even greater care is necessary if the engine has been rebored or a new crankshaft has been installed. In the case of a rebore, the engine will have to be broken in as if the machine were new. This means greater use of the transmission and a restraining hand on the throttle until at least 500 miles (800 km) have been covered. There's no point in keeping to any set speed limit - the main idea is to keep from lugging (labouring) the engine and to gradually increase performance until the 500 mile (800 km) mark is reached. These recommendations can be lessened to an extent when only a new crankshaft is installed. Experience is the best guide, since it's easy to tell when an engine is running freely.

3 If a lubrication failure is suspected, stop the engine immediately and try to find the cause. If an engine is run without oil, even for a short period of time, irreparable damage will occur.

Notes

Chapter 3
Cooling system

Contents

Degrees of difficulty

Easy, suitable for novice with little experience	Fairly easy, suitable for beginner with some experience	Fairly difficult, suitable for competent DIY mechanic	Difficult, suitable for experienced DIY mechanic	Very difficult, suitable for expert DIY or professional

Specifications

General
Coolant type and mixture ratio	See Chapter 1
Radiator cap pressure rating	14 to 18 psi (1.0 to 1.2 bar)
Thermostat rating	
Opening temperature	
600 A and B models	69.5 to 72.5° C (157 to 163 °F)
600 C models	80.5 to 83.5° C (177 to 182° F)
750 models	80 to 84° C (176 to 183° F)
Fully open at:	
600 A and B models	85° C (185° F)
600 C and 750 models	95° C (203° F)
Valve travel (when fully open)	Not less then 8 mm (5/16 in)

Torque specifications
Thermostatic fan switch-to-radiator	8 Nm (69 in-lbs)
Coolant temperature sending unit-to-thermostat housing	8 Nm (69 in-lbs)
Oil cooler hose union bolts	25 Nm (18 ft-lbs)

1 General information

The models covered by this manual are equipped with a liquid cooling system which utilizes a water/antifreeze mixture to carry away excess heat produced during the combustion process (see illustrations). The cylinders are surrounded by water jackets, through which the coolant is circulated by the water pump. The pump is mounted to the left side of the crankcase and is driven by a gear mounted on the secondary shaft. The coolant passes up through a flexible hose and a coolant pipe, which distributes to water around the four cylinders. It flows through the water passages in the cylinder head, through another pipe (or hoses) and into the thermostat housing. The hot coolant then flows down into the radiator (which is mounted on the frame downtubes to take advantage of maximum air flow), where it is cooled by the passing air, through another hose and back to the water pump, where the cycle is repeated.

An electric fan, mounted behind the radiator and automatically controlled by a thermostatic switch, provides a flow of cooling air through the radiator when the motorcycle is not moving. Under certain conditions, the fan may come on even after the engine is stopped, and the ignition switch is off, and may run for several minutes.

The coolant temperature sending unit, threaded into the thermostat housing, senses the temperature of the coolant and controls the coolant temperature gauge on the instrument cluster.

The entire system is sealed and pressurised. The pressure is controlled by a valve which is part of the radiator cap. By pressurising the coolant, the boiling point is raised, which prevents premature boiling of the coolant. An overflow hose, connected between the radiator and reservoir tank, directs coolant to the tank when the radiator cap valve is opened by excessive pressure. The coolant is automatically siphoned back to the radiator as the engine cools.

Many cooling system inspection and service procedures are considered part of routine maintenance and are included in Chapter 1.

On later UK models, the coolant is also used to warm the carburetor bodies via an arrangement of small hoses. The coolant travels from the rear of the cylinder block, through a filter, through the carburetor castings and then rejoins the main cooling system at the water pump. A check valve is fitted above the water pump to ensure the correct flow of coolant.

 Warning: Do not allow antifreeze to come in contact with your skin or painted surfaces of the vehicle. Rinse off spills immediately with plenty of water. Antifreeze is highly toxic if ingested. Never leave antifreeze lying around in an open container or in puddles on the floor; children and pets are attracted by its sweet smell and may drink it. Check with local authorities about disposing of used antifreeze. Many communities have collection centres which will see that antifreeze is disposed of safely.

Caution: Do not remove the radiator cap when the engine and radiator are hot. Scalding hot coolant and steam may be blown out under pressure, which could cause serious injury. To open the radiator cap, remove the rear screw from the right side panel on the inside of the fairing (if equipped). When the engine has cooled, lift up the panel and place a thick rag, like a towel, over the radiator cap; slowly rotate the cap counterclockwise to the first stop. This procedure allows any residual pressure to escape. When the steam has stopped escaping, press down on the cap while turning counterclockwise and remove it

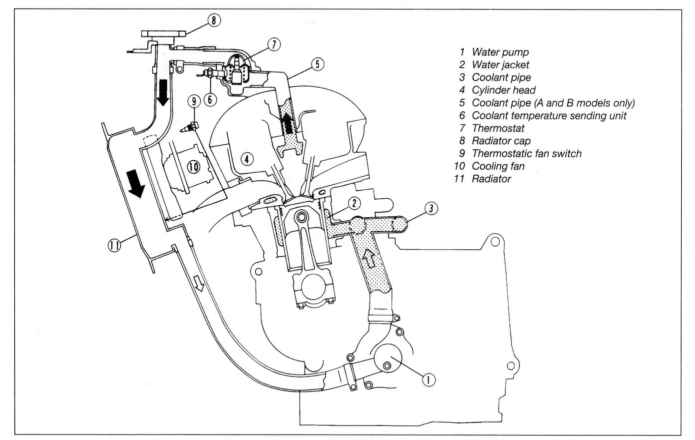

1 Water pump
2 Water jacket
3 Coolant pipe
4 Cylinder head
5 Coolant pipe (A and B models only)
6 Coolant temperature sending unit
7 Thermostat
8 Radiator cap
9 Thermostatic fan switch
10 Cooling fan
11 Radiator

1.1a Coolant flow diagram (600 A and B models shown, 600 C similar)

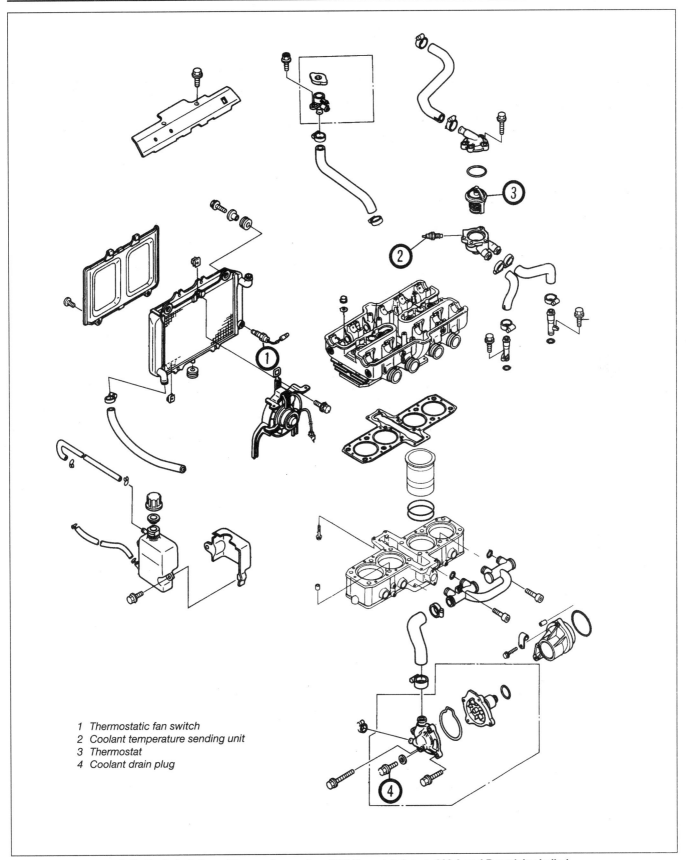

1 Thermostatic fan switch
2 Coolant temperature sending unit
3 Thermostat
4 Coolant drain plug

1.1b Exploded view of the cooling system (600 C model shown, 600 A and B models similar)

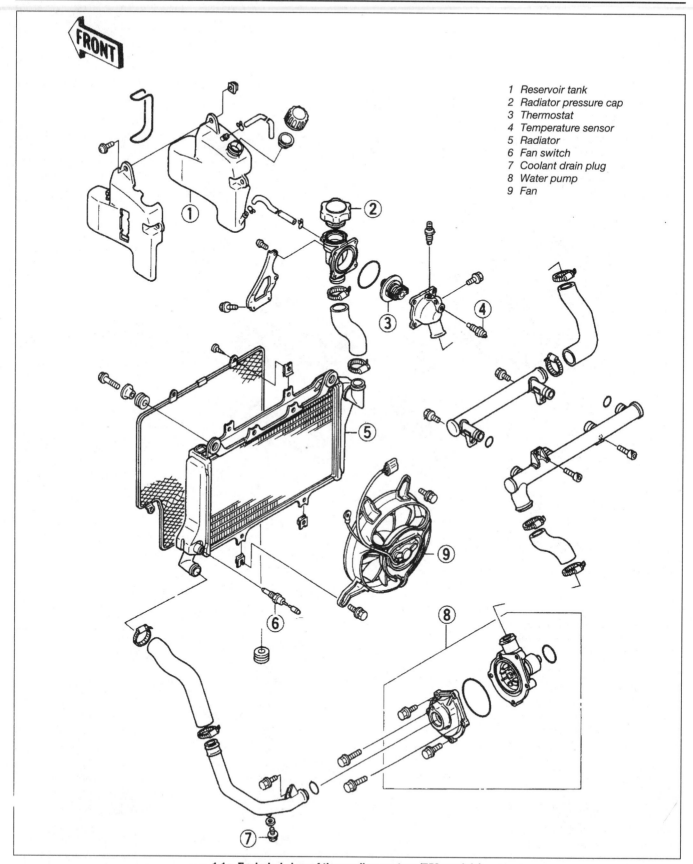

1 Reservoir tank
2 Radiator pressure cap
3 Thermostat
4 Temperature sensor
5 Radiator
6 Fan switch
7 Coolant drain plug
8 Water pump
9 Fan

FRONT

1.1c Exploded view of the cooling system (750 models)

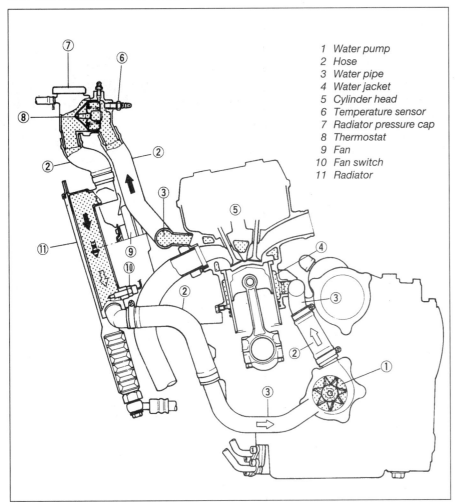

1.1d Coolant flow diagram (750 models)

1 Water pump
2 Hose
3 Water pipe
4 Water jacket
5 Cylinder head
6 Temperature sensor
7 Radiator pressure cap
8 Thermostat
9 Fan
10 Fan switch
11 Radiator

2 Radiator cap - check

1 If problems such as overheating and loss of coolant occur, check the entire system as described in Chapter 1. The radiator cap

opening pressure should be checked by a dealer service department or service station equipped with the special tester required to do the job. If the cap is defective, replace it with a new one.

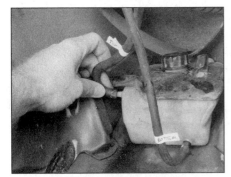

3.2a Label the hoses before disconnecting them from the reservoir (lower fairing-mounted reservoir shown)

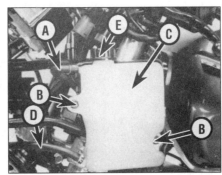

3.2b Coolant reservoir details – 600 C models

A Vent hose
B Mounting bolts
C Reservoir
D Reservoir tank hose
E Cap

3 Coolant reservoir - removal and installation

1 If you're working on a 600 A model or B model, remove the lower fairing (see Chapter 8). If you're working on a 600 C model, remove the upper fairing (see Chapter 8). On 750 models remove the lower and right fairing panels (see Chapter 8).
2 Disconnect the hose(s) from the reservoir **(see illustrations)**. It's a good idea to mark the positions of the hoses so they aren't attached to the wrong fitting when the reservoir is installed.
3 Remove the reservoir retaining screws and detach the reservoir from the lower fairing (early models) or the frame (later models).
4 Installation is the reverse of the removal procedure.

4 Cooling fan and thermostatic fan switch - check and replacement

Check

1 If the engine is overheating and the cooling fan isn't coming on, first remove the seat and check the fuses. If a fuse is blown, check the fan circuit for a short to ground (earth) (see the Wiring diagrams at the end of this book). If the fuses are all good, remove the lower and upper fairings (see Chapter 8) and unplug the fan electrical connector **(see illustration)**. Using two jumper wires, apply battery voltage to the terminals in the fan motor side of the electrical connector. If the fan doesn't work, replace the motor.
2 If the fan does come on, the problem lies in the thermostatic fan switch, the fan relay, or the wiring that connects the components. Remove the jumper wires and plug in the electrical connector to the fan. Unplug the electrical connector to the thermostatic fan switch, attach a jumper wire to the wiring harness side of the electrical connector and ground (earth) the other end of the jumper

4.1 Location of the fan motor electrical connector (arrow)

4.2 Using a jumper wire, ground the fan circuit wiring harness – the fan motor will run if the circuit and motor are okay

1 Electrical connector
2 Thermostatic fan switch

4.6 Remove the fan bracket-to-radiator bolts and separate the bracket from the radiator

4.7 The fan blade assembly is retained to the motor shaft by a single nut

wire **(see illustration)**. If the fan comes on, the circuit to the motor is okay, and the thermostatic fan switch is defective (see Step 10).

3 If the fan still doesn't work, place your hand on the junction block (fuse box). Repeatedly touch the jumper wire to ground (earth) - if you feel a clicking inside the junction block, the relay is probably good. If it's not clicking, check the wiring from the thermostatic fan switch to the junction block. If it is clicking, check the wiring from the junction block to the fan motor. If the wiring checks out okay, the fan relay is most likely the problem, in which case the junction block must be replaced (the relays aren't replaceable individually). Refer to Chapter 9 for further junction block checks.

Replacement

Fan motor

⚠️ **Warning: The engine must be completely cool before beginning this procedure.**

4 Disconnect the cable from the negative terminal of the battery.
5 Remove the radiator (see Section 8).
6 Remove the three (600 models) or four (750

models) bolts securing the fan bracket to the radiator **(see illustration)**, noting which bolt the ground (earth) wire is attached to. Separate the fan and bracket from the radiator.
7 Remove the nut that retains the fan blades to the fan motor shaft **(see illustration)** and remove the fan blade assembly from the motor. Note that on 750 models, the fan and motor unit are replaced as an assembly, rather than separately.
8 Remove the screws that secure the fan motor to the bracket **(see illustration)** and detach the motor from the bracket.
9 Installation is the reverse of the removal procedure. Be sure to reinstall the ground (earth) wire under the bracket-to-radiator bolt.

Thermostatic fan switch

⚠️ **Warning: The engine must be completely cool before beginning this procedure.**

10 Prepare the new switch by wrapping the threads with Teflon tape or by coating the threads with RTV sealant.
11 Unscrew the switch from the radiator (see illustration 4.2 for switch location) and quickly install the new switch, tightening it to the torque listed in this Chapter's Specifications.

12 Plug in the electrical connector to the switch.
13 Check and, if necessary, add coolant to the system (see Daily (pre-ride) checks).

5 Coolant temperature gauge and sending unit - check and replacement

Check

1 If the engine has been overheating but the coolant temperature gauge hasn't been indicating a hotter than normal condition, begin with a check of the coolant level (see Daily (pre-ride) checks). If it's low, add the recommended type of coolant and be sure to locate the source of the leak.
2 Remove the seat and the fuel tank (see Chapter 4). Locate the coolant temperature sending unit, which is screwed into the thermostat housing. Unplug the electrical connector from the sending unit, turn the ignition key to the Run position (don't crank the engine over) and note the temperature gauge - it should read Cold.
3 With the ignition key still in the Run position, connect one end of a jumper wire to the sending unit wire and ground (earth) the other end **(see illustration)**. The needle on the

4.8 Remove the three screws that secure the motor to the bracket, then detach the motor

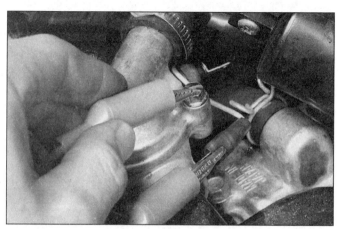

5.3 When the wire to the sending unit is grounded, the needle on the gauge should move all the way past the Hot mark

temperature gauge should swing over past the Hot mark.

Caution: Don't ground(earth) the wire any longer than necessary or the gauge may be damaged.

4 If the gauge passes both of these tests, but doesn't operate correctly under normal riding conditions, the temperature sending unit is defective and must be replaced.

5 If the gauge didn't respond to the tests properly, either the wire to the gauge is bad or the gauge itself is defective.

Replacement

Sending unit

 Warning: The engine must be completely cool before beginning this procedure.

6 Prepare the new sending unit by wrapping the threads with Teflon tape or by coating the threads with RTV sealant.

7 Unscrew the sending unit from the thermostat housing and quickly install the new unit, tightening it to the torque listed in this Chapter's Specifications.

8 Reconnect the electrical connector to the sending unit. Check and, if necessary, add coolant to the system (see *Daily (pre-ride) checks*).

Coolant temperature gauge

9 Refer to Chapter 9 for the coolant temperature gauge replacement procedure.

6 Thermostat - removal, check and installation

 Warning: The engine must be completely cool before beginning this procedure.

Removal

1 If the thermostat is functioning properly, the coolant temperature gauge should rise to the normal operating temperature quickly and then stay there, only rising above the normal position occasionally when the engine gets

unusually hot. If the engine does not reach normal operating temperature quickly, or if it overheats, the thermostat should be removed and checked, or replaced with a new one.

2 Refer to Chapter 1 and drain the cooling system. On 600 models, remove the seat and the fuel tank (see Chapter 4). On 750 models, remove the upper fairing (see Chapter 8).

3 If you're working on a 600 C or 750 model, unplug the electrical connectors and remove the wiring harness bracket **(see illustration)**.

4 On 600 models, remove the three bolts securing the thermostat cover **(see illustration)** and remove the cover (it's not necessary to disconnect the hose from the cover). Withdraw the thermostat from the housing and remove the O-ring from the cover **(see illustrations)**.

5 On 750 models, remove the four bolts which retain the filler neck to the thermostat housing and mounting bracket **(see illustration 1.1d)**. Move the filler neck away from the thermostat housing, lift out the thermostat and remove the O-ring.

Check

6 Remove any coolant deposits, then visually check the thermostat for corrosion, cracks and other damage. If it was open when it was removed, it is defective. Check the O-ring for cracks and other damage.

7 To check the thermostat operation, submerge it in a container of water along with a thermometer.

 Warning: Antifreeze is poisonous. Don't use a cooking pan. The thermostat should be suspended so it does not touch the container.

8 Gradually heat the water in the container with a hotplate or stove and check the temperature when the thermostat first starts to open.

9 Continue heating the water and check the temperature when the thermostat is fully open.

10 Lift the fully open thermostat out of the water and measure the distance the valve has opened.

11 Compare the opening temperature, the fully open temperature and the valve travel to

the values listed in this Chapter's Specifications.

12 If these specifications are not met, or if the thermostat does not open while the water is heated, replace it with a new one.

Installation

13 Install the thermostat into the housing with the spring pointing down **(see illustration 6.4b)**.

14 Install a new O-ring in the groove in the thermostat cover.

15 Place the cover on the housing and install the bolts, tightening them securely.

16 The remainder of installation is the reverse of the removal procedure. Fill the cooling system with the recommended coolant (see *Daily (pre-ride) checks*).

7 Thermostat housing - removal and installation

 Warning: The engine must be completely cool before beginning this procedure.

Removal

1 Refer to Steps 2 and 3 of Section 6.
2 Detach the hose(s) from the housing.
3 On 600 A and B models, remove the bolt securing the thermostat housing to the upper

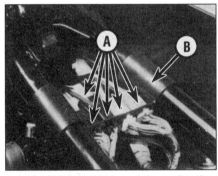

6.3 On 600 C and 750 models, unplug the electrical connectors (A) and remove the wiring harness bracket (B) from between the frame tubes

6.4a Remove the thermostat cover bolts (arrows)

6.4b Remove the thermostat from the housing, noting how it is installed

6.4c Remove the O-ring from the groove in the thermostat cover

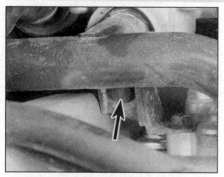

7.3 On early models remove the bolt that retains the thermostat housing to the coolant pipe (arrow)

7.6 Be sure to install a new O-ring on the coolant pipe before installing the thermostat housing

8.6 Remove the bolts that attach the baffle plate to the fan bracket (left side shown)

coolant pipe **(see illustration)**. On 600 C models remove the thermostat housing mounting bolts. On 750 models, the thermostat housing and filler neck can be removed together with the mounting bracket, once the ground (earth) wire has been detached and the two mounting bracket bolts removed.

4 Mark and disconnect any wires or hoses that may interfere with the removal of the thermostat housing.

5 Remove the thermostat housing. On 600 A and B models slide the housing forward, off the coolant pipe. Remove the O-ring from the coolant pipe.

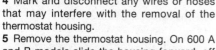

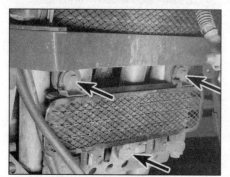

8.8 Remove the three oil cooler mounting bolts (arrows)

Installation

6 On 600 A and B models, install a new O-ring to the coolant pipe **(see illustration)**. Lubricate the O-ring with a little engine oil.
7 Place the thermostat housing in position and install the bolt(s), tightening them securely.
8 Connect the hose(s) to the thermostat housing.
9 The remainder of installation is the reverse of the removal procedure. Fill the cooling system with the recommended coolant (see *Daily (pre-ride) checks*).

8 Radiator - removal and installation

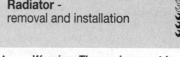

> **Warning: The engine must be completely cool before beginning this procedure.**

Removal

1 Set the bike on its centrestand. Disconnect the cable from the negative terminal of the battery.
2 Remove the upper and lower fairings (see Chapter 8).
3 Drain the coolant (see Chapter 1).
4 Unplug the electrical connectors for the fan motor and the thermostatic fan switch **(see illustrations 4.1 and 4.2)**.

5 Loosen the hose clamps on both radiator hoses (one on each side of the radiator). Detach the hoses. On some models it may be easier to detach the left side hose at the water pump, instead of the radiator.
6 Remove the baffle plate bolts **(see illustration)**.
7 On 600 C models, unplug the electrical connectors from the horn and detach the horn wires from the guide on the radiator. Detach any other wiring that may interfere with radiator removal.
8 On 600 A and B models, remove the oil cooler mounting bolts **(see illustration)**.
9 Remove the radiator mounting bolts **(see illustration)**. On 750 models, note that the radiator screen must be removed for access to the lower mounting bolts.
10 On 600 A and B models, remove the bolts from each side of the frame stay **(see illustration)**.
11 Lift the top of the radiator forward and up, then remove the fairing stay (if equipped) along with the radiator **(see illustration)**. Be sure to check the rubber bushing in the fairing stay and replace it if it's worn.
12 If the radiator is to be repaired, pressure checked or replaced, detach the cooling fan (see Section 4).
13 Carefully examine the radiator for evidence of leaks and damage. It is recommended that any necessary repairs be

8.9 Remove the radiator mounting bolts (arrows)

8.10 Remove the fairing stay bolts (arrows) from both sides

8.11 Lift the radiator up with the fairing stay and carefully guide it out

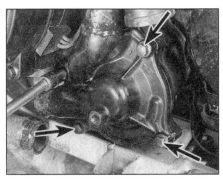

9.6 Detach the hoses then remove the bolts that secure the water pump cover to the pump body

9.7 If you can wiggle the impeller or pull it in-and-out, the pump is defective

9.9 If the cover O-ring is cracked or deteriorated, replace it – make sure the new one seats in the groove fully

performed by a reputable radiator repair shop.

14 If the radiator is clogged, or if large amounts of rust or scale have formed, the repair shop will also do a thorough cleaning job.

15 Make sure the spaces between the cooling tubes and fins are clear. If necessary, use compressed air or running water to remove anything that may be clogging them. If the fins are bent or flattened, straighten them very carefully with a small screwdriver.

Installation

16 Installation is the reverse of the removal procedure. Be sure to replace the hoses if they are deteriorated, and refill the cooling system with the recommended coolant (see *Daily (pre-ride) checks*).

---

9 Water pump - check, removal and installation

> ⚠ **Warning: The engine must be completely cool before beginning this procedure.**

Note: *The water pump on these models can't be overhauled - it must be replaced as a unit.*

Check

1 Visually check around the area of the water pump for coolant leaks. Try to determine if the

leak is simply the result of a loose hose clamp or deteriorated hose.

2 Set the bike on its centrestand.

3 Remove the lower fairing (see Chapter 8).

4 Drain the engine coolant following the procedure in Chapter 1.

5 Loosen the hose clamps and detach the hoses from the water pump cover.

6 Remove the cover bolts **(see illustration)** and separate the cover from the water pump body.

7 Try to wiggle the water pump impeller back-and-forth and in-and-out **(see illustration)**. If you can feel movement, the water pump must be replaced.

8 Check the impeller blades for corrosion. If they are heavily corroded, replace the water pump and flush the system thoroughly (it would also be a good idea to check the internal condition of the radiator).

9 If the cause of the leak was just a defective cover O-ring, remove the old O-ring **(see illustration)** and install a new one.

> **HAYNES HiNT** *Check for signs of leakage from the drainage hole on the underside of the pump body. If either seal fails in the water pump, coolant or oil will leak from the drainage hole, thus preventing oil and coolant mixing.*

Removal

10 Drain the coolant and remove the hoses from the pump (if the cover hasn't already been removed). On 750 models, remove the clutch slave cylinder (see Chapter 2B) and sprocket cover (see Chapter 6) for access.

11 Drain the engine oil (see Chapter 1).

12 Remove any bolts still retaining the pump. Pull the pump straight out to remove it.

13 Check the drainage hole in the sleeve of the pump, just below the pump body **(see illustration)**. If there is coolant residue around it, the water pump is defective (see *Haynes Hint*).

14 If the original water pump is to be installed, check the O-ring on the pump sleeve **(see illustration 9.13)**. If it's cracked or otherwise deteriorated, replace it with a new one.

Installation

15 Installation is basically the reverse of the removal procedure. Before installing the pump, smear a little engine oil on the sleeve O-ring. Be sure to tighten the pump cover bolts securely. Fill the cooling system with the recommended coolant and the crankcase with the specified type and amount of engine oil (see *Daily (pre-ride) checks*).

---

10 Coolant pipe(s) - removal and installation

> ⚠ **Warning: The engine must be completely cool before beginning this procedure.**

1 Place the bike on its centrestand.

2 Remove the lower fairing (see Chapter 8).

3 Drain the engine coolant (see Chapter 1).

Upper coolant pipe (600 A and B models only)

4 Remove the thermostat housing (see Section 7).

5 Remove the coolant pipe-to-cylinder head bolts **(see illustration)** and separate the pipe from the cylinder head.

6 Remove the O-rings from the ends of the pipe and install new ones. If one of the ends doesn't have an O-ring, be sure to retrieve it

9.13 Check the drainage hole in the sleeve for coolant residue – also check the O-ring for deterioration

10.5 Location of the coolant pipe-to-cylinder head bolts (arrows) – the bolt on the left branch of the pipe isn't visible in this photo

10.21 Remove the bolts retaining the lower coolant pipe to the engine (left side shown)

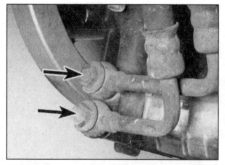

11.3 Remove the union bolts (arrows)

11.4 Remove the cooler mounting bolts and the bracket-to-frame bolts (arrows)

from the thermostat housing or one of the holes in the cylinder head.

7 Check the holes in the head and in the thermostat housing for corrosion, and remove all traces of corrosion if any exists.

8 Install new O-rings on the pipe ends, lubricating them with a little clean engine oil.

9 Install the ends of the coolant pipe into the holes in the cylinder head. Install the mounting bolts, tightening them securely.

10 Install the thermostat housing (see Section 7).

11 Fill the cooling system with the recommended coolant (see Chapter 1) and check for leaks.

12 The remainder of installation is the reverse of the removal procedure.

Upper coolant pipe (750 models)

13 Detach the thermostat hose from the right end of the coolant pipe. Where fitted on UK models, also detach the carburettor warmer hose.

14 Remove the two coolant pipe-to-engine bolts. Pull the pipe out of the water jacket. Make sure both O-rings come out with the pipe - if not, be sure to retrieve them.

15 Check the pipe ends and the holes in the water jacket for corrosion, and remove all traces of corrosion if any exists.

16 Install new O-rings on the pipe ends, lubricating them with a little clean engine oil.

17 Install the ends of the coolant pipe into the holes in the water jacket. Install the mounting bolts, tightening them securely.

18 Fill the cooling system with the recommended coolant (see Chapter 1) and check for leaks.

19 The remainder of installation is the reverse of the removal procedure.

Lower coolant pipe (all models)

20 Detach the hose from the left side of the coolant pipe.

21 Remove the two coolant pipe-to-engine bolts **(see illustration)**.

22 Pull the pipe out of the water jacket. Make sure both O-rings come out with the pipe - if not, be sure to retrieve them.

23 Check the pipe ends and the holes in the water jacket for corrosion, and remove all traces of corrosion if any exists.

24 Install new O-rings on the pipe ends, lubricating them with a little clean engine oil.

25 Install the ends of the coolant pipe into the holes in the water jacket. Install the mounting bolts, tightening them securely.

26 Fill the cooling system with the recommended coolant (see Chapter 1) and check for leaks.

27 The remainder of installation is the reverse of the removal procedure.

11 Oil cooler - removal and installation

Warning: The engine must be completely cool before beginning this procedure.

600 models

1 Set the bike on its centrestand and drain the engine oil (see Chapter 1).

2 Remove the lower fairing (see Chapter 8).

3 Place a drain pan under the front of the crankcase and remove the oil cooler hose-to-crankcase union bolts **(see illustration)**. Retrieve the sealing washers.

4 Remove the oil cooler mounting bolts and the bracket-to-frame bolts **(see illustration)**.

5 Installation is the reverse of removal. Be sure to use new sealing washers if the old ones were leaking or damaged, and fill the crankcase with the recommended type and amount of oil (see Chapter 1). Tighten the union bolts to the torque listed in this Chapter's Specifications.

750 models

6 Set the bike on its centrestand and drain the engine oil (see Chapter 1).

7 Remove the lower fairing (see Chapter 8).

8 Remove the radiator (see Section 8).

9 Place a drain pan under the front of the crankcase and remove the oil line mounting bolts **(see illustration)**. If you're going to detach the oil lines from the cooler, loosen the union bolts at the oil cooler now.

10 Remove the oil cooler mounting bolt and take the cooler out.

11 Installation is the reverse of the removal steps. Use new sealing washers if the oil lines were disconnected from the cooler. Tighten the union bolts to the torque listed in this Chapter's Specifications.

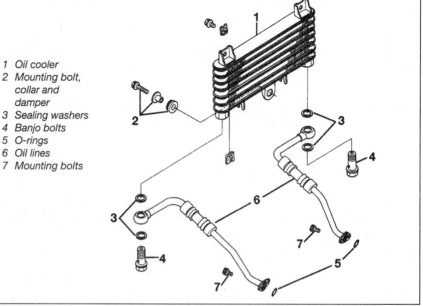

1 Oil cooler
2 Mounting bolt, collar and damper
3 Sealing washers
4 Banjo bolts
5 O-rings
6 Oil lines
7 Mounting bolts

11.9 Oil cooler details (750 models)

Chapter 4
Fuel and exhaust systems

Contents

Degrees of difficulty

| Easy, suitable for novice with little experience | | Fairly easy, suitable for beginner with some experience | | Fairly difficult, suitable for competent DIY mechanic | | Difficult, suitable for experienced DIY mechanic | | Very difficult, suitable for expert DIY or professional | |

Specifications

General

Fuel tank capacity	
600 models	18 lit (4.8 US gal, 4.0 Imp gal)
750 models	21 lit (5.5 US gal, 4.6 Imp gal)
Fuel type	Unleaded gasoline (petrol), 91 octane minimum (RON). Low-lead fuel can be used subject to local regulations.
Carburettor type	
600 models	Keihin CVK32
750 models	Keihin CVK34

Jet sizes

600 A1, A2, A3, A4, B1 models

Main jet	
Cyls. 1 and 4	
California models	108
All others	105
Cyls. 2 and 3	108
Main air jet	100
Jet needle	N27L
Pilot jet	38
Pilot air jet	145
Pilot screw setting	2 turns out
Choke jet	
California models	42
All others	45

Jet sizes (continued)

600 A5 models

Main jet	102 (all cylinders)
Main air jet	100
Jet needle	N52N
Pilot jet	35
Pilot air jet	160
Pilot screw setting	2-1/8 to 2-1/4 turns out

Choke jet

California models	52
All others	45

600 C1 models

Main jet	105
Main air jet	100

Jet needle

California models	N52T
All others	N52Q
Pilot jet	35

Pilot air jet

California models	160
All others	150
Pilot screw setting	2 turns out

Choke jet

California models	48
All others	52

600 C2, C3, C4, C5, C6, C7, C8, C9, C10 models

Main jet

C2 models	105
C3, C4, C5, C6, C7, C8, C9, C10 models	102
Main air jet	100

Jet needle

California models	N52T
All others	N52Q
Pilot jet	35

Pilot air jet

California models	160
All others	150
Pilot screw setting	1-3/4 to 2 turns out

Choke jet

California models	48
All others	52

750 models

Main jet

California models	118
All others	112
Main air jet	Not specified
Jet needle	N53L
Pilot jet	Not specified
Pilot air jet	Not specified

Pilot screw setting

US models	Preset
UK models	2 turns out
Choke jet	Not specified

Carburettor adjustments

Float height	17 mm
Fuel level	0.5 ± 1 mm above the bottom edge of the carburettor body
Choke cable freeplay	2 to 3 mm (1/8 in)
Throttle cable freeplay	see Chapter 1

1 General information

The fuel system consists of the fuel tank, the fuel tap and filter, the carburettors and the connecting lines, hoses and control cables.

The carburettors used on these motorcycles are four constant vacuum Keihins with butterfly-type throttle valves. For cold starting, an enrichment circuit is actuated by a cable and the choke lever mounted on the left handlebar.

The exhaust system is a four-into-two design with a crossover pipe.

Many of the fuel system service procedures are considered routine maintenance items and for that reason are included in Chapter 1.

2 Fuel tank - removal and installation

Warning: Gasoline (petrol) is extremely flammable, so take extra precautions when you work on any part of the fuel system. Don't smoke or allow open flames or bare light bulbs near the work area, and don't work in a garage where a natural gas-type appliance (such as a water heater or clothes dryer) is present. If you spill any fuel on your skin, rinse it off immediately with soap and water. When you perform any kind of work on the fuel system, wear safety glasses and have a class B type fire extinguisher on hand.

1 The fuel tank is held in place at the forward end by two cups, one on each side of the tank, which slide over two rubber dampers on the frame. The rear of the tank is fastened to a bracket by two bolts and rubber insulators, which fit through a flange projecting from the tank.

2 Remove the seat and disconnect the cable from the negative terminal of the battery. On 600 A and B models, remove the side covers. On 600 C models, remove the screw securing the fuel tap knob, then remove the knee grip

covers (see Chapter 8). On 750 models, remove the two bolts from each side of the bike which retain the upper fairing to the tank, then remove the knee grip covers (see Chapter 8).

3 Mark and disconnect the breather hose and, on California models, the evaporative emission control system hoses from the rear of the tank.

4 Disconnect the electrical connector for the fuel gauge sending unit **(see illustration)**.

5 Remove the two bolts securing the rear of the tank to the bracket **(see illustration)**.

6 Turn the fuel tap to the On or Reserve position (600 models). On all models, lift the rear of the tank up, slide back the hose clamps and pull the fuel and vacuum lines **(see illustration)** off the fuel tap.

7 Slide the tank to the rear to disengage the front of the tank from the rubber dampers, then carefully lift the tank away from the machine.

8 Before installing the tank, check the condition of the rubber mounting dampers - if they're hardened, cracked, or show any other signs of deterioration, replace them.

9 When replacing the tank, reverse the above procedure. Make sure the tank seats properly and does not pinch any control cables or wires.

> **HAYNES HINT** *If difficulty is encountered when trying to slide the tank cups onto the dampers, a small amount of light oil should be used to lubricate them.*

3 Fuel tank - cleaning and repair

1 All repairs to the fuel tank should be carried out by a professional who has experience in this critical and potentially dangerous work. Even after cleaning and flushing of the fuel system, explosive fumes can remain and ignite during repair of the tank.

2 If the fuel tank is removed from the vehicle, it should not be placed in an area where

sparks or open flames could ignite the fumes coming out of the tank. Be especially careful inside garages where a natural gas-type appliance is located, because the pilot light could cause an explosion.

4 Idle fuel/air mixture adjustment - general information

1 Due to the increased emphasis on controlling motorcycle exhaust emissions, certain governmental regulations have been formulated which directly affect the carburetion of this machine. In order to comply with the regulations, the carburettors on some models have a metal sealing plug pressed into the hole over the pilot screw (which controls the idle fuel/air mixture) on each carburettor, so they can't be tampered with. These should only be removed in the event of a complete carburettor overhaul, and even then the screws should be returned to their original settings. The pilot screws on other models are accessible, but the use of an exhaust gas analyser is the only accurate way to adjust the idle fuel/air mixture and be sure the machine doesn't exceed the emissions regulations.

2 If the engine runs extremely rough at idle or continually stalls, and if a carburettor overhaul does not cure the problem, take the motorcycle to a Kawasaki dealer service department or other repair shop equipped with an exhaust gas analyser. They will be able to properly adjust the idle fuel/air mixture to achieve a smooth idle and restore low speed performance.

5 Carburettor overhaul - general information

1 Poor engine performance, hesitation, hard starting, stalling, flooding and backfiring are all signs that major carburettor maintenance may be required.

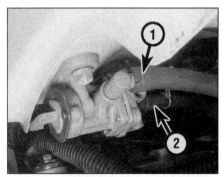

2.6 With the fuel tap in the ON or RES position, disconnect the vacuum hose and the fuel line from the tap fittings

1 Fuel line   2 Vacuum hose

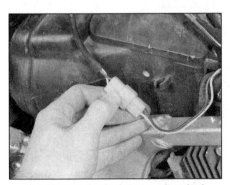

2.4 Unplug the fuel gauge electrical connector

2.5 Disconnect the hoses and remove the fuel tank mounting bolts

6.5 Loosen the clamps on the intake manifold tubes

6.6 Mark and disconnect the vacuum hoses from the carburettors

2 Keep in mind that many so-called carburettor problems are really not carburettor problems at all, but mechanical problems within the engine or ignition system malfunctions. Try to establish for certain that the carburettors are in need of maintenance before beginning a major overhaul.

3 Check the fuel tap filter, the fuel lines, the fuel tank cap vent, the intake manifold hose clamps, the vacuum hoses, the air filter element, the cylinder compression, the spark plugs, and the carburettor synchronisation before assuming that a carburettor overhaul is required.

4 Most carburettor problems are caused by dirt particles, varnish and other deposits which build up in and block the fuel and air passages. Also, in time, gaskets and O-rings shrink or deteriorate and cause fuel and air leaks which lead to poor performance.

5 When the carburettor is overhauled, it is generally disassembled completely and the parts are cleaned thoroughly with a carburettor cleaning solvent and dried with filtered, unlubricated compressed air. The fuel and air passages are also blown through with compressed air to force out any dirt that may have been loosened but not removed by the solvent. Once the cleaning process is complete, the carburettor is reassembled using new gaskets, O-rings and, generally, a new inlet needle valve and seat.

6 Before disassembling the carburettors, make sure you have a carburettor rebuild kit (which will include all necessary O-rings and other parts), some carburettor cleaner, a supply of rags, some means of blowing out the carburettor passages and a clean place to work. It is recommended that only one carburettor be overhauled at a time to avoid mixing up parts.

6 Carburettors - removal and installation

⚠ **Warning:** *Gasoline (petrol) is extremely flammable, so take extra precautions when you work on any part of the fuel system. Don't smoke or allow open flames or bare light bulbs near the work area, and don't work in a garage where a natural gas-type appliance (such as a water heater or clothes dryer) is present. If you spill any fuel on your skin, rinse it off immediately with soap and water. When you perform any kind of work on the fuel system, wear safety glasses and have a class B type fire extinguisher on hand.*

Removal

1 Remove the fuel tank (see Section 2). On 600 C and 750 models, remove the side covers (see Chapter 8, if necessary).

2 On US models, pull the large hose leading to the vacuum switch out of the front of the air filter housing and secure it out of the way.

3 Disconnect the choke cable from the carburettor assembly (see Section 10).

4 Loosen the lockwheel on the throttle cable adjuster at the handlebar and turn the adjuster in all the way.

5 Loosen the clamp screws on the intake manifolds (the rubber tubes that connect the carburettors to the engine) **(see illustration)**.

6 Mark and disconnect the vacuum hoses from the carburettors **(see illustration)**. Disconnect and plug the coolant hoses on UK models so equipped.

7 Slide the spring bands on the ducts from the air filter housings away from the carburettors **(see illustration)**. On 600 A and B models, remove the air filter housing (see Section 11).

8 Pull the carburettor assembly to the rear, clear of the intake manifold tubes **(see illustration)**. Raise the assembly up far enough to disconnect the throttle cables from the throttle pulley **(see illustration)**, then remove the carburettors from the machine.

6.7 Push the spring bands back, toward the air cleaner housing

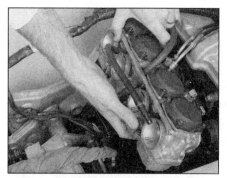

6.8a Gently pull the assembly to the rear, detaching the carburettors from the intake manifold tubes

6.8b Raise the carburettors up, then align the throttle cables with the slots in the throttle pulley and pass the cables through

9 After the carburettors have been removed, stuff clean rags into the intake manifold tubes to prevent the entry of dirt or other objects.

Installation

10 Position the assembly over the intake manifold tubes. Lightly lubricate the ends of the throttle cables with multi-purpose grease and attach them to the throttle pulley. Make sure the accelerator and decelerator cables are in their proper positions.
11 Tilt the front of the assembly down and insert the fronts of the carburettors into the intake manifold tubes. Push the assembly forward and tighten the clamps.
12 Install the air filter housing (see Section 11).
13 Make sure the ducts from the air cleaner housing are seated properly, then slide the spring bands into position.
14 Connect the choke cable to the assembly and adjust it (see Section 10).
15 On US models connect the large hose from the vacuum switch to the air filter housing. On UK models so equipped, connect the coolant lines to the carburettors. Connect the other vacuum hoses that were previously disconnected.
16 Adjust the throttle grip freeplay (see Chapter 1).
17 Install the fuel tank (see Section 2). On 600 models, turn the fuel tap to PRI and check for leaks, then return the tap to the ON position. On 750 models, press the button on the fuel tap several times to prime the carburettors and check for leaks.
18 Check and, if necessary, adjust the idle speed and carburettor synchronisation (see Chapter 1).
19 Install the seat and side covers.

7 Carburettors - disassembly, cleaning and inspection

Warning: Gasoline (petrol) is extremely flammable, so take extra precautions when you work on any part of the fuel system. Don't smoke or allow open flames or bare light bulbs near the work area, and

7.2b The idle adjusting screw holder is retained by two screws

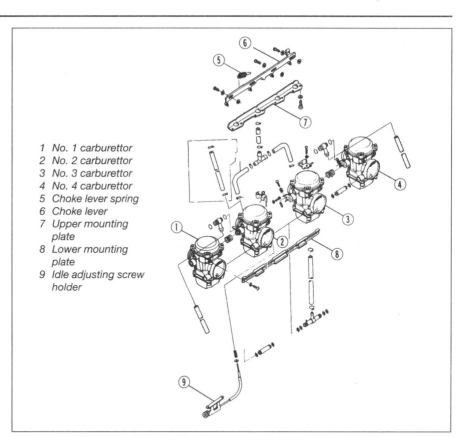

1 No. 1 carburettor
2 No. 2 carburettor
3 No. 3 carburettor
4 No. 4 carburettor
5 Choke lever spring
6 Choke lever
7 Upper mounting plate
8 Lower mounting plate
9 Idle adjusting screw holder

7.2a Exploded view of the carburettors

don't work in a garage where a natural gas-type appliance (such as a water heater or clothes dryer) is present. If you spill any fuel on your skin, rinse it off immediately with soap and water. When you perform any kind of work on the fuel system, wear safety glasses and have a class B type fire extinguisher on hand.

Disassembly

1 Remove the carburettors from the machine as described in Section 6. Set the assembly on a clean working surface. **Note:** Unless the O-rings on the fuel, vent and (if applicable) coolant fittings between the carburettors are leaking, don't detach the carburettors from their mounting brackets. Also, work on one carburettor at a time to avoid getting parts mixed up.
2 If the carburettors must be separated from each other (during a complete overhaul, for example) remove the idle adjusting screw assembly **(see illustrations)**, being careful not to lose the spring and washer on the end of the screw. Disconnect the fuel hoses. Remove the choke lever spring and choke lever by removing the three screws and six plastic washers (two washers per screw, one on each side of the lever) **(see illustration)**, then remove the screws securing the upper and lower mounting plates to the carburettors **(see illustrations)**. Mark the position of each

7.2c If the carburettors are to be separated, disconnect the choke spring, remove the three choke lever screws and detach the lever

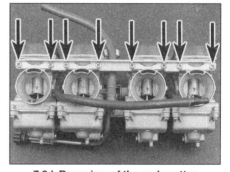

7.2d Rear view of the carburettor assembly, showing the upper mounting plate screws (arrows)

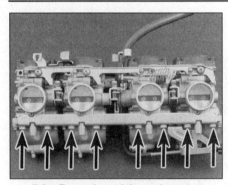

7.2e Front view of the carburettor assembly, showing the lower mounting plate screws (arrows)

carburettor and gently separate them, noting how the throttle linkage is connected, and being careful not to lose any springs or fuel, vent and coolant (where applicable) fittings that are present between the carburettors.

3 Remove the four screws securing the top cover to the carburettor body. Lift the cover off and remove the piston spring **(see illustration)**.

4 Peel the diaphragm away from its groove in the carburettor body, being careful not to tear it. Lift out the diaphragm/piston assembly **(see illustration)**.

5 Remove the piston spring seat and separate the needle from the piston **(see illustrations)**.

6 Remove the four screws retaining the float bowl to the carburettor body, then detach the bowl **(see illustration 7.3)**.

7 Push the float pivot pin out and detach the float (and fuel inlet valve needle) from the carburettor body **(see illustration)**. Detach the valve needle from the float.

8 Unscrew the main jet from the needle jet holder **(see illustration)**.

9 Unscrew the needle jet holder/air bleed pipe **(see illustration)**.

10 Using a wood or plastic tool, push the needle jet out of the carburettor body **(see illustration)**.

11 Using a small, flat-bladed screwdriver, remove the pilot jet **(see illustration)**.

1 Screw
2 Spring washer
3 Washer
4 Top cover
5 Piston spring
6 Spring seat
7 Jet needle
8 Diaphragm/vacuum piston assembly
9 Needle jet
10 Needle jet holder/air bleed pipe
11 Main jet
12 Pilot jet
13 Float assembly
14 Pivot pin
15 Fuel inlet valve needle
16 Clip
17 O-ring
18 Float bowl
19 Drain screw
20 O-ring
21 Screw
22 Spring washer
23 Sealing plug (US models only)
24 Pilot screw
25 Spring
26 Washer
27 O-ring
28 Choke plunger
29 Spring
30 Nut
31 Cap

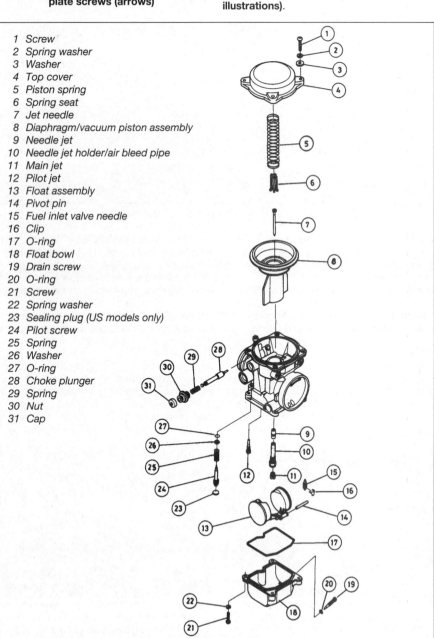

7.3 Carburettor – exploded view

7.4 Remove the diaphragm/piston assembly from the carburettor body

7.5a Remove the vacuum piston spring seat from the piston

7.5b Remove the needle from the piston

7.7 Extract the float pivot pin, then remove the float and valve needle assembly

7.8 Unscrew the main jet from the needle jet holder

7.9 Unscrew the needle jet holder/air bleed pipe

12 The pilot (idle mixture) screw is located in the bottom of the carburettor body **(see illustration)**. On US models, this screw is hidden behind a plug which will have to be removed if the screw is to be taken out. To do this, punch a hole in the plug with an awl or a scribe, then pry it out. On all models, turn the pilot screw in, counting the number of turns until it bottoms lightly. Record that number for use when installing the screw. Now remove the pilot screw along with its spring, washer and O-ring.

13 The choke plunger can be removed by unscrewing the nut that retains it to the carburettor body **(see illustration 7.3)** if the carburettors have been separated from each other (see Step 2).

Cleaning

Caution: Use only a carburettor cleaning solution that is safe for use with plastic parts (be sure to read the label on the container).

14 Submerge the metal components in the carburettor cleaner for approximately thirty minutes (or longer, if the directions recommend it).

15 After the carburettor has soaked long enough for the cleaner to loosen and dissolve most of the varnish and other deposits, use a

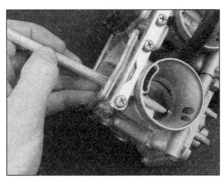

7.10 Working from the top of the carburettor, push the needle jet out with a wood or plastic tool

brush to remove the stubborn deposits. Rinse it again, then dry it with compressed air. Blow out all of the fuel and air passages in the main and upper body.

Caution: Never clean the jets or passages with a piece of wire or a drill bit, as they will be enlarged, causing the fuel and air metering rates to be upset.

Inspection

16 Check the operation of the choke plunger. If it doesn't move smoothly, replace it, along with the return spring.

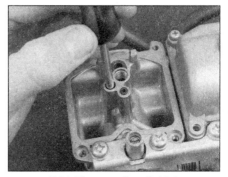

7.11 Unscrew the pilot jet

17 Check the tapered portion of the pilot screw for wear or damage **(see illustration)**. Replace the pilot screw if necessary.

18 Check the carburettor body, float bowl and top cover for cracks, distorted sealing surfaces and other damage. If any defects are found, replace the faulty component, although replacement of the entire carburettor will probably be necessary (check with your parts supplier for the availability of separate components).

19 Check the diaphragm for splits, holes and general deterioration. Holding it up to a light will help to reveal problems of this nature.

7.12 Location of the pilot screw (arrow)

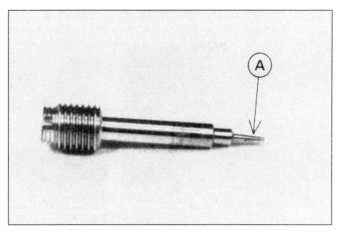

7.17 Check the tapered portion of the pilot screw (A) for wear or damage

20 Insert the vacuum piston in the carburettor body and see that it moves up-and-down smoothly. Check the surface of the piston for wear. If it's worn excessively or doesn't move smoothly in the bore, replace the carburettor.

21 Check the jet needle for straightness by rolling it on a flat surface (such as a piece of glass). Replace it if it's bent or if the tip is worn.

22 Check the tip of the fuel inlet valve needle. If it has grooves or scratches in it, it must be replaced. Push in on the rod in the other end of the needle, then release it - if it doesn't spring back, replace the valve needle (see illustration).

23 Check the O-rings on the float bowl and the drain plug (in the float bowl). Replace them if they're damaged.

24 Operate the throttle shaft to make sure the throttle butterfly valve opens and closes smoothly. If it doesn't, replace the carburettor.

25 Check the floats for damage. This will usually be apparent by the presence of fuel inside one of the floats. If the floats are damaged, they must be replaced.

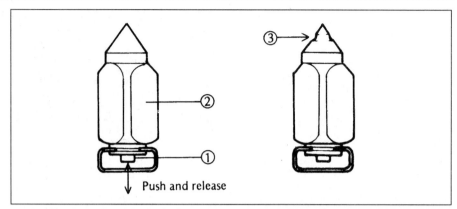

7.22 Check the tip of the fuel inlet valve needle for grooves or scratches – also make sure the rod in the end of the needle pops back out quickly after it's pushed in

1 Rod   2 Valve needle   3 Groove in tip

---

8 Carburettors - reassembly and fuel level adjustment

Caution: When installing the jets, be careful not to over-tighten them they're made of soft material and can strip or shear easily.

Note: *When reassembling the carburettors, be sure to use the new O-rings, gaskets and other parts supplied in the rebuild kit.*

Reassembly

1 If the choke plunger was removed, install it in its bore, followed by its spring and nut. Tighten the nut securely and install the cap.

2 Install the pilot screw (if removed) along with its spring, washer and O-ring, turning it in until it seats lightly. Now, turn the screw out the number of turns that was previously recorded. If you're working on a US model, install a new metal plug in the hole over the

screw. Apply a little bonding agent around the circumference of the plug after it has been seated.

3 Install the pilot jet, tightening it securely.

4 Turn the carburettor body upside-down and install the needle jet into its hole, small diameter first (see illustration).

5 Install the needle jet holder/air bleed pipe, tightening it securely.

6 Install the main jet into the needle jet holder/air bleed pipe, tightening it securely.

7 Drop the jet needle down into its hole in the vacuum piston and install the spring seat over the needle. Make sure the spring seat doesn't cover the hole at the bottom of the vacuum piston - reposition it if necessary.

8 Install the diaphragm/vacuum piston assembly into the carburettor body. Lower the spring into the piston. Seat the bead of the diaphragm into the groove in the top of the carburettor body, making sure the diaphragm isn't distorted or kinked (see illustration). This is not always an easy task. If the diaphragm seems too large in diameter and doesn't want to seat in the groove, place the top cover over the carburettor diaphragm, insert your finger into the throat of the carburettor and push up on the vacuum piston. Push down gently on the top cover - it should drop into place, indicating the diaphragm has seated in its groove.

9 Install the top cover, tightening the screws securely. If you're working on the no. 3 carburettor, don't forget to install the choke cable bracket on the right front corner.

10 Invert the carburettor. Attach the fuel inlet valve needle to the float. Set the float into position in the carburettor, making sure the valve needle seats correctly. Install the float pivot pin. To check the float height, hold the carburettor so the float hangs down, then tilt it back until the valve needle is just seated (the rod in the end of the valve shouldn't be compressed). Measure the distance from the carburettor body to the top of the float (see illustration) and compare your measurement to the float height listed in this Chapter's Specifications. If it isn't as specified, carefully bend the tang that contacts the valve needle up or down until the float height is correct.

11 Install the O-ring into the groove in the float bowl. Place the float bowl on the carburettor and install the screws, tightening them securely.

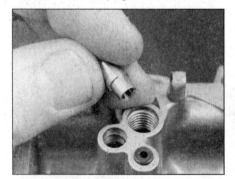

8.4 Install the needle jet, small diameter end first

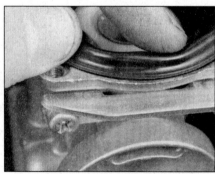

8.8 Make sure the bead of the vacuum diaphragm seats in its groove, and that the diaphragm isn't distorted

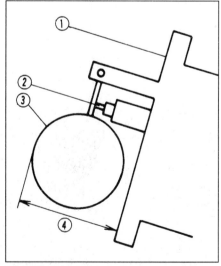

8.10 Float height adjustment details

1 Carburettor body   3 Float
2 Float valve needle rod   4 Float height

12 If the carburettors were separated, install new O-rings on the fuel, vent and (if applicable) coolant fittings. Lubricate the O-rings on the fittings with a light film of oil and install them into their respective holes, making sure they seat completely **(see illustration 7.3)**.

13 Position the coil springs between the carburettors, gently push the carburettors together, then make sure the throttle linkages are correctly engaged. Check the fuel, vent and coolant (if applicable) fittings to make sure they engage properly also.

14 Install the upper and lower mounting plates and install the screws, but don't tighten them completely yet. Set the carburettors on a sheet of glass, then align them with a straightedge placed along the edges of the bores. When the centrelines of the carburettors are all in horizontal and vertical alignment, tighten the mounting plate screws securely.

15 Install the choke lever, making sure it engages correctly with all the choke plungers. Position a plastic washer on each side of the choke lever **(see illustration)** and install the screws, tightening them securely. Install the lever return spring, then make sure the choke mechanism operates smoothly.

16 Install the throttle linkage springs **(see illustration)**. Visually synchronise the throttle butterfly valves, turning the adjusting screws on the throttle linkage, if necessary, to equalise the clearance between the butterfly valve and throttle bore of each carburettor. Check to ensure the throttle operates smoothly.

Fuel level adjustment

 Warning: Gasoline (petrol) is extremely flammable, so take extra precautions when you work on any part of the fuel system. Don't smoke or allow open flames or bare light bulbs near the work area, and don't work in a garage where a natural gas-type appliance (such as a water heater or clothes dryer) is present. If you spill any fuel on your skin, rinse it off immediately with soap and water. When you perform any kind of work on the fuel system, wear safety glasses and have a class B type fire extinguisher on hand.

17 Lightly clamp the carburettor assembly in the jaws of a vice. Make sure the vise jaws are lined with wood. Set the fuel tank next to the vise, but at an elevation that is higher than the carburettors (resting on a box, for example). Connect a hose from the fuel tap to the fuel inlet fitting on the carburettor assembly - make sure the fuel tap is in the ON or RES position.

18 Attach Kawasaki service tool no. 57001-1017 to the drain fitting on the bottom of one of the carburettor float bowls (all four will be checked) **(see illustration)**. This is a clear plastic tube graduated in millimeters. An alternative is to use a length of clear plastic tubing and an accurate ruler. Hold the graduated tube (or the free end of the clear

8.15 Make sure to install a plastic washer on each side of the choke lever when installing the screws

plastic tube) against the carburettor body, as shown in the accompanying illustration. If the Kawasaki tool is being used, raise the zero mark to a point several millimeters above the bottom edge of the carburettor main body. If a piece of clear plastic tubing is being used, make a mark on the tubing at a point several millimeters above the bottom edge of the carburettor main body.

19 Unscrew the drain screw at the bottom of the float bowl a couple of turns, then turn the fuel tap to the PRI position (600 models) or press the button on the tap (750 models) - fuel will flow into the tube. Wait for the fuel level to stabilise, then slowly lower the tube until the zero mark is level with the bottom edge of the carburettor body. **Note:** *Don't lower the zero mark below the bottom edge of the carburettor then bring it back up - the reading won't be accurate.*

8.16 Install the throttle linkage springs

20 Measure the distance between the mark and top of the fuel level in the tube or gauge. This distance is the fuel level - write it down on a piece of paper, screw in the drain screw, turn the fuel tap to the On or Reserve position, then move on to the next carburettor and check it the same way.

21 Compare your fuel level readings to the value listed in this Chapter's Specifications. If the fuel level in any carburettor is not correct, remove the float bowl and bend the tang up or down (see Step 10), as necessary, then recheck the fuel level. **Note:** *Bending the tang up increases the float height and lowers the fuel level - bending it down decreases the float height and raises the fuel level.*

22 After the fuel level for each carburettor has been adjusted, install the carburettor assembly (see Section 6).

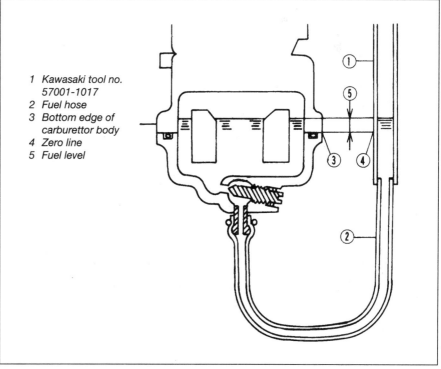

1 Kawasaki tool no. 57001-1017
2 Fuel hose
3 Bottom edge of carburettor body
4 Zero line
5 Fuel level

8.18 Checking the fuel level in a carburettor

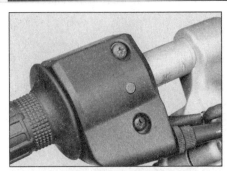

9.4 Remove the screws from the cable/switch housing and detach the front half of the housing

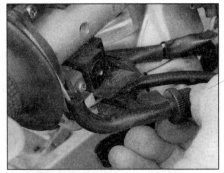

9.5 Detach the accelerator cable and guide from the rear half of the housing

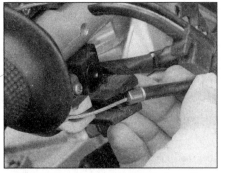

9.6 Pull the decelerator cable from the housing

9 Throttle cables and grip - removal, installation and adjustment

Throttle cables

Removal

1 Remove the fuel tank (see Section 2).
2 Detach the front brake master cylinder from the handlebar and position it out of the way (see Chapter 7).
3 Loosen the accelerator cable lockwheel and screw the cable adjuster in.
4 Remove the cable/switch housing screws **(see illustration)** and remove the front half of the housing.

9.7 Align the cables with the slots in the throttle grip pulley, then slide the cable ends out

5 Detach the accelerator cable and guide from the cable/switch housing **(see illustration)**.
6 Pull the decelerator cable from the rear half of the cable/switch housing **(see illustration)**.
7 Remove the rear half of the cable/switch housing from the handlebar. Detach both cables from the throttle grip pulley **(see illustration)**.
8 Detach the decelerator cable from the throttle pulley at the carburettor assembly **(see illustration)**.
9 Rotate the throttle pulley at the carburettor and detach the accelerator cable from the pulley **(see illustration)**.
10 Remove the cables, noting how they are routed.

Installation

11 Route the cables into place. Make sure they don't interfere with any other components and aren't kinked or bent sharply.
12 Lubricate the end of the accelerator cable with multi-purpose grease and connect it to the throttle pulley at the carburettor. Pass the inner cable through the slot in the bracket, then seat the cable housing in the bracket.
13 Repeat the previous step to connect the decelerator cable.
14 Route the decelerator cable around the backside of the handlebar and connect it to the rear hole in the throttle grip pulley.
15 From the front side of the handlebar, connect the accelerator cable to the forward hole in the throttle grip pulley.

16 Attach the rear half of the cable/switch housing to the handlebar, seating the decelerator cable in the groove in the housing as it's installed.
17 Hold the rear half of the cable/switch housing to the handlebar, then push the accelerator cable guide into place, making sure the notched portion is correctly engaged with the housing.
18 Install the front half of the cable/switch housing, making sure the locating pin engages with the hole in the handlebar. If necessary, rotate the housing back and forth, until the locating pin drops into the hole and the housing halves mate together. Install the screws and tighten them securely.
19 Install the master cylinder (see Chapter 7).

Adjustment

20 Follow the procedure outlined in Chapter 1 to adjust the cables.
21 Turn the handlebars back and forth to make sure the cables don't cause the steering to bind.
22 Install the fuel tank.

Throttle grip

Removal

23 Follow Steps 2 through 7 to detach the upper ends of the throttle cables from the throttle grip pulley.
24 Remove the screw retaining the handlebar end **(see illustration)** and remove the end. Slide the throttle grip off the handlebar.

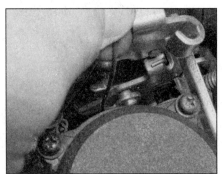

9.8 Detach the decelerator cable from the carburettor throttle pulley

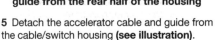

9.9 Rotate the throttle pulley at the carburettor assembly to full throttle, then detach the cable from the pulley

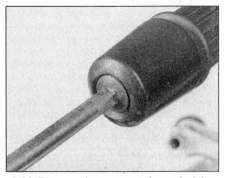

9.24 Remove the screw at the end of the handlebar, then slide the end and throttle grip off

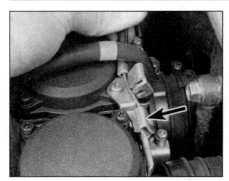

10.2 Pull back on the cable casing and remove it from its bracket, then detach the cable end from the choke lever (arrow)

10.3 Remove the screws from the choke cable/switch housing, then pull the front half of the housing off the handlebar

10.5 Make sure the choke cable guide seats in the housing properly

Installation

25 Clean the handlebar and apply a light coat of multi-purpose grease.
26 If it is necessary to replace the rubber grip itself, slice the old grip lengthwise with a sharp knife and remove it. Lubricate the new grip and handlebar with soapy water or a commercially available grip cement (which acts as a lubricant and then dries, bonding the rubber to the handlebar or twist grip) and push the grip on.
27 Push the twist grip on and install the handlebar end. Apply a non-hardening thread locking compound to the screw and tighten it securely.
28 Attach the cables following Steps 14 through 19, then adjust the cables (see Chapter 1).

10 Choke cable - removal, installation and adjustment

Removal

1 Remove the seat and fuel tank (see Section 2).
2 Pull the choke cable casing away from its mounting bracket at the carburettor

assembly, pass the inner cable through the opening in the bracket (see illustration). Detach the cable end from the choke lever by the no. 4 carburettor.
3 Remove the two screws securing the choke cable/switch housing halves to the left handlebar (see illustration). Pull the front half of the housing off and separate the choke cable from the lever.
4 Remove the cable, noting how it's routed.

Installation

5 Route the cable into position. Connect the upper end of the cable to the choke lever. Make sure the cable guide seats properly in the housing (see illustration). Place the housing up against the handlebar, making sure the pin in the housing fits into the hole in the handlebar. Install the screws, tightening them securely.
6 Connect the lower end of the cable to the choke lever. Pull back on the cable casing and connect it to the bracket on the no. 3 carburettor.

Adjustment

7 Check the freeplay at the lever on the handlebar. It should move about two to three millimeters (1/8-inch).
8 If the freeplay isn't as specified, loosen the cable adjusting locknut (see illustration) and

turn the adjusting nut in or out, as necessary, until the freeplay at the lever is correct. Tighten the locknut.
9 Install the fuel tank and all of the other components that were previously removed.

11 Air filter housing - removal and installation

1 Remove the seat. On 600 C and 750 models, remove the knee grip covers.
2 Remove the side covers (see Chapter 8).
3 Remove the fuel tank (see Section 2).
4 Remove the fuel tank rear mount.
5 On 600 A and B models, disconnect the cables from the battery (negative cable first), then remove the battery.
6 On 600 A and B models, remove the voltage regulator/rectifier (see Chapter 9) and its bracket.
7 On 600 A and B models, remove the left-side mounting bolt (see illustration).
8 On 600 A and B models, remove the right-side mounting bolt (see illustration).
9 On 600 A and B models, remove the rear mounting bolts (see illustration).
10 Slide the spring bands that attach the ducts to the carburettors toward the air filter housing.

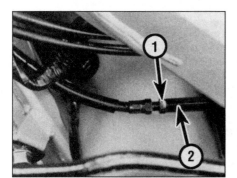

10.8 To adjust the choke cable, loosen the locknut (1) and turn the adjusting nut (2) in or out until the freeplay is correct

11.7 Once the voltage regulator/rectifier has been detached from its bracket, the left-side mounting bolt (arrow) for the air filter housing can be removed (A and B models)

11.8 The right-side mounting bolt for the air filter housing (arrow) is located in front of the igniter unit (A and B models)

11.9 Location of the rear mounting bolts (arrows) for the air filter housing (A and B models)

11.12 The engine breather hose is located under the air filter housing and is secured to the engine by a hose clamp (C model shown, A and B models similar)

1 Breather hose   2 Clamp

11 On 600 C and 750 models, remove the carburettor assembly (see Section 6).

12 Using a pair of pliers, squeeze the hose clamp on the crankcase breather hose (underneath the air filter housing) and detach the hose from the crankcase **(see illustration)**.

13 Detach the evaporative emission control system hose from the air cleaner housing (if equipped).

14 On 600 A and B models, lift the air filter housing up and out of the frame **(see illustration)**.

15 On 600 C and 750 models, remove the air ducts from the front of the air filter housing **(see illustration)**. Lift up the air filter housing far enough to disengage the lug on the bottom of the housing from the hole in the frame **(see illustration)**. Slide the housing forward and remove it out the left or right side of the machine.

16 Installation is the reverse of removal.

11.14 Lift the housing out of the frame, front side first (A and B models)

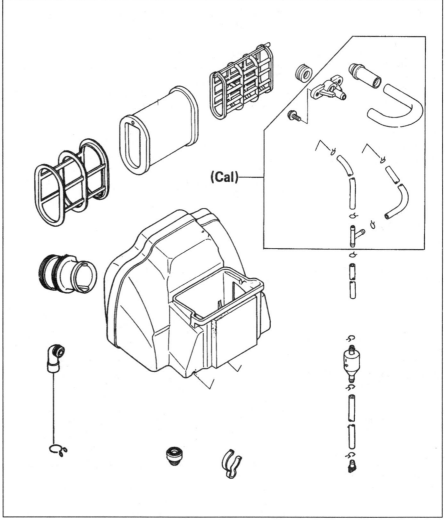

11.15a Air cleaner details (750 models)

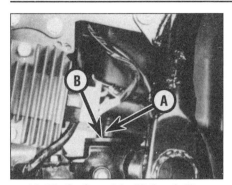

11.15b On C models, lift the air filter housing up to detach the lug (A) from the hole in the frame (B), slide the housing forward and remove it out from the side of the bike

12.5 Remove the nuts and slide the exhaust pipe holders away from the cylinder head

12.6 Remove the muffler mounting bolts from the passenger footpeg bracket on each side

12 Exhaust system - removal and installation

1 Remove the lower fairing (see Chapter 8).
2 Drain the coolant (see Chapter 1) then remove the radiator (see Chapter 3).

3 On 750 models, remove the oil cooler (see Chapter 3).
4 Loosen the clamps securing the connecting pipe to the right and left side exhaust pipes (it's located under the engine).
5 Remove the exhaust pipe holder nuts and slide the holders off the mounting studs **(see illustration)**. If the split keepers didn't come off with the holders, remove them from the pipes.

6 Remove the muffler mounting bolts at the footpeg brackets **(see illustration)**.
7 Pull the exhaust system forward, separate the right side pipes from the left side pipes and remove the system from the machine.
8 Installation is the reverse of removal, but be sure to install a new gasket at the connecting pipe.

Notes

Chapter 5
Ignition system

Contents

Degrees of difficulty

Easy, suitable for novice with little experience	Fairly easy, suitable for beginner with some experience	Fairly difficult, suitable for competent DIY mechanic 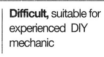	Difficult, suitable for experienced DIY mechanic	Very difficult, suitable for expert DIY or professional

Specifications

Ignition coil
Primary resistance
  600 A and B models and 750 models .... 1.8 to 2.8 ohms
  600 C models .... 2.61 to 3.19 ohms
Secondary resistance
  600 A and B models and 750 models .... 10 to 16 K ohms
  600 C models .... 13.5 to 16.5 K ohms
Arcing distance .... 7 mm (1/4 in) or more

Pickup coil
Resistance
  600 models .... 360 to 440 ohms
  750 models .... 390 to 590 ohms
Air gap - 750 models .... 0.5 mm (0.02 in)

Ignition timing
600 A and B models
  California models .... From 7.5° BTDC @ 1250 rpm to 35° BTDC @ 10,000 rpm
  All other models .... From 12.5° BTDC @ 1050 rpm to 40° BTDC @ 10,000 rpm
600 C models
  California models .... From 7.5° BTDC @ 1300 rpm to 35° BTDC @ 10,000 rpm
  All other models .... From 10° BTDC @ 1050 rpm to 32.5° BTDC @ 10,000 rpm
750 models
  California models .... From 5° BTDC @ 1250 rpm to 35° BTDC @ 3500 rpm
  All other models .... From 10° BTDC @ 1000 rpm to 35° BTDC @ 3500 rpm

Torque specifications
Pickup coil cover bolts - 600 models .... 5 Nm (43 in-lbs)
Timing rotor Allen bolt .... 25 Nm (18 ft-lbs)

1 General information

This motorcycle is equipped with a battery operated, fully transistorised, breakerless ignition system. The system consists of the following components:

Pickup coils
IC igniter unit
Battery and fuse
Ignition coils
Spark plugs
Stop and main (key) switches
Primary and secondary circuit wiring

The transistorised ignition system functions on the same principle as a conventional DC ignition system with the pickup unit and igniter performing the tasks normally associated with the breaker points and mechanical advance system. As a result, adjustment and maintenance of ignition components is eliminated (with the exception of spark plug replacement).

Because of their nature, the individual ignition system components can be checked but not repaired. If ignition system troubles occur, and the faulty component can be isolated, the only cure for the problem is to replace the part with a new one. Keep in mind that most electrical parts, once purchased, can't be returned. To avoid unnecessary expense, make very sure the faulty component has been positively identified before buying a replacement part.

2 Ignition system - check

 Warning: Because of the very high voltage generated by the ignition system, extreme care should be taken when these checks are performed.

1 If the ignition system is the suspected cause of poor engine performance or failure to start, a number of checks can be made to isolate the problem.
2 Make sure the ignition stop switch is in the Run or On position.

Engine will not start

3 Remove the fuel tank (see Chapter 4). Disconnect one of the spark plug wires, connect the wire to a spare spark plug and lay the plug on the engine with the threads contacting the engine. If necessary, hold the spark plug with an insulated tool **(see illustration)**. Crank the engine over and make sure a well-defined, blue spark occurs between the spark plug electrodes.

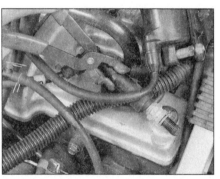
2.3 With the wire attached, ground (earth) a spark plug to the engine and operate the starter - bright blue sparks should be visible

 Warning: Don't remove one of the spark plugs from the engine to perform this check - atomised fuel being pumped out of the open spark plug hole could ignite, causing severe injury!

4 If no spark occurs, the following checks should be made:
5 Unscrew a spark plug cap from a plug wire and check the cap resistance with an ohmmeter **(see illustration)**. If the resistance is infinite, replace it with a new one. Repeat this check on the remaining plug caps.
6 Make sure all electrical connectors are clean and tight. Check all wires for shorts, opens and correct installation.
7 Check the battery voltage with a voltmeter and the specific gravity with an hydrometer (see Chapter 1 and *Fault Finding Equipment* in the Reference section). If the voltage is less than 12-volts or if the specific gravity is low, recharge the battery.
8 Check the ignition fuse and the fuse connections. If the fuse is blown, replace it with a new one; if the connections are loose or corroded, clean or repair them.
9 Refer to Section 3 and check the ignition coil primary and secondary resistance.
10 Refer to Section 4 and check the pickup coil resistance.
11 If the preceding checks produce positive results but there is still no spark at the plug,

2.13 A simple spark gap testing fixture can be made from a block of wood, a large alligator clip, two nails, a screw and a piece of wire

2.5 Unscrew the spark plug caps from the plug wires and measure their resistance with an ohmmeter

remove the IC igniter and carry out the checks described in Section 5.

Engine starts but misfires

12 If the engine starts but misfires, make the following checks before deciding that the ignition system is at fault.
13 The ignition system must be able to produce a spark across a seven millimeter (1/4-inch) gap (minimum). A simple test fixture **(see illustration)** can be constructed to make sure the minimum spark gap can be jumped. Make sure the fixture electrodes are positioned seven millimeters apart.
14 Connect one of the spark plug wires to the protruding test fixture electrode, then attach the fixture's alligator clip to a good engine ground (earth) **(see illustration)**.
15 Crank the engine over (it will probably start and run on the remaining cylinders) and see if well-defined, blue sparks occur between the test fixture electrodes. If the minimum spark gap test is positive, the ignition coil for that cylinder (and its companion cylinder) is functioning properly. Repeat the check on one of the spark plug wires that is connected to the other coil. If the spark will not jump the gap during either test, or if it is weak (orange colored), refer to Paragraphs 5 through 11 of this Section and perform the component checks described.

2.14 Connect the tester to a good ground (earth) and attach one of the spark plug wires - when the engine is cranked, sparks should jump the gap between the nails

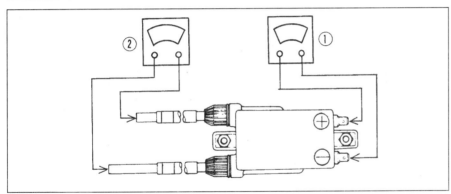

3.4 To check the resistance of the primary windings, connect the leads of the ohmmeter to the primary terminals (1) - to check the resistance of the secondary windings, attach the leads of the ohmmeter to the spark plug wires of that coil (2)

3 Ignition coils - check, removal and installation

Check

1 In order to determine conclusively that the ignition coils are defective, they should be tested by a Kawasaki dealer service department which is equipped with the special electrical tester required for this check.

2 However, the coils can be checked visually (for cracks and other damage) and the primary and secondary coil resistances can be measured with an ohmmeter. If the coils are undamaged, and if the resistances are as specified, they are probably capable of proper operation.

3 To check the coils for physical damage, they must be removed (see Step 9). To check the resistances, simply remove the fuel tank (see Chapter 4), unplug the primary circuit electrical connectors from the coil(s) and remove the spark plug wires from the plugs that are connected to the coil being checked.

Mark the locations of all wires before disconnecting them.

4 To check the coil primary resistance, attach one ohmmeter lead to one of the primary terminals and the other ohmmeter lead to the other primary terminal **(see illustration)**.

5 Place the ohmmeter selector switch in the R x 1 position and compare the measured resistance to the value listed in this Chapter's Specifications.

6 If the coil primary resistance is as specified, check the coil secondary resistance by disconnecting the meter leads from the primary terminals and attaching them to the spark plug wire terminals **(see illustration 3.4)**.

7 Place the ohmmeter selector switch in the R x 100 position and compare the measured resistance to the values listed in this Chapter's Specifications.

8 If the resistances are not as specified, unscrew the spark plug wire retainers from the coil, detach the wires and check the resistance again. If it is now within specifications, one or both of the wires are bad. If it's still not as specified, the coil is probably defective and should be replaced with a new one.

Removal and installation

9 To remove the coils, refer to Chapter 4 and remove the fuel tank, then disconnect the spark plug wires from the plugs. After labeling them with tape to aid in reinstallation, unplug the coil primary circuit electrical connectors.

10 Support the coil with one hand and remove the coil mounting nuts **(see illustrations)**, then withdraw the coil from its bracket. If necessary, detach the bracket from the frame - it's secured by two bolts.

11 Installation is the reverse of removal. If a new coil is being installed, unscrew the spark plug wire terminals from the coil, pull the wires out and transfer them to the new coil. Make sure the primary circuit electrical connectors are attached to the proper terminals. Just in case you forgot to mark the wires, the black and red wires connect to the no. 1 and 4 ignition coil (red to positive, black to negative) and the red and green wires attach to the no. 2 and 3 coil (red to positive, green to negative).

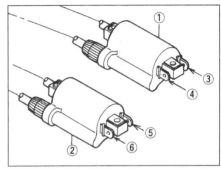

3.10a Ignition coil connection details - 750 models

1 Coil for cylinders 1 and 4
2 Coil for cylinders 2 and 3
3 Red wire
4 Black wire
5 Red wire
6 Green wire

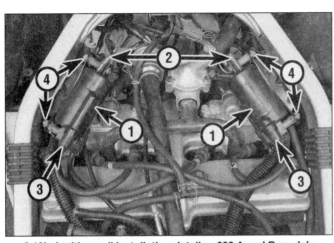

3.10b Ignition coil installation details - 600 A and B models

1 Ignition coils
2 Primary wires
3 Spark plug wire caps
4 Coil mounting nuts

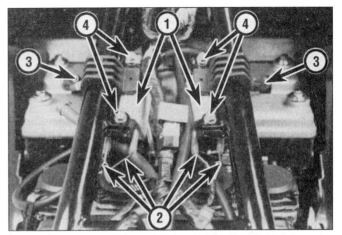

3.10c Ignition coil installation details - 600 C models

1 Ignition coils
2 Primary wires
3 Spark plug wire caps
4 Coil mounting nuts

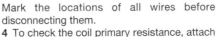

4.2 Depress the tang and unplug the electrical connector for the pickup coils

4.7 Remove the screws that secure the pickup coil cover

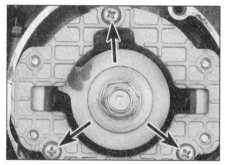

4.8a Remove the screws (arrows) and detach the pickup coil mounting plate - 600 models

4.8b Detach the wiring harness from the clamps (arrows) below the clutch housing on 600 models

4 Pickup coils - check, removal and installation

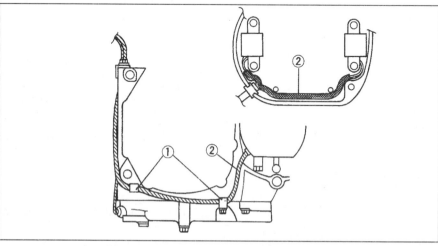

4.8c Wiring harness routing and clamps on 750 models

1 Wiring harness retainers   2 Wiring harness

Check

1 Remove the right side cover.
2 Disconnect the electrical connector for the pickup coil (see illustration).
3 Probe each pair of terminals in the pickup coil connector with an ohmmeter (the black and yellow wire terminals, then the blue and black/white wire terminals) and compare the resistance reading with the value listed in this Chapter's Specifications.
4 Set the ohmmeter on the highest resistance range. Measure the resistance between a good ground (earth) and each terminal in the electrical connector. The meter should read infinity.
5 If the pickup coils fail either of the above tests, they must be replaced.

Removal

6 On 750 models, remove the lower fairing for access (see Chapter 8).
7 Remove the screws that secure the pickup coil cover to the engine case (see illustration) and detach the cover from the engine.
8 On 600 models, unscrew the pickup coil mounting plate screws and remove the pickup coils with the plate (see illustration). On 750 models, remove the two screws which retain each pickup coil. Detach the wiring harness

from the clamps beneath the clutch housing (see illustrations).

Installation

9 If the new pickup coils don't come with a mounting plate, remove the screws and detach the pickup coils from the plate, noting how they're installed (see illustration). Place the new pickup coils on the plate and install the screws, tightening them securely.
10 On 600 models, position the pickup coil mounting plate on the engine and install the screws, tightening them securely. Be sure to

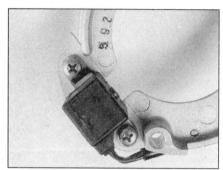

4.9 Each pickup coil is secured to the mounting plate with two screws - 600 models

seat the grommet on the wiring harness into the notch in the case.
11 On 750 models, install the pickup coils but only secure their screws finger-tight. Rotate the ignition timing rotor, using a spanner on its centre hexagon, so that the air gap between the rotor tip and pickup coil raised trigger can be measured with a feeler gauge (see illustration). The gap should be

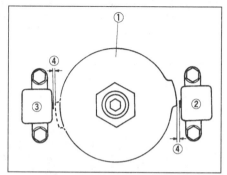

4.11 Set the pickup coil air gaps to specifications - 750 models

1 Timing rotor
2 Pickup coil for cylinders 1 and 4
3 Pickup coil for cylinders 2 and 3
4 Air gap

as given in the Specifications at the beginning of this Chapter. Set the pickup coil position so that the air gap is correct, then tighten the coil screws. Do the same with the other pickup coil. Seat the wiring grommet into the notch in the crankcase and make sure the wiring between the pickup coils is routed correctly **(see illustration 4.8c)**.

12 Install the pickup coil cover, making sure the notch in the cover is positioned on the lower side **(see illustration)**. Tighten the cover bolts to the torque listed in this Chapter's Specifications.

13 Attach the harness to the clamps on the bottom of the clutch housing **(see illustration 4.8b or c)**.

14 Plug in the electrical connector and install the side cover.

5 IC igniter - removal, check and installation

Removal

1 Remove the right side cover.

2 Pull the igniter from its bracket **(see illustration)**. Unplug the electrical connector.

4.12 The notch in the pickup coil cover serves as a drain for moisture - make sure it's positioned on the bottom

Check

3 If a check of all the other ignition system components does not reveal the cause of an ignition fault, the igniter unit should be removed from the bike and tested in accordance with the accompanying table **(see illustrations)**. Note that different test meters may give slightly different readings from the Kawasaki tester (Pt. No. 57001-983), so confirm the diagnosis by taking the suspect unit to a Kawasaki dealer before buying a replacement.

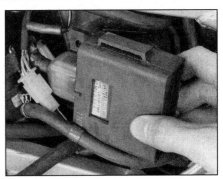

5.2 To remove the igniter, simply pull it from its mounting bracket and detach the electrical connector

Installation

4 If a new igniter is being installed, slide the rubber cover off the old one and transfer it to the new one.

5 Plug in the electrical connector and push the igniter into its bracket, making sure the slots of the rubber cover are engaged properly with the tabs on the bracket.

6 Install the side cover.

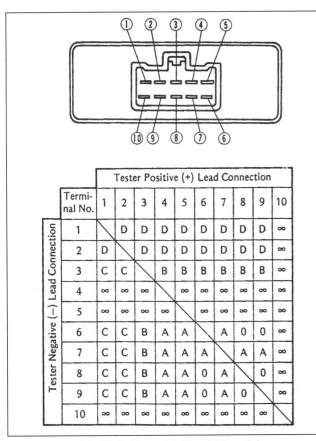

	Tester Positive (+) Lead Connection									
Terminal No.	1	2	3	4	5	6	7	8	9	10
1		D	D	D	D	D	D	D	D	∞
2	D		D	D	D	D	D	D	D	∞
3	C	C		B	B	B	B	B	B	∞
4	∞	∞	∞		∞	∞	∞	∞	∞	∞
5	∞	∞	∞	∞		∞	∞	∞	∞	∞
6	C	C	B	A	A		A	0	0	∞
7	C	C	B	A	A	A		A	A	∞
8	C	C	B	A	A	0	A		0	∞
9	C	C	B	A	A	0	A	0		∞
10	∞	∞	∞	∞	∞	∞	∞	∞	∞	

(Row label: Tester Negative (−) Lead Connection)

5.3a IC igniter test table - 600 models

0 Zero ohms (continuity)
A 0.3 to 4.2 K ohms
B 6.6 to 21.4 K ohms
C 25 to 75 K ohms
D 125 to 375 K ohms
∞ Infinity (no continuity)

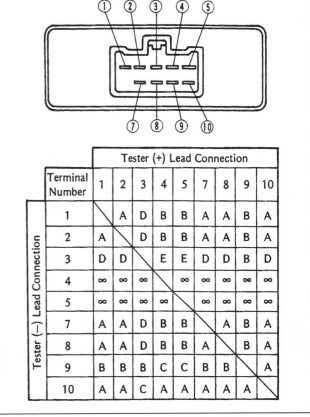

	Tester (+) Lead Connection								
Terminal Number	1	2	3	4	5	7	8	9	10
1		A	D	B	B	A	A	B	A
2	A		D	B	B	A	A	B	A
3	D	D		E	E	D	D	B	D
4	∞	∞	∞		∞	∞	∞	∞	∞
5	∞	∞	∞	∞		∞	∞	∞	∞
7	A	A	D	B	B		A	B	A
8	A	A	D	B	B	A		B	A
9	B	B	B	C	C	B	B		A
10	A	A	C	A	A	A	A	A	

(Row label: Tester (−) Lead Connection)

5.3b IC igniter test table - 750 models

A 2 to 6 K ohms
B 5 to 11 K ohms
C 9 to 20 K ohms
D 15 to 28 K ohms
E 25 to 55 K ohms
∞ Infinity (no continuity)

Notes

Chapter 6
Frame, suspension and final drive

Contents

Degrees of difficulty

Easy, suitable for novice with little experience 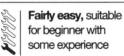	Fairly easy, suitable for beginner with some experience	Fairly difficult, suitable for competent DIY mechanic	Difficult, suitable for experienced DIY mechanic	Very difficult, suitable for expert DIY or professional

Specifications

Forks

Fork oil type, quantity and level	see Chapter 1
Spring free length	
600 A and B models	
Standard	478 mm (18.82 in)
Minimum	468 mm (18.43 in)
600 C1 to C5 (UK) and C1 to C6 (US) models	
Standard	481 mm (18.94 in)
Minimum	472 mm (18.58 in)
600 C6, C7 (UK) and C7 to C10 (US) models	
Standard	494.8 mm (19.48 in)
Minimum	485 mm (19.09 in)
750 models	
Standard	413 mm (16.26 in)
Minimum	405 mm (15.94 in)

Rear sprocket

Runout (maximum)	0.5 mm (0.020 in)
Diameter - 600 models	
Standard	187.02 to 187.52 mm (7.368 to 7.388 in)
Minimum	186.7 mm (7.355 in)

Torque specifications

Frame rear section mounting bolts (600 C and 750 models only)
Lower bolts	44 Nm (33 ft-lbs)
Upper bolts	19 Nm (14 ft-lbs)
Frame downtube bolts	24 Nm (18 ft-lbs)
Handlebar-to-upper triple clamp bolts	23 Nm (16.5 ft-lbs)
Handlebar tube-to-mount bolt (600 C and 750 models only)	23 Nm (16.5 ft-lbs)
Brake plunger bolts (600 A and B models only)	7 Nm (61 in-lbs)
Anti-dive unit-to-fork leg bolts (600 A and B models only)	7 Nm (61 in-lbs)
ESCS-to-fork leg bolts (600 C and 750 models only)	7 Nm (61 in-lbs)

Damper rod bolt
600 A and B models	30 Nm (22 ft-lbs)
600 C models	39 Nm (29 ft-lbs)
750 models	21 Nm (15 ft-lbs)
Fork top plug	22 Nm (16.5 ft-lbs)
Steering stem nut	39 Nm (29 ft-lbs)
Fork clamp bolts	21 Nm (15 ft-lbs)

Rear shock absorber mounting bolts/nuts
600 A and B models	53 Nm (39 ft-lbs)
600 C models	49 Nm (36 ft-lbs)
750 models	59 Nm (43 ft-lbs)

Tie-rod-to-rocker arm bolt/nut
600 A and B models	53 Nm (39 ft-lbs)
600 C models	49 Nm (36 ft-lbs)
750 models	59 Nm (43 ft-lbs)

Tie-rod-to-swingarm bolts/nuts
600 A and B models	53 Nm (39 ft-lbs)
600 C models	49 Nm (36 ft-lbs)
750 models	59 Nm (43 ft-lbs)

Rocker arm pivot shaft nut
600 A and B models	53 Nm (39 ft-lbs)
600 C models	49 Nm (36 ft-lbs)
750 models	59 Nm (43 ft-lbs)

Swingarm pivot shaft nut
600 A and B models and 750 models	88 Nm (65 ft-lbs)
600 C models	94 Nm (69 ft-lbs)
Engine sprocket cover bolts - 600 models	9 Nm (78 in-lbs)
Rear sprocket-to-wheel coupling nuts	73 Nm (54 ft-lbs)
Engine sprocket holding plate bolts - 600 models	10 Nm (87 in-lbs)
Engine sprocket nut - 750 models	98 Nm (72 ft-lbs)

1 General information

The machines covered by this manual use two different frame designs. On 600 A and B models, the frame is of the full cradle design, constructed of square tubing; steel for the A models and aluminum alloy for the B models. The downtubes on these models are detachable, which allows for easy engine removal. The B models have integral frame members where the removable plastic side stays are on A models.

The frame on 600 C and 750 models is fabricated from round steel tubing and is of the diamond design. The downtubes aren't detachable, but the rear section of the frame is.

The front forks are of the conventional coil spring, hydraulically damped telescopic type. On 600 A and B models they are air-assisted.

The forks on 600 C and 750 models are designed to run at atmospheric pressure.

A and B models use a variable damping system, controlled by a knob at the lower end of each fork leg, on the AVDS units. A brake valve mounted to the top of these units firms-up the damping characteristics of the forks during braking, which reduces fork compression. On 600 C and 750 models, all this is controlled electrically by the Electric Suspension Control System (ESCS); note that this system is not fitted to the 600 C6, C7 UK models and 600 C7, C8, C9, C10 US models.

The rear suspension is Kawasaki's Uni-Trak design, which consists of a single shock absorber, a rocker arm, two tie-rods and a square-section aluminum swingarm. The shock absorber has four damping settings and is also air adjustable.

The final drive uses an endless chain (which means it doesn't have a master link). A rubber damper is installed between the rear wheel coupling and the wheel.

2 Frame - inspection and repair

1 The frame should not require attention unless accident damage has occurred. In most cases, frame replacement is the only satisfactory remedy for such damage. A few frame specialists have the jigs and other equipment necessary for straightening the frame to the required standard of accuracy, but even then there is no simple way of assessing to what extent the frame may have been overstressed.

2 After the machine has accumulated a lot of miles, the frame should be examined closely for signs of cracking or splitting at the welded joints. Rust corrosion can also cause weakness at these joints. Loose engine mount bolts can cause ovaling or fracturing of the mounting tabs. Minor damage can often be repaired by welding, depending on the extent and nature of the damage.

3 Remember that a frame which is out of alignment will cause handling problems. If misalignment is suspected as the result of an accident, it will be necessary to strip the machine completely so the frame can be thoroughly checked.

3 Frame rear section (600 C and 750 models only) - removal and installation

600 C models

1 Remove the rider's seat.
2 Remove the knee grip covers and the side covers (see Chapter 8).
3 Remove the passenger seat and tailpiece (see Chapter 8).
4 Detach the air valve (see Chapter 1), the starter solenoid and voltage regulator/rectifier (see Chapter 9) and the rear brake fluid reservoir (see Chapter 7) from the frame.
5 Remove the rear section of the rear fender (see Chapter 8).
6 Detach any wiring harness clamps or other components which may interfere with removal of the frame rear section.
7 Unscrew the bolts and detach the frame rear section.
8 If necessary, unbolt the passenger footpeg brackets, the seat lock and the helmet lock from the frame section.
9 Installation is the reverse of the removal procedure. Be sure to tighten the bolts to the torque listed in this Chapter's Specifications.

750 models

10 Remove the seat, knee grip covers and side covers.
11 Remove the air cleaner housing cover.
12 Remove the battery (see Chapter 9).
13 Disconnect the wiring connectors for the frame rear section on the right side of the motorcycle, the left side, just forward of the starter relay and at the rear inside the tailpiece.
14 Unbolt the rear master cylinder (but leave the fluid line connected).
15 Support the exhaust system and remove

the muffler mounting bolts on both sides of the motorcycle.
16 Unbolt the rear frame section and remove it together with the rear fender.
17 Installation is the reverse of the removal steps.

4 Footpegs and brackets - removal and installation

Rider's left side

1 If it's only necessary to detach the footpeg from the bracket, pry the C-clip off the pivot pin (see illustration), slide out the pin and detach the footpeg from the bracket. Be careful not to lose the spring. Installation is the reverse of removal, but be sure to install the spring correctly.
2 If it's necessary to remove the entire bracket from the frame, mark the relationship of the shift lever to the shift shaft, then remove the clamp bolt. Slide the lever off the shaft.
3 Unscrew the shift lever pivot bolt and the bracket-to-frame bolt and separate the footpeg and bracket from the frame.
4 Installation is basically the reverse of removal. Apply a thin coat of grease to the shift pedal pivot bolt, and be sure to line up the matchmarks on the shift lever and shift shaft. The link rod should be parallel to the shift pedal.

Rider's right side

5 If it's only necessary to detach the footpeg from the bracket, pry the C-clip off the pivot pin (see illustration 4.1), slide out the pin and detach the footpeg from the bracket. Be careful not to lose the spring. Installation is the reverse of removal, but be sure to install the spring correctly.
6 If it's necessary to remove the entire bracket from the frame, unplug the electrical connector for the rear brake light switch.
7 Remove the cotter pin from the clevis pin that attaches the brake pedal to the master cylinder, then remove the clevis pin (see Chapter 9).
8 Unbolt the master cylinder from the bracket.

9 Remove the Allen-head bolts that secure the bracket to the frame, then detach the footpeg and bracket.
10 Installation is the reverse of removal.

Passenger footpegs and brackets (either side)

11 If it's only necessary to detach the footpeg from the bracket, pry the C-clip off the pivot pin (see illustration 4.1), slide out the pin and detach the footpeg from the bracket. Be careful not to lose the spring. Installation is the reverse of removal, but be sure to install the spring correctly.
12 If it's necessary to remove the entire bracket, unscrew the two Allen head bolts and detach the bracket from the frame.
13 Installation is the reverse of removal.

5 Side and centrestand - maintenance

1 The centrestand pivots on two bolts attached to the frame. Periodically, remove the pivot bolts and grease them thoroughly to avoid excessive wear.
2 Make sure the return spring is in good condition. A broken or weak spring is an obvious safety hazard.
3 The sidestand is attached to a bracket bolted to the frame. An extension spring anchored to the bracket ensures that the stand is held in the retracted position.
4 Make sure the pivot bolt is tight and the extension spring is in good condition and not overstretched. An accident is almost certain to occur if the stand extends while the machine is in motion.

6 Handlebars - removal and installation

1 The handlebars are individual assemblies that slip over the tops of the fork tubes, each being retained to the steering head by two Allen-head bolts. If the handlebars must be removed for access to other components, such as the forks or the steering head, simply remove the bolts and slip the handlebar(s) off the fork tubes (see illustration). It's not necessary to disconnect the cables, wires or hoses, but it is a good idea to support the assembly with a piece of wire or rope, to avoid unnecessary strain on the cables, wires and (on the right side) the brake hose.
2 If the handlebars are to be removed completely, refer to Chapter 9 for the master cylinder removal procedure, Chapter 4 for the throttle grip removal procedure and Chapter 9 for the switch removal procedure.
3 Check the handlebars for cracks and distortion and replace them if any undesirable conditions are found. When installing the handlebars, tighten the bolts to the torque listed in this Chapter's Specifications.

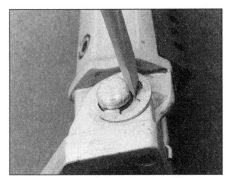

4.1 Pry off the C-clip and push the pivot pin out to detach the footpeg from the bracket

6.1 To detach a handlebar, remove the two bolts and slide the assembly off the fork tube - if it's stuck, wiggle it back and forth while pulling up

7.9 On A and B models, unbolt the brake plungers from the anti-dive units - don't disconnect the brake lines

7.11 If the forks are going to be disassembled, loosen the upper clamp bolt, then loosen the top plug with a 1/2 inch drive ratchet

7.12a Loosen the fork lower clamp bolts

7 Forks -
removal and installation

Removal

1 Set the bike on its centrestand.
2 Remove the upper fairing (see Chapter 8).
3 If you're working on a 600 A or B model, release the air pressure from the forks.
4 Remove the front fender (see Chapter 8).
5 Remove the wheel (see Chapter 7).
6 Remove the handlebars (see Section 6). Support them so the cables, wires and brake hose aren't strained or kinked.
7 Unbolt the brake calipers (see Chapter 7) and hang them with pieces of wire or rope. On 600 A and B models, unbolt the junction blocks from the fork tubes.
8 If you're removing the right side fork on a 600 C or 750 model, detach the ESCS electrical connector from the ESCS unit.
9 If you're working on a 600 A or B model, unbolt the brake plungers from the anti-dive units (see illustration).
10 Remove any wiring harness clamps or straps from the fork tubes.
11 If the forks will be disassembled after removal, loosen the upper triple clamp bolts, then loosen the top plugs (see illustration).

12 Loosen the fork upper and lower triple clamp bolts (see illustration), then slide the fork tubes down slightly, using a twisting motion, then retighten the lower clamp bolts. Remove the retaining rings from their grooves (see illustration).
13 Loosen the lower clamp bolts again and slide the fork tube down (one at a time) and out of the steering stem. If you're working on a 600 A or B model, hold the sleeve of the air connecting pipe stationary so the pipe doesn't bend or break (see illustration).
14 If you're working on a 600 A or B model, stick a piece of tape over the air holes to prevent oil loss.

Installation

15 Slide each fork leg into the lower triple clamp. On 600 A and B models, install the retaining rings, then lubricate the O-rings in the sleeves of the air connecting pipe assembly with the recommended fork oil (see Chapter 1). Install the the air connecting pipe assembly over the tops of the fork tubes.
16 Slide the fork legs up, installing the tops of the tubes into the upper triple clamp. On 600 A and B models, align the air holes in the fork tubes with the pipes on the sleeves. Push the forks up until the sleeves of the air connecting pipe assembly contact the retaining rings. On 600 C models, push the

forks into the upper clamp until 16.0 to 17.5 mm (0.63 to 0.69 in) of the fork tubes protrude past the upper surface of the upper clamp (see illustration). On 750 models, push the forks into the upper clamp until 15 mm (0.59 inch) of the fork tubes protrude past the upper surface of the upper clamp.
17 The remainder of installation is the reverse of the removal procedure. Be sure to tighten the clamp bolts and the brake plunger bolts (600 A and B models) to the torque listed in this Chapter's Specifications. Tighten the caliper mounting bolts to the torque listed in the Chapter 7 Specifications.
18 If you're working on a 600 A or B model, charge the forks with the recommended air pressure (see Chapter 1).
19 Pump the front brake lever several times to bring the pads into contact with the discs.

8 Forks - disassembly,
inspection and reassembly

Disassembly

1 Remove the forks following the procedure in Section 7. Work on one fork leg at a time to avoid mixing up the parts.

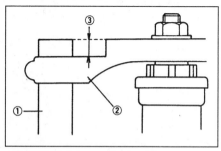

7.16 Fork tube protrusion above upper clamp - 600 C and 750 models

1 Fork tube
2 Upper clamp
3 16 to 17.5 mm (0.63 to 0.69 inch) -
   600 C models
   15 mm (0.59 inch) - 750 models

7.12b Slide the forks down a little, pull the sleeves of the air connecting tube up and, using a small screwdriver, pry the retaining rings out of their grooves (A and B models only)

7.13 Hold the sleeve of the air connecting tube while pulling the fork tube down, so the air connecting tube doesn't break (A and B models only)

2 Remove the axle clamp bolts. If you're working on the left fork leg of a 600 A or B model, remove the axle nut. If you're working on the right fork leg on a 600 C or 750 model, remove the axle clamp bolt.

3 Remove the top plug and washer (it should have been loosened before the forks were removed), spacer and spring guide (C models only) and spring **(see illustrations)**. When removing the plug, exert downward pressure on it to counteract the pressure from the fork spring - this will help to prevent damage to the threads.

4 Invert the fork assembly over a container and allow the oil to drain out. Be careful not to

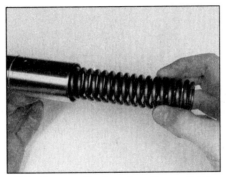

8.3a Remove the top plug, then pull out the fork spring

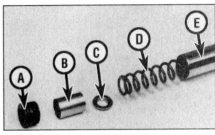

8.3b On 600 C and 750 models, a spacer and spring guide must also be removed

A Top plug   D Spring
B Spacer   E Inner fork tube
C Spring guide

1   Top plug
2   O-ring
3   Upper clamp
4   Air connecting pipe
5   O-rings
6   Washer
7   Fork spring
8   Fork tube (inner)
9   Inner guide bushing
10 Steering stem/lower clamp
    assembly
11 Travel Control Valve
12 Brake plunger assembly
13 Damper rod
14 Teflon ring
15 Rebound spring
16 Washers
17 Damper rod base
18 Dust seal
19 Retaining ring
20 Washer
21 Oil seal
22 Washer
23 Outer guide bushing
24 Fork tube (outer)
25 Copper washer
26 Allen-head bolt
27 Anti-dive unit
28 Drain screw
29 Gasket
30 Axle clamp bolt

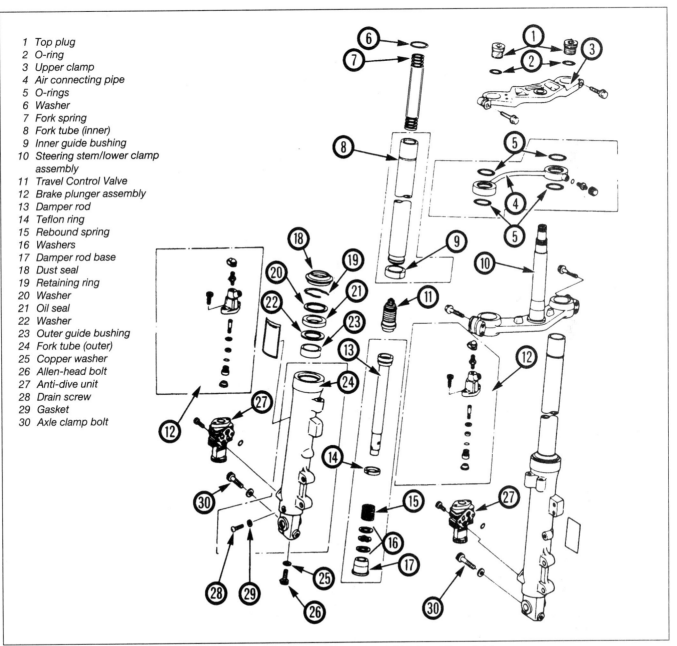

8.3c Details of the forks and steering head (600 A and B models)

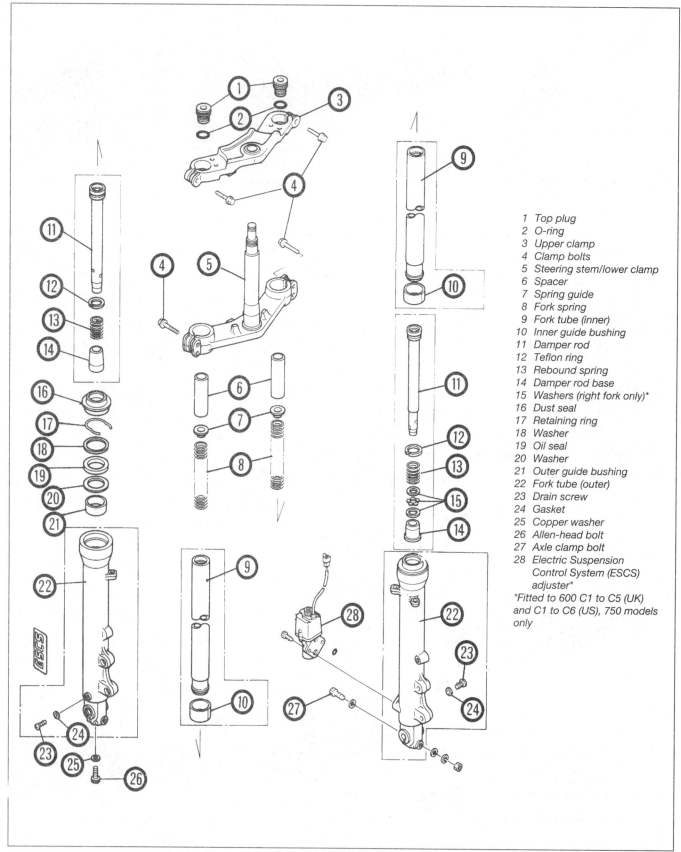

1 Top plug
2 O-ring
3 Upper clamp
4 Clamp bolts
5 Steering stem/lower clamp
6 Spacer
7 Spring guide
8 Fork spring
9 Fork tube (inner)
10 Inner guide bushing
11 Damper rod
12 Teflon ring
13 Rebound spring
14 Damper rod base
15 Washers (right fork only)*
16 Dust seal
17 Retaining ring
18 Washer
19 Oil seal
20 Washer
21 Outer guide bushing
22 Fork tube (outer)
23 Drain screw
24 Gasket
25 Copper washer
26 Allen-head bolt
27 Axle clamp bolt
28 Electric Suspension
   Control System (ESCS)
   adjuster*

*Fitted to 600 C1 to C5 (UK)
and C1 to C6 (US), 750 models
only

8.3d Details of the forks and steering head (600 C and 750 models)

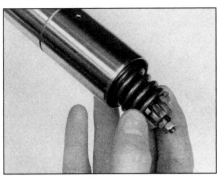

8.4 Remove the TCV from the inner tube

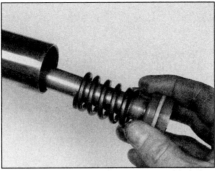

8.6 Remove the damper rod and rebound spring from the inner tube

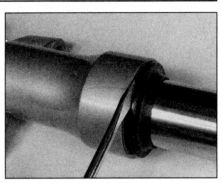

8.7 Pry the dust seal out of the outer tube with a small screwdriver

let the Travel Control Valve (TCV) fall out (600 A and B models only). Gently tap the fork leg with a rubber mallet until the valve comes out - be ready to catch it **(see illustration)**.

5 Prevent the damper rod from turning using a holding handle (Kawasaki tool no. 57001-183) and adapter (Kawasaki tool no. 57001-1057) engaged in the head of the damper rod. Unscrew the Allen bolt at the bottom of the outer tube and retrieve the copper washer.

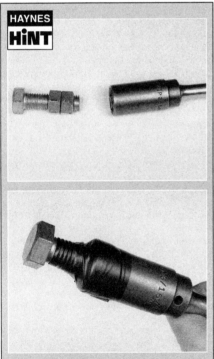

HAYNES HiNT

If you don't have access to these special tools, you can fabricate your own using a bolt with a 23 mm head, two nuts, a socket (to fit on the nuts), a long extension and a ratchet. Thread the two nuts onto the bolt and tighten them against each other. Insert the assembly into the socket and tape it into place. Now, insert the tool into the fork tube and engage the bolt head into the hex recess in the damper rod.

6 Pull out the damper rod and the rebound spring **(see illustration)**. Don't remove the Teflon ring from the damper rod unless a new one will be installed.

7 Pry the dust seal from the outer tube **(see illustration)**.

8 Pry the retaining ring from its groove in the outer tube **(see illustration)**. Remove the ring and the washer that's present underneath it.

9 Hold the inner tube and, using a dead-blow plastic mallet, tap the outer tube off (be careful not to let it fall) **(see illustration)**. An alternative method is to hold the outer tube and yank the inner tube upward, repeatedly (like a slide hammer), until the seal and outer tube guide bushing pop loose.

10 Slide the seal, washer and outer tube guide bushing from the inner tube.

11 Invert the outer tube and remove the damper rod base and the washers **(see illustration)**. **Note:** *On 600 C1 to C5 (UK) , C1 to C6 (US) and all 750 models, only the right fork leg has washers.*

Inspection

12 Clean all parts in solvent and blow them dry with compressed air, if available. Check the inner and outer fork tubes, the guide bushings and the damper rod for score marks, scratches, flaking of the chrome and excessive or abnormal wear. Look for dents in the tubes and replace them if any are found. Check the fork seal seat for nicks, gouges and scratches. If damage is evident, leaks will

8.9 If you don't have the special tool, use a plastic dead-blow hammer to tap the outer tube from the inner tube

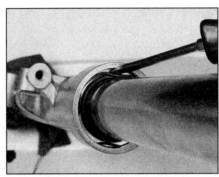

8.8 Pry out the retaining ring

occur around the seal-to-outer tube junction. Replace worn or defective parts with new ones.

13 Have the fork inner tube checked for runout at a dealer service department or other repair shop.

Caution: If it is bent, it should not be straightened; replace it with a new one.

14 Measure the overall length of the long spring and check it for cracks and other damage. Compare the length to the minimum length listed in this Chapter's Specifications. If it's defective or sagged, replace both fork springs with new ones. Never replace only one spring.

15 Check the TCV for any signs of damage and replace it if necessary (600 A and B models only). Don't attempt to disassemble it.

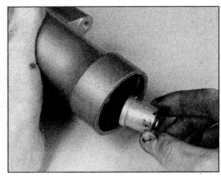

8.11 Retrieve the damper rod base and washers from the outer tube

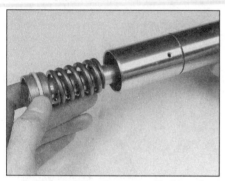

8.17 Install the damper rod/rebound spring assembly into the inner tube

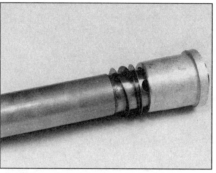

8.18 Install the damper rod washers and base on the end of the damper rod - the flat washer should be between the two spring washers

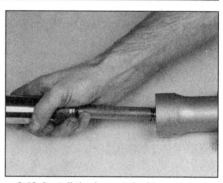

8.19 Install the inner tube/damper rod assembly into the outer tube

Reassembly

16 If it's necessary to replace the inner guide bushing (the one that won't come off that's on the bottom of the inner tube), pry it apart at the slit and slide it off. Make sure the new one seats properly.

17 Place the rebound spring over the damper rod, slide the rod assembly into the inner fork tube until it protrudes from the lower end of the tube **(see illustration)**.

18 Install the damper rod washers and base onto the end of the damper rod **(see illustration)**. **Note:** Remember, only the right fork leg has washers on 600 C1 to C5 (UK) , C1 to C6 (US) and all 750 models.

19 Insert the inner tube/damper rod assembly into the outer tube **(see illustration)** until the Allen-head bolt (with copper washer) can be threaded into the damper rod from the lower end of the outer tube. **Note:** Apply a non-permanent thread locking compound to

the threads of the bolt. Keep the two tubes fairly horizontal so the damper rod base and washers don't fall off. Using the tool described in Step 5, hold the damper rod and tighten the Allen-head bolt to the torque listed in this Chapter's Specifications.

20 Slide the outer guide bushing down the inner tube. The slit in the bushing must point to the left or right - not to the front or rear. Using a special bushing driver (Kawasaki tool no. 57001-1104 or equivalent for 600 A and B models, or no. 57001-1219 or equivalent for 600 C and 750 models) and a used guide bushing placed on top of the guide bushing being installed, drive the bushing into place until it is fully seated **(see illustration)**. If you don't have access to one of these tools, it is highly recommended that you take the assembly to a Kawasaki dealer service department or other motorcycle repair shop to have this done. It is possible, however, to drive the bushing into place using a section of tubing and an old guide bushing **(see illustration)**. Wrap tape around the ends of the tubing to prevent it from scratching the fork tube.

21 Slide the washer down the inner tube, into position over the guide bushing.

22 Lubricate the lips and the outer diameter of the fork seal with the recommended fork oil (see Chapter 1) and slide it down the inner tube, with

the lips facing down **(see illustration)**. Drive the seal into place with a special seal driver (Kawasaki tool no. 57001-1104 or equivalent for 600 A and B models, or no. 57001-1219 or equivalent for 600 C and 750 models). If you don't have access to one of these, it is recommended that you take the assembly to a Kawasaki dealer service department or other motorcycle repair shop to have the seal driven in. If you are very careful, the seal can be driven in with a hammer and a drift punch. Work around the circumference of the seal, tapping gently on the outer edge of the seal until it's seated. Be careful - if you distort the seal, you'll have to disassemble the fork again and end up taking it to a dealer anyway!

23 Install the washer and the retaining ring, making sure the ring is completely seated in its groove.

24 Install the dust seal, making sure it seats completely.

25 Install the drain screw and a new gasket, if it was removed.

26 Install the TCV, with the nuts at the top (A and B models only).

27 On 600 A and B models, extend the fork fully and add the recommended type and amount of fork oil (see Chapter 1). On 600 C and 750 models, this should be done with the forks fully compressed.

28 Install the fork spring, with the closer-wound springs at the top. On 600 C and 750 models, install the spring guide and spacer.

29 Check the O-ring on the top plug - if it's deteriorated, replace it. Lubricate the O-ring

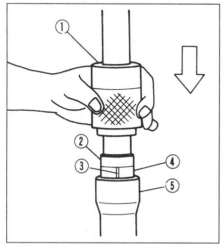

8.20a Here's the proper way to drive the outer guide bushing into place

1 Tool no. 57001-1104 (this is the tool for A and B models; tool for C models and 750 is similar)
2 Used guide bushing
3 Slit - this must face to the left or right
4 New guide bushing
5 Outer tube

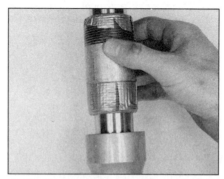

8.20b If you don't have the proper tool, a section of pipe can be used the same way the special tool would be used - as a slide hammer (be sure to tape the ends of the pipe so it doesn't scratch the fork tube)

8.22 Install the oil seal

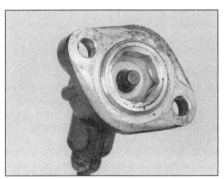

9.3 When the brake lever is squeezed, the plunger should extend approximately 2 mm

9.7 Insert a 14 mm nut into the recessed hex in the seal case, place a wrench on the nut and unscrew the case from the housing

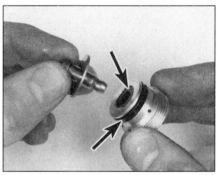

9.8 Pull the plunger and washer out of the seal case, then remove the O-rings (arrows)

with fork oil and install the top plug. Tighten it to the torque listed in this Chapter's Specifications after the fork has been installed.

30 Install the fork by following the procedure outlined in Section 7. If you won't be installing the fork right away, store it in an upright position and stick a piece of tape over the air hole at the top, to prevent leakage.

9 Brake plunger unit (600 A and B models only) - check and overhaul

Check

1 Unbolt the brake plunger from the anti-dive unit (see illustration 7.9). Don't disconnect the brake line.
2 Unbolt the junction block from the fork leg (so it doesn't get bent).
3 Gently squeeze the front brake lever and watch the plunger - it should come out about 2 mm (see illustration).
4 Let go of the brake lever and push on the plunger with your finger - it should retract without having to apply much pressure.
5 If the plunger doesn't perform as described, proceed to the next step.

Overhaul

6 Disconnect the brake line from the plunger unit. You may have to bolt it to the anti-dive unit if the fitting is very tight. Also, use a flare nut wrench, if available, to prevent rounding off the fitting.
7 Remove the rubber cap and unscrew the seal case from the plunger housing (see illustration).
8 Remove the O-rings from the seal case (see illustration).
9 Check the bore of the plunger housing for corrosion and scoring. If either of these conditions are found, replace the brake plunger assembly.
10 If the bore is okay, reassemble the valve

using a new seal ring and O-ring. Lubricate the components with clean brake fluid and tighten the seal case securely.
11 Installation is the reverse of removal, but be sure to tighten the mounting bolts to the torque listed in this Chapter's Specifications and bleed the front brakes following the procedure in Chapter 7.

10 Anti-dive valve assembly (600 A and B models only) - check, removal and installation

Check

1 The forks must be removed from the machine to check the operation of the anti-dive units. Remove the handlebar (see Section 6), loosen the top plug, then refer to Section 7 and remove one fork leg at a time.
2 Remove the top plug and pull out the fork spring.
3 Hold the fork leg upright and compress it (be careful not to let any oil spill out). The stroke should be smooth and shouldn't offer too much resistance.
4 Extend the fork. Depress the rod on the anti-dive valve with your thumb and compress the fork again (see illustration). The fork

should offer more resistance than the first time.
5 If the valve doesn't change the damping characteristics of the fork when depressed, replace it. Don't attempt to disassemble the valve - it isn't serviceable.

Removal and installation

6 Relieve the air pressure in the forks and drain the fork oil (see Chapter 1).
7 Unbolt the brake plunger from the anti-dive unit (see illustration 7.9).
8 Remove the two Allen-head bolts and detach the anti-dive unit from the fork leg.
9 If you're installing the original unit, check the condition of the O-rings. If they are cracked, hardened or show any signs of general deterioration, replace them (see illustration).
10 Lubricate the O-rings with clean fork oil. Position the unit on the fork leg, apply a non-hardening thread locking compound to the threads of the bolts and tighten the bolts to the torque listed in this Chapter's Specifications.
11 Install the fork (if removed) following the procedure described in Section 7. Fill the fork with the recommended type and amount of fork oil, then charge the fork with the recommended amount of air pressure (see Chapter 1).

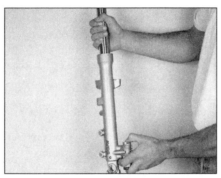

10.4 The fork should be harder to compress when the rod on the anti-dive unit is depressed

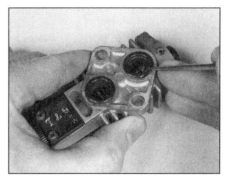

10.9 Be sure to check the O-rings on the anti-dive unit and replace them if necessary

11.6 The ESCS relay (A) is mounted on a bracket directly behind the steering head

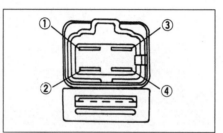

11.7 ESCS relay terminal identification

1   Brown wire terminal
2   Brown/red wire terminal
3   Black/yellow wire terminal
4   Blue/red wire terminal

11 Electric Suspension Control System (ESCS) - check and replacement

Note: *ESCS is not fitted to the 600 C6, C7 (UK) and 600 C7, C8, C9, C10 (US) models.*

Circuit check

1 Turn the ignition switch to the On position.
2 Squeeze the front brake lever and listen for a click coming from the ESCS unit, which is mounted under the fuel tank.
3 If no click is heard, check the front brake light switch (see Chapter 9), the ESCS relay and the ESCS unit.
4 If a click is heard, depress the rear brake pedal - you shouldn't hear a click. If you do, check the ESCS diode. If no click is heard, operation is normal.

ESCS relay

5 Remove the fuel tank (see Chapter 4).
6 Remove the ESCS relay from its bracket **(see illustration)** and unplug the electrical connector.
7 Using an ohmmeter, check the resistance between the terminal for the blue/red wire and the terminal for the black/yellow wire **(see illustration)**. The reading should be 90 to 100 ohms.

8 Using two jumper wires, connect battery voltage to the two terminals that were checked in Step 7. Set the ohmmeter on the R x 1 range and connect the leads across the remaining terminals (the ones for the brown/red wire and the solid brown wire). This should cause the relay to click.
9 If the relay fails either test, replace it.

ESCS unit

10 Remove the fuel tank, if you haven't already done so (see Chapter 4).
11 Unplug the electrical connector from the ESCS unit.
12 Using jumper wires, apply battery voltage to the terminals of the ESCS unit. The wire from the positive terminal of the battery must be connected to the white wire terminal, and the wire from the negative terminal must be connected to the black wire terminal. When the connection is made, the ESCS unit should click. If it doesn't, replace it.

ESCS diode

13 Remove the fuel tank, if you haven't already done so (see Chapter 4).
14 Remove the ESCS diode **(see illustration)** and unplug the electrical connector.
15 Using an ohmmeter set on the R x 10 or R x 100 scale, check the resistance across the terminals of the diode, reverse the ohmmeter

leads and check the resistance again. The resistance of the diode should be low in one direction and more than ten times as much in the other direction. If not, replace the diode.

ESCS adjuster

16 Remove the front wheel (see Chapter 7).
17 Remove the right handlebar (see Section 6) and unscrew the top plug. Pull out the spacer, spring guide and the fork spring.
18 Compress the fork leg slowly (be careful not to spill any fork oil - it would be a good idea to cover the fuel tank and bodywork before doing this). The fork shouldn't offer much resistance. Allow the fork to extend.
19 Turn the ignition to the On position. Squeeze the brake lever and compress the fork again - it should be much harder to compress than before.
20 If it isn't harder to compress, and all of the ESCS components check out okay, replace the ESCS adjuster. Follow Steps 6, 8, 9, 10 and 11 of Section 10 for the replacement procedure.

12 Steering head bearings - replacement

1 If the steering head bearing check/adjustment (see Chapter 1) does not remedy excessive play or roughness in the steering head bearings, the entire front end must be disassembled and the bearings and races replaced with new ones.
2 Refer to Chapter 4 and remove the fuel tank. Refer to Section 7 and remove the front forks. On 750 models, unbolt the brake hose joint from the steering head; there is no need to disconnect any of the hydraulic lines, but make sure they are supported so that no strain is placed on the single hose from the master cylinder.
3 Remove the steering stem nut **(see illustration)**, then lift off the upper triple clamp (sometimes called the fork bridge or crown).

11.14 Location of the ESCS diode (A)

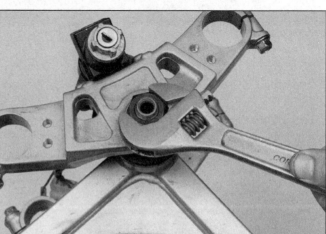

12.3 Unscrew the steering stem nut and lift off the upper triple clamp

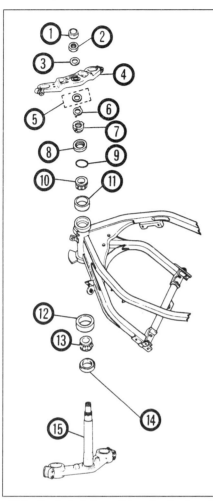

12.4b Unscrew the locknut with an adjustable spanner wrench

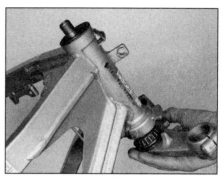

12.5 Pull the steering stem/lower triple clamp out - if it's stuck, tap gently on the steering stem with a soft-face hammer

12.4a Exploded view of the steering head

1 Cap	8 Stem cap
2 Steering stem nut	9 O-ring
3 Washer	10 Upper bearing
4 Upper triple clamp	11 Upper race
5 Washer (West Germany, Norway and Sweden models only)	12 Lower race
	13 Lower bearing
	14 Grease seal
	15 Steering stem/lower triple clamp
6 Lockwasher	
7 Locknut	

4 Remove the lockwasher from the stem locknut **(see illustration)**. Using an adjustable spanner wrench, remove the stem locknut/stem cap assembly **(see illustration)** while supporting the steering head from the bottom. Lift off the nut and race cover.
5 Remove the steering stem and lower triple clamp assembly **(see illustration)**. If it's stuck, gently tap on the top of the steering stem with a plastic mallet or a hammer and a wood block.
6 Remove the upper bearing and O-ring **(see illustration)**.
7 Clean all the parts with solvent and dry them thoroughly, using compressed air, if available. If you do use compressed air, don't let the bearings spin as they're dried - it could ruin them. Wipe the old grease out of the frame steering head and bearing races.
8 Examine the races in the steering head for cracks, dents, and pits. If even the slightest amount of wear or damage is evident, the races should be replaced with new ones.
9 To remove the races, drive them out of the steering head with Kawasaki tool no. 57001-1107 or equivalent **(see illustration)**. A slide hammer with the proper internal-jaw puller will

also work. When installing the races, use Kawasaki press shaft no. 57001-1075 and drivers no. 57001-1106 and 57001-1076 **(see illustration)**, or tap them gently into place with a hammer and punch or a large socket. Do not strike the bearing surface or the race will be damaged.

> **HAYNES HINT** *New races will be easier to install if left overnight in a refrigerator. This will cause them to contract and slip into place in the frame with very little effort.*

10 Check the bearings for wear. Look for cracks, dents, and pits in the races and flat spots on the bearings. Replace any defective parts with new ones. If a new bearing is required, replace both of them as a set.
11 To remove the lower bearing from the steering stem, use a bearing puller (Kawasaki tool no. 57001-158 or equivalent, combined with adapter no. 57001-318 and

12.6 Lift out the O-ring and the upper bearing

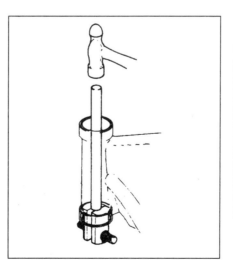

12.9a Driving out the lower bearing race with the special Kawasaki tool

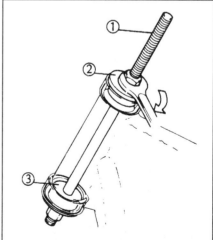

12.9b Using the special Kawasaki tool to press the outer races into the frame

1 Driver press shaft (tool no. 57001-1075)
2 Driver (tool no. 57001-1106)
3 Driver (tool no. 57001-1076)

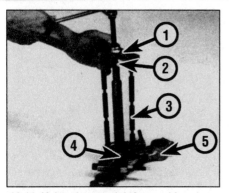

12.11 Using the special Kawasaki tools to pull the lower bearing off the steering stem

1   Bearing puller (tool no. 57001-158)
2   Adapter (tool no. 57001-317)
3   Pole (tool no. 57001-1190)
4   Bearing inner race
5   Lower triple clamp

pole no. 57001-1190) **(see illustration)**. Don't remove this bearing unless it, or the grease seal underneath, must be replaced.
12 Check the grease seal under the lower bearing and replace it with a new one if necessary.
13 Inspect the steering stem/lower triple clamp for cracks and other damage. Do not attempt to repair any steering components. Replace them with new parts if defects are found.
14 Check the rubber portion of the stem locknut/stem cap assembly - if it's worn or deteriorated, replace it.
15 Pack the bearings with high-quality grease (preferably a moly-based grease) **(see illustration)**. Coat the outer races with grease also.
16 Install the grease seal and lower bearing onto the steering stem. Drive the lower bearing onto the steering stem using Kawasaki stem bearing driver no. 57001-137 and adapter no. 57001-1074 **(see illustration)**. If you don't have access to these tools, a section of pipe with a diameter the same as the inner race of the bearing can be used. Drive the bearing on until it is fully seated.
17 Insert the steering stem/lower triple clamp

13.8 Loosen the shock absorber upper nut (arrow) - 600 shown

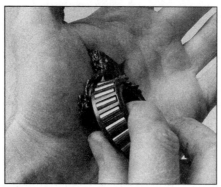

12.15 Work the grease completely into the rollers

into the frame head. Install the upper bearing, O-ring and the stem locknut/stem cap assembly. Using the adjustable spanner, tighten the locknut while moving the lower triple clamp back and forth. Continue to tighten the nut (to approximately 40 Nm / 30 ft-lbs) until the steering head becomes tight, then back off until there is some play in the bearings. Now, turn the locknut until there is no more play, and tighten the nut just a fraction of a turn from that point (not too much, though, or the steering will be too tight). Make sure the steering head turns smoothly.
18 Install the lockwasher on the locknut, then install the upper triple clamp on the steering stem. Install the washer and nut, tightening the nut to the torque listed in this Chapter's Specifications.
19 The remainder of installation is the reverse of removal.

13 Rear shock absorber - removal and installation

1 Set the bike on its centrestand.
2 Remove the side covers.
3 Unscrew the nut and detach the air valve from its bracket.
4 Unscrew the damper adjusting knob.
5 On 600 models, remove the IC igniter from its bracket and position it out of the way.

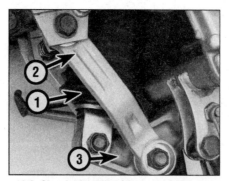

13.9 Shock absorber (1), tie-rods (2) and rocker arm (3)

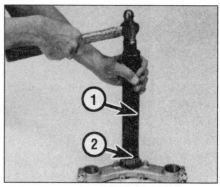

12.16 Driving the lower bearing onto the steering stem with the special Kawasaki tools (a piece of tubing, with a diameter the same as the inner race of the bearing can be used instead)

1   Stem bearing driver (tool no. 57001-137)
2   Adapter (tool no. 57001-1074)

Remove the screw and detach the bracket from the frame.
6 Loosen the locknut and unscrew the damper adjusting rod from the shock absorber.
7 On 750 models, remove the rear frame section (see Section 3).
8 Loosen the shock absorber upper nut **(see illustration)**. Don't remove it yet.
9 Remove the shock absorber lower nut and bolt and the tie-rod lower nut and bolt **(see illustration)**.
10 Remove the upper nut and bolt. Pull the tie-rods back and lower the shock absorber to the ground.
11 Installation is the reverse of the removal procedure. Tighten the shock absorber and tie-rod nuts to the torque values listed in this Chapter's Specifications.

14 Rear suspension linkage (Uni-Trak) - removal, check and installation

Tie-rod(s)

1 Position the bike on its centrestand.
2 Remove the tie-rod lower nut and bolt **(see illustrations)**.
3 Remove the nut(s) and bolt(s) that secure the tie-rods to the swingarm and remove the tie-rod(s).
4 Push the sleeve out of the tie-rod **(see illustration)**. Check the O-rings for cracking and general deterioration and replace them if necessary. Check the bushing in the tie-rod and the outer surface of the sleeve for wear, replacing them if necessary. **Note:** *A bushing driver will be required if the bushing is to be replaced.*
5 Apply a thin coat of moly-based grease to the bushing and sleeve and install the sleeve in the bushing. Press the O-rings into place around the sleeve.

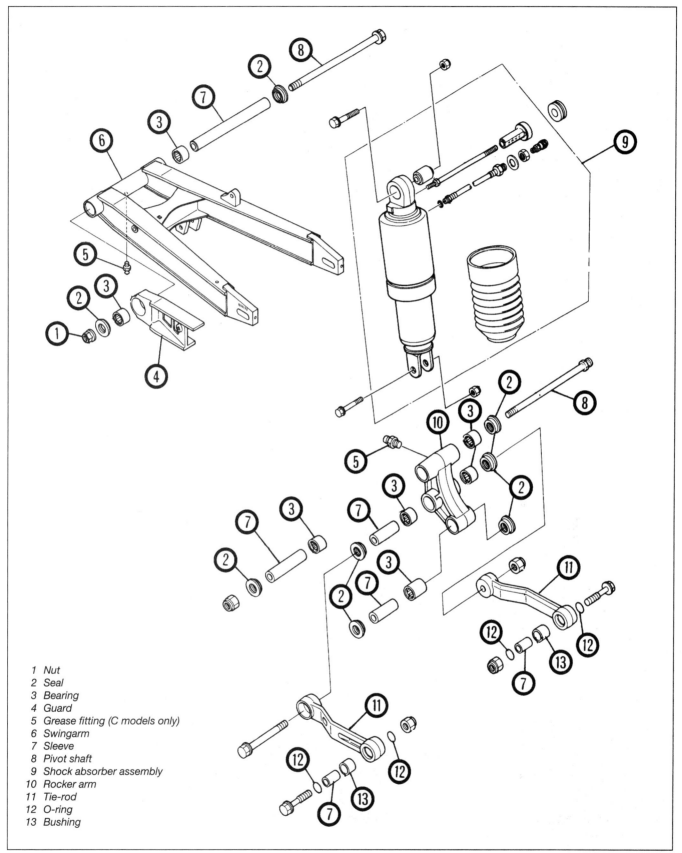

1 Nut
2 Seal
3 Bearing
4 Guard
5 Grease fitting (C models only)
6 Swingarm
7 Sleeve
8 Pivot shaft
9 Shock absorber assembly
10 Rocker arm
11 Tie-rod
12 O-ring
13 Bushing

14.2a Rear suspension - 600 models

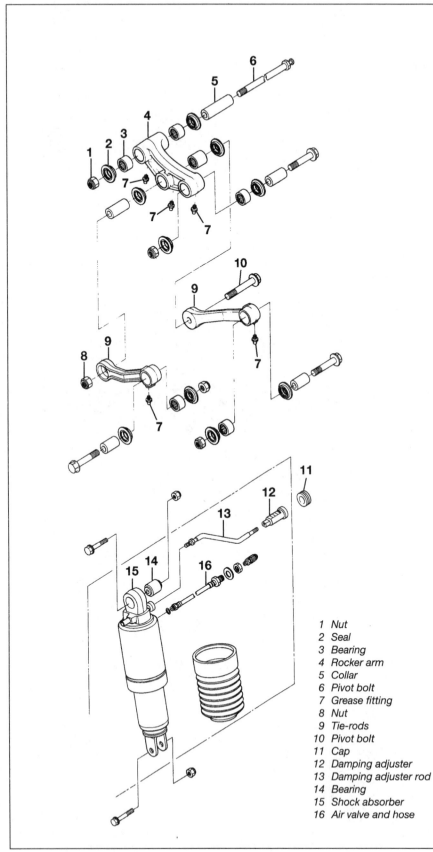

1 Nut
2 Seal
3 Bearing
4 Rocker arm
5 Collar
6 Pivot bolt
7 Grease fitting
8 Nut
9 Tie-rods
10 Pivot bolt
11 Cap
12 Damping adjuster
13 Damping adjuster rod
14 Bearing
15 Shock absorber
16 Air valve and hose

14.2b Rear shock absorber and suspension linkage - 750 models

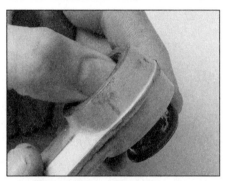

14.4 Push the sleeve out of the end of the tie-rod and remove the O-rings

6 Installation is the reverse of removal. Tighten the bolts to the torque listed in this Chapter's Specifications.

Rocker arm

7 Remove the lower fairing (see Chapter 8).
8 Support the bike with a floor jack, with a piece of wood on the jack head, positioned underneath the engine. It would be a good idea to have an assistant steady the bike during this procedure.
9 Retract the centrestand and disconnect the spring.
10 Disconnect the shock absorber and the tie-rods from the rocker arm **(see illustration 13.9)**. Support the swingarm with a rope or piece of wire so it doesn't fall.
11 Remove the nut and pull out the rocker arm pivot bolt **(see illustration)**. Detach the rocker arm from the frame.

14.11 Remove the nut (arrow) then push the rocker arm pivot shaft out

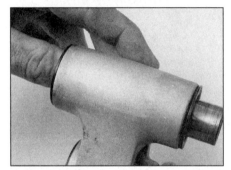

14.12a To check the condition of the needle bearings, push the sleeves out of the rocker arm

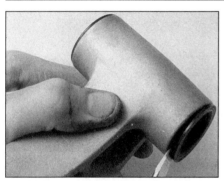

14.12b Pry out the grease seals for access to the bearings

14.13a To replace the bearings, knock them out with a hammer and punch . . .

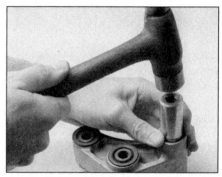

14.13b . . . then drive them in with a socket that just fits into the bore of the rocker arm

12 Push the sleeves out of the rocker arm **(see illustration)**. Check the needle bearings for dryness and discoloration. If necessary, pry out the grease seal **(see illustration)**, clean the bearings with solvent, dry and repack them with moly-based grease.

13 If the bearings are deteriorated, drive them out of the rocker arm with a hammer and punch **(see illustration)**. Install new bearing sets by driving them in with a hammer and a socket of the appropriate size **(see illustration)**.

14 Coat the bearings with grease, install the grease seals and slide the bushings into place.

15 The remainder of installation is the reverse of the removal procedure. Tighten the nut of the rocker arm pivot bolt, the shock absorber-to-rocker arm bolt/nut and the tie-rod-to-rocker arm bolt/nut to the torque values listed in this Chapter's Specifications.

15 Swingarm bearings - check

1 Refer to Chapter 7 and remove the rear wheel, then refer to Section 13 and remove the rear shock absorber.

2 Grasp the rear of the swingarm with one hand and place your other hand at the junction of the swingarm and the frame. Try to move the rear of the swingarm from side-to-side. Any wear (play) in the bearings should be felt as movement between the swingarm and the frame at the front. The swingarm will actually be felt to move forward and backward at the front (not from side-to-side). If any play is noted, the bearings should be replaced with new ones (see Section 17).

3 Next, move the swingarm up and down through its full travel. It should move freely, without any binding or rough spots. If it does not move freely, refer to Section 17 for servicing procedures.

16 Swingarm - removal and installation

1 Raise the bike and set it on its centrestand.

2 Remove the rear wheel (see Chapter 7).

3 Detach the torque arm from the swingarm **(see illustration)**. Support the rear brake caliper and torque arm with a piece of rope or wire don't let them hang by the brake hose.

4 Remove the swingarm pivot nut **(see illustration)**. Don't remove the pivot shaft yet.

5 Detach the tie-rods and the shock absorber from the rocker arm **(see illustration 13.8)**. Support the swingarm while doing this.

6 Support the swingarm and pull the pivot shaft out. Remove the swingarm and support the drive chain so that it does not pick up dirt from the ground. If necessary, remove the bolts and detach the tie-rods from the swingarm.

7 Check the pivot bearings in the swingarm for dryness or deterioration. If they're in need of lubrication or replacement, refer to Section 17.

8 Installation is the reverse of the removal procedure, noting the following points:

 a) Remember to loop the drive chain over the swingarm before installation.

 b) Be sure the bearing seals are in position before installing the pivot shaft.

 c) Tighten the pivot shaft nut and the shock absorber and tie-rod lower mounting bolts/nuts to the torque values listed in this Chapter's Specifications.

 d) Adjust the drive chain as described in Chapter 1.

17 Swingarm bearings - replacement

1 Remove the swingarm (see Section 16).

2 Slide the sleeve out **(see illustration)**.

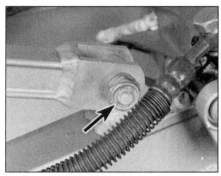

16.3 Remove the torque arm-to-swingarm nut and bolt, then support the brake caliper and torque arm with wire or rope

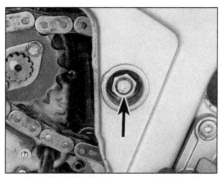

16.4 The swingarm pivot shaft nut is located on the left side of the frame - pry out the plug for access

17.2 Slide the sleeve out of the front of the swingarm

17.3 Pry the seals out with a screwdriver

18.1 Mark the relationship of the shift lever to the shift shaft so it can be reinstalled in the same position

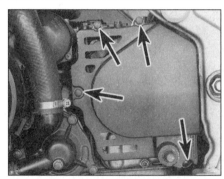

18.2 Remove the engine sprocket cover bolts (arrows) and slide the cover off

3 Pry out the seals **(see illustration)**.
4 Refer to Section 14, Steps 12, 13 and 14 for the bearing service procedures.

18 Drive chain - removal, cleaning and installation

Removal

Note: *The original equipment drive chain fitted to all models is an endless chain. Removal requires the removal of the swingarm as detailed below, or if the necessary chain breaking and joining tool is available, the chain can be separated and rejoined (see Tools and Workshop Tips in the Reference section).*

 Warning: NEVER install a drive chain which uses a clip-type master (split) link.

1 Mark the relationship of the shift lever to the shift shaft **(see illustration)**. Remove the shift lever pinch bolt and slide the lever off the shaft.
2 On 600 models, remove the bolts securing the engine sprocket cover to the engine case **(see illustration)**, then slide the sprocket cover off.
3 On 750 models, remove the clutch slave cylinder (see Chapter 2B). Remove the screws

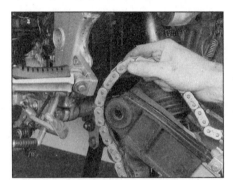

18.5 Pull the swingarm back and pass the chain between the frame and the swingarm

securing the engine sprocket cover to the crankcase and the alternator lower mounting bracket. Remove the engine sprocket cover, taking note of the dowel positions. Remove the engine sprocket (see Section 19).
4 Remove the rear wheel (see Chapter 7), and on 600 models, lift the chain off the engine sprocket.
5 Detach the swingarm from the frame by following the first few Steps of Section 16. Pull the swingarm back far enough to allow the chain to slip between the frame and the front of the swingarm **(see illustration)**.

Cleaning

6 Soak the chain in kerosene (paraffin) or diesel fuel for approximately five or six minutes. Remove the chain, wipe it off then blow dry it with compressed air immediately.

Caution: Don't use gasoline (petrol), solvent or other cleaning fluids. Don't use high-pressure water. Remove the chain, wipe it off, then blow dry it with compressed air immediately. The entire process shouldn't take longer than ten minutes - if it does, the O-rings in the chain rollers could be damaged.

Installation

7 Installation is the reverse of the removal procedure. Tighten the suspension fasteners and the engine sprocket cover bolts to the torque values listed in this Chapter's

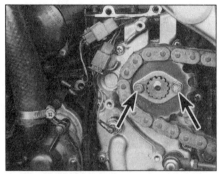

19.5 Remove the holding plate bolts and slide the sprocket and chain off the shaft

Specifications. Tighten the rear axle nut to the torque listed in the Chapter 7 Specifications.
8 Connect the shift lever to the shift shaft, lining up the marks. If it's installed correctly, the link rod should be parallel to the shift pedal.
9 Lubricate the chain following the procedure described in Chapter 1.

19 Sprockets - check and replacement

1 Set the bike on its centrestand.
2 Whenever the drive chain is inspected, the sprockets should be inspected also. If you are replacing the chain, replace the sprockets as well. Likewise, if the sprockets are in need of replacement, install a new chain also.

Engine sprocket

3 Remove the engine sprocket cover following the procedure outlined in the previous Section.
4 Check the wear pattern on the sprocket (see Chapter 1). If the sprocket teeth are worn excessively, replace the chain and sprockets as a set.
5 To remove the engine sprocket on 600 models, place the transmission in first gear, have an assistant apply the rear brake and loosen the holding plate bolts **(see illustration)**. Remove the holding plate and pull the engine sprocket and chain off the shaft, then separate the sprocket from the chain; if necessary slacken the rear axle nut and fully back off the drive chain adjusters to allow the chain to be slipped off of the sprocket.
6 To install the engine sprocket, engage it with the chain, making sure the IN mark, or recess, faces the engine case. Install the holding plate, apply a non-hardening thread locking compound to the bolts, then tighten the bolts to the torque listed in this Chapter's Specifications.
7 To remove the engine sprocket on 750 models, first bend back the lockwasher tab

19.7a Bend back the lockwasher (arrow) . . .

19.7b . . . and insert a rod through the sprocket teeth into the cover slot to prevent the sprocket from turning

19.11 Check the runout of the rear sprocket with a dial indicator

(see illustration). Insert a rod between two sprocket teeth into the slot in the external shift mechanism cover to hold the sprocket from turning and loosen the nut (see illustration). Remove the nut and lockwasher and pull the engine sprocket and chain off the shaft, then separate the sprocket from the chain; if necessary slacken the rear axle nut and fully back off the drive chain adjusters to allow the chain to be slipped off of the sprocket.

8 To install the engine sprocket, engage it with the chain and slip the sprocket on the shaft. If the lockwasher is damaged or weakened, replace it with a new one. Install the lockwasher and nut, using the method employed on removal to prevent the sprocket rotating as the nut is tightened to the torque listed in this Chapter's Specifications. Bend a portion of the lockwasher up against on of the nut flats to lock it in position.

9 On all models, install the engine sprocket

cover and shift lever (see Section 18). Adjust the drive chain freeplay (see Chapter 1).

Rear sprocket

10 Check the wear pattern on the sprocket (see Chapter 1). If the sprocket teeth are worn excessively, replace the chain and sprockets as a set.

11 Attach a dial indicator to the swingarm, with the plunger of the indicator touching the sprocket near its outer diameter (see illustration). Turn the wheel and measure the runout. If the runout exceeds the maximum runout listed in this Chapter's Specifications, replace the rear sprocket. As stated before, it's a good idea to replace the chain and the sprockets as a set. However, if the components are relatively new or in good condition, but the sprocket is warped, you may be able to get away with just replacing the rear sprocket.

12 Remove the rear wheel (see Chapter 7).

13 On 600 models, use vernier calipers to measure the diameter of the rear sprocket (see illustration). If the diameter of the sprocket is less than the minimum diameter listed in this Chapter's Specifications, replace the chain and sprockets.

14 To remove the rear sprocket, unscrew the nuts holding it to the wheel coupling and lift the sprocket off. When installing the sprocket, apply a non-hardening thread locking compound to the threads of the studs. Tighten the nuts to the torque listed in this Chapter's Specifications. Also, check the condition of the rubber damper under the rear wheel coupling (see Section 20).

15 Install the rear wheel (see Chapter 7) and adjust the drive chain freeplay (see Chapter 1).

20 Rear wheel coupling/rubber damper - check and replacement

1 Remove the rear wheel (see Chapter 7).

2 Lift the collar and rear sprocket/rear wheel coupling from the wheel (see Chapter 7, illustrations 13.4a and 13.4b).

3 Lift the rubber damper from the wheel (see illustration) and check it for cracks, hardening and general deterioration. Replace it with a new one if necessary.

4 Checking and replacement procedures for the coupling bearing are similar to those described for the wheel bearings. Refer to Chapter 7, Section 13.

5 Installation is the reverse of the removal procedure.

19.13 Measure the diameter of the rear sprocket (A) and replace it, the engine sprocket and the chain if it has worn past the minimum diameter

20.3 Replace the rubber damper if it shows signs of deterioration

Notes

Chapter 7
Brakes, wheels and tyres

Contents

Degrees of difficulty

Easy, suitable for novice with little experience 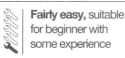	Fairly easy, suitable for beginner with some experience	Fairly difficult, suitable for competent DIY mechanic	Difficult, suitable for experienced DIY mechanic	Very difficult, suitable for expert DIY or professional 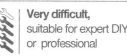

Specifications

Brakes

Brake fluid type . . . . DOT 4
Brake disc minimum thickness . . . . See Chapter 1
Disc thickness
  600 A and B models and 750 models
    Front
      Standard . . . . 4.8 to 5.1 mm (0.189 to 0.2 in)
      Minimum* . . . . 4.5 mm (0.177 in)
    Rear
      Standard . . . . 5.8 to 6.1 mm (0.228 to 0.240 in)
      Minimum* . . . . 5.5 mm (0.217 in)
  600 C models
    Front
      Standard . . . . 4.3 to 4.6 mm (0.170 to 0.181 in)
      Minimum* . . . . 4.0 mm (0.157 in)
    Rear
      Standard . . . . 5.8 to 6.1 mm (0.228 to 0.240 in)
      Minimum* . . . . 5.5 mm (0.217 in)
*Refer to marks stamped into the disc (they supersede information printed here)
Disc runout (maximum, front and rear, all models) . . . . 0.3 mm (0.012 in)

Wheels and tyres

Wheel runout
  Axial (side-to-side) . . . . 0.5 mm (0.020 in)
  Radial (out-of-round) . . . . 0.8 mm (0.031 in)
Rear axle runout . . . . 0.2 mm (0.007 in)
Tyre pressures . . . . See Chapter 1

Wheels and tyres (continued)

Tyre sizes

600 A and B models

Front 110/90V16

Rear 130/90V16

600 C models

Front 110/80V16

Rear 130/90V16

750 models - US/Canada F1 and F2 models

Front 110/90 V16

Rear .... 140/70 V18 or 140/70 VB18

750 models - All UK models, US/Canada F3 and F4 models

Front 110/90 V16/V250

Rear .... 140/70 V18/V250, 140/70 VR18, 140/70 VB18 or 140/70 V18

Torque specifications

Front caliper mounting bolts .... 32 Nm (24 ft-lbs)

Rear caliper mounting bolts

600 models .... 32 Nm (24 ft-lbs)

750 models .... 34 Nm (25 ft-lbs)

Banjo fitting bolts

600 models .... 25 Nm (18 ft-lbs)

750 models .... 29 Nm (22 ft-lbs)

Brake disc-to-wheel bolts .... 23 Nm (16.5 ft-lbs)

Master cylinder mounting bolts

Front .... 8.8 Nm (78 in-lbs)

Rear .... 23 Nm (16.5 ft-lbs)

Front axle .... 88 Nm (65 ft-lbs)

Front axle clamp bolt .... 21 Nm (15 ft-lbs)

Rear axle nut .... 110 Nm (80 ft-lbs)

---

1 General information

The models covered by this manual are equipped with hydraulic disc brakes on the front and rear. All 600 A and B models employ single piston calipers, while 600 C and 750 models use dual piston calipers.

All models are equipped with cast aluminum wheels, which require very little maintenance and allow tubeless tyres to be used.

Caution: Disc brake components rarely require disassembly. Do not disassemble components unless absolutely necessary. If any hydraulic brake line connection in the system is loosened, the entire system should be disassembled, drained, cleaned and then properly filled and bled upon reassembly. Do not use solvents on internal brake components. Solvents will cause seals to swell and distort. Use only clean brake fluid or alcohol for cleaning. Use care when working with brake fluid as it can injure your eyes and it will damage painted surfaces and plastic parts.

2 Brake caliper - removal, overhaul and installation

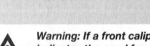

Warning: If a front caliper indicates the need for an overhaul (usually due to leaking

fluid or sticky operation), BOTH front calipers should be overhauled and all old brake fluid flushed from the system. Also, the dust created by the brake system may contain asbestos, which is harmful to your health. Never blow it out with compressed air and don't inhale any of it. An approved filtering mask should be worn when working on the brakes. Do not, under any circumstances, use petroleum-based solvents to clean brake parts. Use brake cleaner or denatured alcohol only!

Note: *If you are removing the caliper only to replace or inspect the brake pads, don't disconnect the hose from the caliper.*

Removal

1 Place the bike on its centrestand.

2.3a Unscrew the banjo fitting bolt, . . .

2 If you're removing the left front caliper, disconnect the lower end of the speedometer cable from the hub.

3 Note: *Remember, if you're just removing the caliper to inspect or replace the pads, ignore this step.* Disconnect the brake hose from the caliper. Remove the brake hose banjo fitting bolt (except on models with a threaded fitting) and separate the hose from the caliper **(see illustration)**. Discard the sealing washers. If a threaded fitting (tube nut) is used instead of a banjo fitting, use a flare-nut wrench to unscrew it. If equipped, remove the hose support bolt from the caliper bracket. Plug the end of the hose or wrap a plastic bag tightly around it to prevent excessive fluid loss and contamination. Unscrew the caliper mounting bolts **(see illustration)**.

2.3b . . . then remove the caliper mounting bolts

4 Lift off the caliper, being careful not to strain or twist the brake hose. If you're removing the rear caliper, detach the brake hose from the clip on the swingarm.

Overhaul

5 Remove the brake pads and anti-rattle spring from the caliper (see Section 3, if necessary). Clean the exterior of the caliper with denatured alcohol or brake system cleaner.

6 Remove the caliper bracket and the slider pin boots from the caliper **(see illustrations)**.

7 Place a few rags between the piston(s) and the caliper frame to act as a cushion, then use

2.6a Pull the bracket out of the caliper . . .

2.6b . . . and remove the slider pin boots - grab the end with a pair of needle-nose pliers, twist, then push the boot through the hole

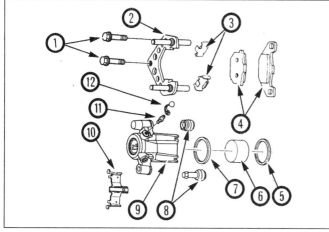

2.6c Single piston front caliper - exploded view

1 *Caliper mounting bolts*	5 *Dust seal*	9 *Caliper body*
2 *Caliper bracket*	6 *Piston*	10 *Anti-rattle spring*
3 *Pad support clips*	7 *Piston seal*	11 *Bleeder valve*
4 *Brake pads*	8 *Slider pin boots*	12 *Bleeder valve cap*

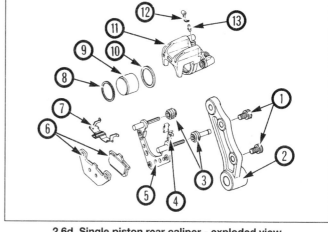

2.6d Single piston rear caliper - exploded view

1 *Caliper mounting bolts*	5 *Caliper bracket*	10 *Piston seal*
2 *Caliper holder*	6 *Brake pads*	11 *Caliper body*
3 *Slider pin boots*	7 *Anti-rattle spring*	12 *Bleeder valve cap*
4 *Pad support clips*	8 *Dust seal*	13 *Bleeder valve*
	9 *Piston*	

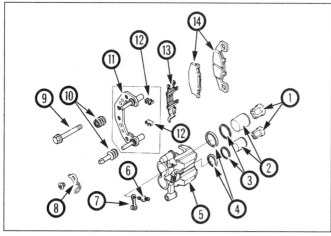

2.6e Dual piston front caliper - exploded view

1 *Inserts*	7 *Bleeder valve cap*	10 *Slider pin boots*
2 *Pistons*	8 *Speedometer cable guide (left side only)*	11 *Caliper bracket*
3 *Dust seals*		12 *Pad support clips*
4 *Piston seals*	9 *Caliper mounting bolt*	13 *Anti-rattle spring*
5 *Caliper body*		14 *Brake pads*
6 *Bleeder valve*		

2.6f Dual piston rear caliper - exploded view

1 *Pad support clips*	6 *Inserts*	10 *Caliper body*
2 *Caliper bracket*	7 *Pistons*	11 *Bleeder valve caps*
3 *Brake pads*	8 *Piston seals*	
4 *Anti-rattle spring*	9 *Slider pin boots*	12 *Bleeder valves*
5 *Dust seals*		

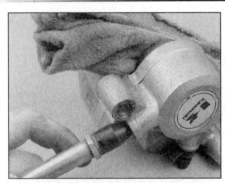

2.7 With a rag placed between the piston and caliper frame, use compressed air to ease the piston out of the bore

2.9 The dust seal should be removed with a plastic or wooden tool to avoid damage to the bore and seal groove (a pencil works well) - remove the piston seal the same way

2.15 Bottom the piston in the caliper bore - make sure it goes in straight

compressed air, directed into the fluid inlet, to remove the piston(s) **(see illustration)**. Use only enough air pressure to ease the piston(s) out of the bore. If a piston is blown out, even with the cushion in place, it may be damaged.

 Warning: Never place your fingers in front of the piston in an attempt to catch or protect it when applying compressed air, as serious injury could occur.

8 If compressed air isn't available, reconnect the caliper to the brake hose and pump the brake lever or pedal until the piston(s) is/are free.
9 Using a wood or plastic tool, remove the dust seal(s) **(see illustration)**. Metal tools may cause bore damage.
10 Using a wood or plastic tool, remove the piston seal(s) from the groove in the caliper bore.
11 Clean the piston(s) and the bore(s) with denatured alcohol, clean brake fluid or brake system cleaner and blow dry them with filtered, unlubricated compressed air. Inspect the surfaces of the piston(s) for nicks and burrs and loss of plating. Check the caliper bore(s), too. If surface defects are present, the caliper must be replaced. If the caliper is in bad shape, the master cylinder should also be checked.
12 Temporarily reinstall the caliper bracket. Make sure it slides smoothly in-and-out of the caliper. If it doesn't, check the slider pins for burrs or excessive wear. Also check the slider pin bores in the caliper for wear and scoring.

Replace the caliper bracket, the caliper, or both if necessary.
13 Lubricate the piston seal(s) with clean brake fluid and install it in its groove in the caliper bore. Make sure it isn't twisted and seats completely.
14 Lubricate the dust seal(s) with clean brake fluid and install it in its groove, making sure it seats correctly.
15 Lubricate the piston(s) with clean brake fluid and install it into the caliper bore. Using your thumbs, push the piston all the way in **(see illustration)**, making sure it doesn't get cocked in the bore.
16 Install the slider pin boots **(see illustration)**.
17 Apply a thin coat of PBC (poly butyl cuprysil) grease, or silicone grease designed for high-temperature brake applications, to the slider pins on the caliper bracket **(see illustration)**. Install the caliper bracket to the caliper and seat the boots over the lips on the bracket.

Installation

18 Install the anti-rattle spring and the brake pads (see Section 3).
19 Install the caliper, tightening the mounting bolts to the torque listed in this Chapter's Specifications.
20 Connect the brake hose to the caliper, using new sealing washers on each side of

the fitting. Tighten the banjo fitting bolt to the torque listed in this Chapter's Specifications. If a threaded fitting (tube nut) is used instead of a banjo bolt, tighten it securely with a flare-nut wrench.
21 Reconnect the speedometer cable.
22 Fill the master cylinder with the recommended brake fluid (see *Daily (pre-ride) checks*) and bleed the system (see Section 8). Check for leaks.
23 Check the operation of the brakes carefully before riding the motorcycle.

3 Brake pads - replacement

 Warning: When replacing the front brake pads always replace the pads in BOTH calipers - never just on one side. Also, the dust created by the brake system may contain asbestos, which is harmful to your health. Never blow it out with compressed air and don't inhale any of it. An approved filtering mask should be worn when working on the brakes.

1 Set the bike on its centrestand.
2 Remove the caliper following the procedure in Section 2, but don't disconnect the brake hose.
3 Remove the inner brake pad from the caliper bracket **(see illustration)**.

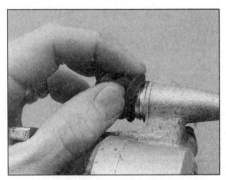

2.16 Install the slider pin boots

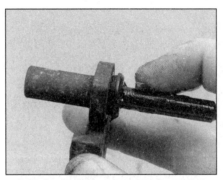

2.17 Apply a thin coat of the specified grease to the slider pins on the caliper bracket

3.3 Pull the inner pad (arrow) from the caliper bracket

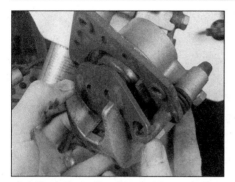

3.4 Push the caliper bracket towards the piston and remove the outer pad

3.5 Remove the anti-rattle spring and check it for distortion, replacing it if necessary

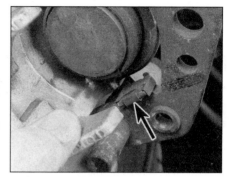

3.6 Also check the pad support clips (arrow) on the caliper bracket - replace them if they're damaged

4 Push the caliper bracket in (toward the piston) until the pins on the bracket clear the holes in the pad backing plate, then remove the outer pad **(see illustration)**.

5 Remove the anti-rattle spring **(see illustration)**. If it appears damaged, replace it.

6 Check the pad support clips on the caliper bracket **(see illustration)**. If they are missing or distorted, replace them.

7 Check the condition of the brake disc(s) (see Section 4). If they are in need of machining or replacement, follow the procedure in that Section to remove them. If they are okay, deglaze them with sandpaper or emery cloth, using a swirling motion.

8 Remove the cap from the master cylinder reservoir and siphon out some fluid. Push the piston into the caliper as far as possible, while checking the master cylinder reservoir to make sure it doesn't overflow. If you can't depress the piston with thumb pressure, try using a C-clamp. If the piston sticks, remove the caliper and overhaul it as described in Section 2.

9 Install the anti-rattle spring in the caliper.

10 Install the outer pad in the caliper and pull the caliper bracket out, so the pins on the bracket engage with the holes in the pad backing plate.

11 Install the inner pad in the caliper bracket.

12 Install the caliper, tightening the mounting

bolts to the torque listed in this Chapter's Specifications.

13 Refill the master cylinder reservoir (see *Daily (pre-ride) checks*) and install the diaphragm and cap.

14 Operate the brake lever or pedal several times to bring the pads into contact with the disc. Check the operation of the brakes carefully before riding the motorcycle.

4 Brake disc(s) - inspection, removal and installation

Inspection

1 Set the bike on its centrestand.

2 Visually inspect the surface of the disc(s) for score marks and other damage. Light scratches are normal after use and won't affect brake operation, but deep grooves and heavy score marks will reduce braking efficiency and accelerate pad wear. If the discs are badly grooved they must be machined or replaced.

3 To check disc runout, mount a dial indicator to a fork leg or the swingarm, with the plunger on the indicator touching the surface of the disc about 1/2-inch from the outer edge **(see illustration)**. Slowly turn the wheel (if you're checking the front discs, have an assistant sit

on the seat to raise the front wheel off the ground) and watch the indicator needle, comparing your reading with the limit listed in this Chapter's Specifications. If the runout is greater than allowed, check the hub bearings for play (see Chapter 1). If the bearings are worn, replace them and repeat this check. If the disc runout is still excessive, it will have to be replaced.

4 The disc must not be machined or allowed to wear down to a thickness less than the minimum allowable thickness, listed in this Chapter's Specifications. The thickness of the disc can be checked with a micrometer **(see illustration)**. If the the thickness of the disc is less than the minimum allowable, it must be replaced. The minimum thickness is also stamped into the disc **(see illustration)**.

Removal

5 Remove the wheel (see Section 11 for front wheel removal or Section 12 for rear wheel removal).

Caution: Don't lay the wheel down and allow it to rest on one of the discs - the disc could become warped. Set the wheel on wood blocks so the disc doesn't support the weight of the wheel.

6 Mark the relationship of the disc to the wheel, so it can be installed in the same position. Remove the Allen-head bolts that

4.3 Use a dial indicator to check disc runout - if the reading exceeds the specified limit, the disc will have to be replaced

4.4a Use a micrometer to measure the thickness of the disc at several points

4.4b The minimum allowable thickness is stamped into the disc

4.6 Loosen the disc retaining bolts a little at a time to prevent distortion

5.4 Use a six-point box-end wrench to remove the banjo bolt from the master cylinder - be prepared for spillage

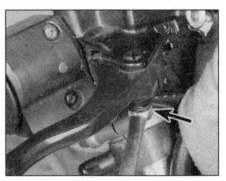

5.5 Remove the locknut then unscrew the brake lever pivot bolt

retain the disc to the wheel **(see illustration)**. Loosen the bolts a little at a time, in a criss-cross pattern, to avoid distorting the disc.

7 Take note of any paper shims that may be present where the disc mates to the wheel. If there are any, mark their position and be sure to include them when installing the disc.

Installation

8 Position the disc on the wheel, aligning the previously applied matchmarks (if you're reinstalling the original disc). Make sure the arrow (stamped on the disc) marking the direction of rotation is pointing in the proper direction.

9 Apply a non-hardening thread locking compound to the threads of the bolts. Install the bolts, tightening them a little at a time, in a

5.6 Remove the master cylinder mounting bolts

criss-cross pattern, until the torque listed in this Chapter's Specifications is reached. Clean off all grease from the brake disc(s) using acetone or brake system cleaner.

10 Install the wheel.

11 Operate the brake lever or pedal several times to bring the pads into contact with the disc. Check the operation of the brakes carefully before riding the motorcycle.

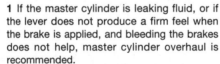

5 Front brake master cylinder - removal, overhaul and installation

1 If the master cylinder is leaking fluid, or if the lever does not produce a firm feel when the brake is applied, and bleeding the brakes does not help, master cylinder overhaul is recommended.

2 Before disassembling the master cylinder, read through the entire procedure and make sure that you have the correct rebuild kit. Also, you will need some new, clean brake fluid of the recommended type, some clean rags and internal snap-ring pliers. **Note:** *To prevent damage to the paint from brake fluid, always cover the fuel tank when working on the master cylinder.*

Caution: Disassembly, overhaul and reassembly of the brake master cylinder must be done in a spotlessly clean work area to avoid contamination and possible failure of the brake hydraulic system components.

Removal

3 Loosen, but do not remove, the screws holding the reservoir cover in place.

4 Pull back the rubber boot, loosen the banjo fitting bolt **(see illustration)** and separate the brake hose from the master cylinder. Wrap the end of the hose in a clean rag and suspend the hose in an upright position or bend it down carefully and place the open end in a clean container. The objective is to prevent excess loss of brake fluid, fluid spills and system contamination.

5 Remove the locknut from the underside of the lever pivot bolt, then unscrew the bolt **(see illustration)**.

6 Remove the master cylinder mounting bolts **(see illustration)** and separate the master cylinder from the handlebar.

Caution: Do not tip the master cylinder upside down or brake fluid will run out.

7 Disconnect the electrical connectors from the brake light switch **(see illustration)**.

Overhaul

8 Detach the top cover and the rubber diaphragm, then drain the brake fluid into a suitable container. Wipe any remaining fluid out of the reservoir with a clean rag.

9 Carefully remove the rubber dust boot from the end of the piston **(see illustration)**.

10 Using snap-ring pliers, remove the snap-ring **(see illustrations)** and slide out the piston, the cup seals and the spring. Lay the

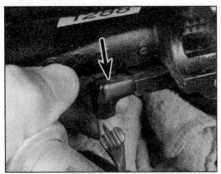

5.7 Unplug the two electrical connectors from the brake light switch

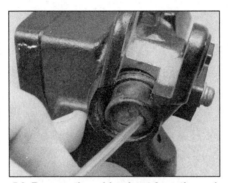

5.9 Remove the rubber boot from the end of the master cylinder piston, . . .

5.10a . . . then depress the piston and remove the snap-ring with a pair of snap-ring pliers

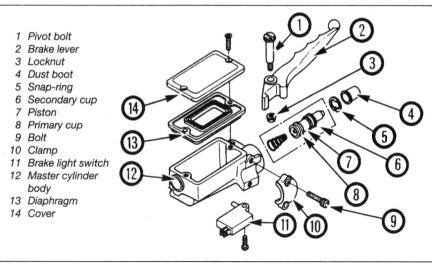

1 Pivot bolt
2 Brake lever
3 Locknut
4 Dust boot
5 Snap-ring
6 Secondary cup
7 Piston
8 Primary cup
9 Bolt
10 Clamp
11 Brake light switch
12 Master cylinder body
13 Diaphragm
14 Cover

5.10b Exploded view of the master cylinder

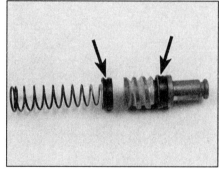

5.12 Make sure the lips of the piston cups (arrows) are facing away from the lever end of the piston

parts out in the proper order to prevent confusion during reassembly.

11 Clean all of the parts with brake system cleaner (available at auto parts stores), isopropyl alcohol or clean brake fluid. Check the master cylinder bore for corrosion, scratches, nicks and score marks. If damage is evident, the master cylinder must be replaced with a new one. If the master cylinder is in poor condition, then the calipers should be checked as well.

Caution: Do not, under any circumstances, use a petroleum-based solvent to clean brake parts. If compressed air is available, use it to dry the parts thoroughly (make sure it's filtered and unlubricated).

12 Remove the old cup seals from the piston and spring and install the new ones. Make sure the lips face away from the lever end of the piston **(see illustration)**. If a new piston is included in the rebuild kit, use it regardless of the condition of the old one.

13 Before reassembling the master cylinder, soak the piston and the rubber cup seals in clean brake fluid for ten or fifteen minutes. Lubricate the master cylinder bore with clean brake fluid, then carefully insert the piston and related parts in the reverse order of disassembly. Make sure the lips on the cup seals do not turn inside out when they are slipped into the bore.

14 Depress the piston, then install the snap-ring (make sure the snap-ring is properly seated in the groove with the sharp edge facing out). Install the rubber dust boot (make sure the lip is seated properly in the piston groove).

Installation

15 Attach the master cylinder to the handlebar and tighten the bolts to the torque listed in this Chapter's Specifications. The arrow and the word "up" on the master cylinder clamp should be pointing up and readable. Install the brake lever and tighten

the pivot bolt locknut.

16 Connect the brake hose to the master cylinder, using new sealing washers. Tighten the banjo fitting bolt to the torque listed in this Chapter's Specifications. Refer to Section 8 and bleed the air from the system.

6 Rear brake master cylinder - removal, overhaul and installation

1 If the master cylinder is leaking fluid, or if the pedal does not produce a firm feel when the brake is applied, and bleeding the brakes does not help, master cylinder overhaul is recommended.

2 Before disassembling the master cylinder, read through the entire procedure and make sure that you have the correct rebuild kit. Also, you will need some new, clean brake fluid of the recommended type, some clean rags and internal snap-ring pliers.

Caution: Disassembly, overhaul and reassembly of the brake master cylinder must be done in a spotlessly clean work area to avoid contamination and possible failure of the brake hydraulic system components.

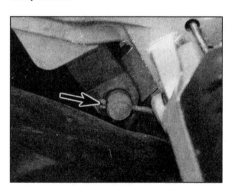

6.4 Remove the cotter pin / split pin (arrow) from the clevis pin

Removal

3 Set the bike on its centrestand. Remove the right side cover (see Chapter 8 if necessary).

4 Remove the cotter pin (split pin) from the clevis pin on the master cylinder pushrod **(see illustration)**. Remove the clevis pin.

5 Have a container and some rags ready to catch spilling brake fluid. Using a pair of pliers, slide the clamp up the fluid feed hose and detach the hose from the master cylinder **(see illustration)**. Direct the end of the hose into the container, unscrew the cap on the master cylinder reservoir and allow the fluid to drain.

6 Using a six-point box-end wrench, unscrew the banjo fitting bolt from the top of the master cylinder **(see illustration 6.5)**. Discard the sealing washers on either side of the fitting.

7 Remove the two master cylinder mounting bolts and detach the cylinder from the bracket.

Overhaul

8 Using a pair of snap-ring pliers, remove the snap-ring from the fluid inlet fitting **(see**

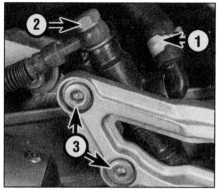

6.5 Mounting details of the rear brake master cylinder

1 Fluid feed hose
2 Banjo fitting bolt
3 Master cylinder mounting bolts

6.8a Remove the snap-ring that secures the fluid inlet fitting

6.8b Remove the O-ring from the bore and discard it

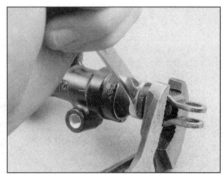

6.9a Hold the clevis with a pair of pliers and loosen the locknut

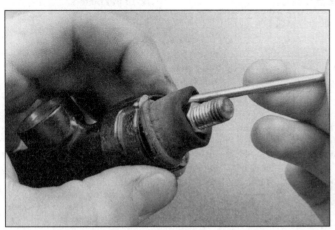

6.9b Remove the dust boot from the pushrod

6.10a Depress the pushrod and remove the snap-ring from the cylinder bore

illustration) and detach the fitting from the master cylinder. Remove the O-ring from the bore (see illustration).

9 Hold the clevis with a pair of pliers and loosen the locknut (see illustration). Carefully remove the rubber dust boot from the pushrod (see illustration).

10 Depress the pushrod and, using snap-ring pliers, remove the snap-ring (see illustrations). Slide out the piston, the cup seal and spring. Lay the parts out in the proper order to prevent confusion during reassembly.

11 Clean all of the parts with brake system cleaner (available at auto parts stores), isopropyl alcohol or clean brake fluid. Check the master cylinder bore for corrosion, scratches, nicks and score marks. If damage is evident, the master cylinder must be replaced with a new one. If the master cylinder is in poor condition, then the caliper should be checked as well.

Caution: Do not, under any circumstances, use a petroleum-based solvent to clean brake parts. If compressed air is available, use it to dry the parts thoroughly (make sure it's filtered and unlubricated).

12 Remove the old cup seals from the piston and spring and install the new ones. Make sure the lips face away from the pushrod end of the piston (see illustration). If a new piston is included in the rebuild kit, use it regardless of the condition of the old one.

13 Before reassembling the master cylinder, soak the piston and the rubber cup seals in clean brake fluid for ten or fifteen minutes. Lubricate the master cylinder bore with clean brake fluid, then carefully insert the parts in the reverse order of disassembly. Make sure

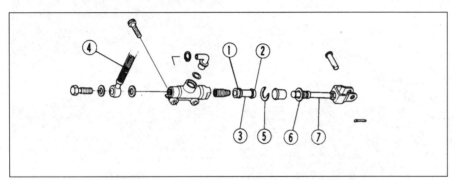

6.10b Exploded view of the rear brake master cylinder

1 Primary cup	3 Piston	5 Retainer	7 Pushrod
2 Secondary cup	4 Brake hose	6 Piston stop	

6.12 Make sure the lips of the cups (arrow) face away from the pushrod end of the piston

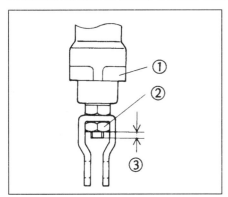

6.15 Allow 3.5 to 5.5 mm of the pushrod to protrude past the adjusting nut, then tighten the locknut

1 Master cylinder
2 Adjusting nut
3 Rod position

the lips on the cup seals do not turn inside out when they are slipped into the bore.

14 Lubricate the end of the pushrod with PBC (poly butyl cuprysil) grease, or silicone grease designed for brake applications, and install the pushrod and stop washer into the cylinder bore. Depress the pushrod, then install the snap-ring (make sure the snap-ring is properly seated in the groove with the sharp edge facing out). Install the rubber dust boot (make sure the lip is seated properly in the groove in the piston stop nut).

15 Install the clevis to the end of the pushrod, leaving about 3.5 to 5.5 mm of the end of the pushrod exposed past the adjusting nut **(see illustration)**, then tighten the locknut. This will ensure the brake pedal will be positioned correctly.

16 Install the feed hose fitting, using a new O-ring. Install the snap-ring, making sure it seats properly in its groove.

Installation

17 Position the master cylinder on the frame and install the bolts, tightening them to the torque listed in this Chapter's Specifications.

18 Connect the banjo fitting to the top of the master cylinder, using new sealing washers on each side of the fitting. Tighten the banjo fitting bolt to the torque listed in this Chapter's Specifications.

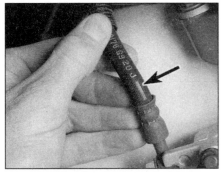

7.2 Flex the brake hoses and check for cracks, bulges and leaking fluid

19 Connect the fluid feed hose to the inlet fitting and install the hose clamp.

20 Connect the clevis to the brake pedal and secure the clevis pin with a new cotter pin (split pin).

21 Fill the fluid reservoir with the specified fluid (see *Daily (pre-ride) checks*) and bleed the system following the procedure in Section 8. Install the side cover.

22 Check the position of the brake pedal (see Chapter 1) and adjust it if necessary. Check the operation of the brakes carefully before riding the motorcycle.

---

7 Brake hoses and lines - inspection and replacement

Inspection

1 Once a week, or if the motorcycle is used less frequently, before every ride, check the condition of the brake hoses and the metal lines that feed the brake fluid to the anti-dive units on the front forks.

2 Twist and flex the rubber hoses **(see illustration)** while looking for cracks, bulges and seeping fluid. Check extra carefully around the areas where the hoses connect with the banjo fittings, as these are common areas for hose failure.

3 Inspect the metal pipes connected to the anti-dive units on the front forks. If the plating on the lines is chipped or scratched, the pipes may rust. If the pipes are rusted, scratched or cracked, replace them. Note that the metal pipes must be replaced at the interval in Chapter 1.

Replacement

4 Most brake hoses have banjo fittings on each end of the hose **(see illustration)**. Cover the surrounding area with plenty of rags and unscrew the banjo bolts on either end of the hose. If a threaded fitting is used instead of a banjo bolt, use a flare nut wrench to loosen it. Detach the hose from any clips that may be present and remove the hose.

5 Position the new hose, making sure it isn't twisted or otherwise strained, between the two components. Make sure the metal tube portion of the banjo fitting is located between the casting protrusions on the component it's connected to, if equipped. Install the banjo bolts, using new sealing washers on both sides of the fittings, and tighten them to the torque listed in this Chapter's Specifications.

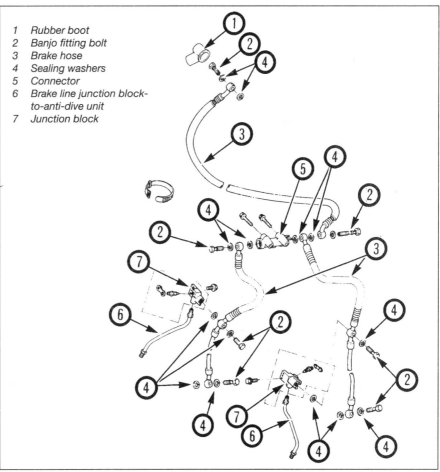

1 Rubber boot
2 Banjo fitting bolt
3 Brake hose
4 Sealing washers
5 Connector
6 Brake line junction block-to-anti-dive unit
7 Junction block

7.4 Front brake hose installation details (typical)

If a threaded fitting is used instead of a banjo bolt, tighten it securely, again using a flare nut wrench.

6 Flush the old brake fluid from the system, refill the system with the recommended fluid (see *Daily (pre-ride) checks*) and bleed the air from the system (see Section 8). Check the operation of the brakes carefully before riding the motorcycle.

8 Brake system - bleeding

1 Bleeding the brake is simply the process of removing all the air bubbles from the brake fluid reservoir, the lines and the brake caliper. Bleeding is necessary whenever a brake system hydraulic connection is loosened, when a component or hose is replaced, or when the master cylinder or caliper is overhauled. Leaks in the system may also allow air to enter, but leaking brake fluid will reveal their presence and warn you of the need for repair.

2 To bleed the brake, you will need some new, clean brake fluid of the recommended type (see Chapter 1), a length of clear vinyl or plastic tubing, a small container partially filled with clean brake fluid, some rags and a wrench to fit the brake caliper bleeder valve.

3 Cover the fuel tank and other painted components to prevent damage in the event that brake fluid is spilled.

4 Remove the reservoir cap or cover and slowly pump the brake lever or pedal a few times, until no air bubbles can be seen floating up from the holes at the bottom of the reservoir. Doing this bleeds the air from the master cylinder end of the line. Reinstall the reservoir cap or cover.

5 Attach one end of the clear vinyl or plastic tubing to the brake caliper bleeder valve and submerge the other end in the brake fluid in the container.

6 Remove the reservoir cap or cover and check the fluid level. Do not allow the fluid level to drop below the lower mark during the bleeding process.

7 Carefully pump the brake lever or pedal three or four times and hold it while opening the caliper bleeder valve **(see illustration)**. When the valve is opened, brake fluid will flow out of the caliper into the clear tubing and the lever will move toward the handlebar or the pedal will move down.

8 Retighten the bleeder valve, then release the brake lever or pedal gradually. Repeat the process until no air bubbles are visible in the brake fluid leaving the caliper and the lever or pedal is firm when applied. **Note:** *The rear calipers on later models have two bleeder valves - air must be bled from both, one after the other. Remember to add fluid to the reservoir as the level drops.* Use only new, clean brake fluid of the recommended type. Never reuse the fluid lost during bleeding.

9 If you're bleeding the front brakes, repeat this procedure to the other caliper, then bleed both anti-dive units on the lower ends of the forks, followed by the junction blocks about half-way up the forks **(see illustrations)**. **Note:** *Later models don't have bleeder valves on the anti-dive units or the junction blocks.* Be sure to check the fluid level in the master cylinder reservoir frequently.

10 Replace the reservoir cover, wipe up any spilled brake fluid and check the entire system for leaks.

> **HAYNES HiNT** *If bleeding is difficult, it may be necessary to let the brake fluid in the system stabilise for a few hours (it may be aerated). Repeat the bleeding procedure when the tiny bubbles in the system have settled out.*

9 Wheels - inspection and repair

1 Place the motorcycle on the centrestand, then clean the wheels thoroughly to remove mud and dirt that may interfere with the inspection procedure or mask defects. Make

8.7 When bleeding the brakes, a clear piece of tubing is attached to the bleeder valve and submerged in brake fluid - the air bubbles can easily be seen in the tube and container (when no more bubbles appear, the air has been purged from the caliper)

a general check of the wheels and tyres as described in Chapter 1.

2 With the motorcycle on the centrestand and the wheel in the air, attach a dial indicator to the fork slider or the swingarm and position the stem against the side of the rim **(see illustration)**. Spin the wheel slowly and check the side-to-side (axial) runout of the rim, then compare your readings with the value listed in this Chapter's Specifications. In order to accurately check radial runout with the dial indicator, the wheel would have to be removed from the machine and the tyre removed from the wheel. With the axle clamped in a vise, the wheel can be rotated to check the runout.

3 An easier, though slightly less accurate, method is to attach a stiff wire pointer to the fork slider or the swingarm and position the end a fraction of an inch from the wheel (where the wheel and tyre join). If the wheel is true, the distance from the pointer to the rim will be constant as the wheel is rotated. Repeat the procedure to check the runout of the rearwheel. **Note:** *If wheel runout is excessive, refer to the appropriate Section in this Chapter and check the wheel bearings very carefully before replacing the wheel.*

4 The wheels should also be visually

8.9a After bleeding the front calipers on 600 A and B models, bleed the anti-dive units. . .

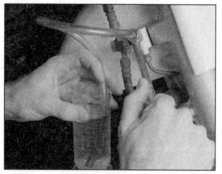

8.9b . . . and the bleeder valves on the junction blocks

9.2 Axial runout of the wheel can be checked with a dial indicator mounted to a fork tube (front) or the swingarm (rear)

inspected for cracks, flat spots on the rim and other damage. Since tubeless tyres are involved, look very closely for dents in the area where the tyre bead contacts the rim. Dents in this area may prevent complete sealing of the tyre against the rim, which leads to deflation of the tyre over a period of time.

5 If damage is evident, or if runout in either direction is excessive, the wheel will have to be replaced with a new one. Never attempt to repair a damaged cast aluminum wheel.

10 Wheels - alignment check

1 Misalignment of the wheels, which may be due to a cocked rear wheel or a bent frame or triple clamps, can cause strange and possibly serious handling problems. If the frame or triple clamps are at fault, repair by a frame specialist or replacement with new parts are the only alternatives.

2 To check the alignment you will need an assistant, a length of string or a perfectly straight piece of wood and a ruler graduated in 1/64 inch increments. A plumb bob or other suitable weight will also be required.

3 Place the motorcycle on the centrestand, then measure the width of both tyres at their widest points. Subtract the smaller measurement from the larger measurement, then divide the difference by two. The result is the amount of offset that should exist between the front and rear tyres on both sides.

4 If a string is used, have your assistant hold one end of it about half way between the floor and the rear axle, touching the rear sidewall of the tire.

5 Run the other end of the string forward and pull it tight so that it is roughly parallel to the floor. Slowly bring the string into contact with the front sidewall of the rear tire, then turn the front wheel until it is parallel with the string. Measure the distance from the front tyre sidewall to the string.

6 Repeat the procedure on the other side of the motorcycle. The distance from the front tyre sidewall to the string should be equal on both sides.

7 As was previously pointed out, a perfectly straight length of wood may be substituted for the string. The procedure is the same.

8 If the distance between the string and tyre is greater on one side, or if the rear wheel appears to be cocked, refer to Chapter 6, Swingarm bearings - check, and make sure the swingarm is tight.

9 If the front-to-back alignment is correct, the wheels still may be out of alignment vertically.

10 Using the plumb bob, or other suitable weight, and a length of string, check the rear wheel to make sure it is vertical. To do this, hold the string against the tyre upper sidewall and allow the weight to settle just off the floor. When the string touches both the upper and lower tyre sidewalls and is perfectly straight, the wheel is vertical. If it is not, place thin spacers under one leg of the centrestand.

11 Once the rear wheel is vertical, check the front wheel in the same manner. If both wheels are not perfectly vertical, the frame and/or major suspension components are bent.

11 Front wheel - removal and installation

Removal

1 Remove the lower portion of the fairing (see Chapter 8). Place the motorcycle on the centrestand, then raise the front wheel off the ground by placing a floor jack, with a wood block on the jack head, under the engine.

2 Disconnect the speedometer cable from the drive unit.

3 Remove one of the brake calipers and support it with a piece of wire. Don't disconnect the brake hose from the caliper.

4 Loosen the right side axle clamp bolt **(see illustration)**. On early models unscrew the axle **(see illustration)**. On later models, the axle nut on the left side must be removed.

5 Support the wheel, then pull out the axle **(see illustration)** and carefully lower the

wheel until the brake disc clears the caliper that is still installed. It will probably be necessary to tilt the wheel to one side to allow removal. Don't lose the spacer that fits into the right side of the hub. If the axle is corroded, remove the corrosion with fine emery cloth. **Note:** *Do not operate the front brake lever with the wheel removed.*

Caution: Don't lay the wheel down and allow it to rest on one of the discs - the disc could become warped. Set the wheel on wood blocks so the disc doesn't support the weight of the wheel.

6 Check the condition of the wheel bearings (see Section 13).

Installation

7 Installation is the reverse of removal. Apply a thin coat of grease to the seal lip, then slide the collar into the right side of the hub. Position the speedometer drive unit in place in the left side of the hub, then slide the wheel into place. Make sure the notches in the speedometer drive housing line up with the lugs in the wheel. If the disc will not slide between the brake pads, remove the wheel and carefully pry them apart with a piece of wood.

8 Slip the axle into place, then tighten the axle (early models) or the axle nut (later models) to the torque listed in this Chapter's Specifications. Tighten the right side axle clamp bolt to the torque listed in this Chapter's Specifications.

9 Apply the front brake, pump the forks up and down several times and check for binding and proper brake operation.

12 Rear wheel - removal and installation

Removal

1 Set the bike on its centrestand.

2 Remove the chain guard (see Chapter 6 if necessary).

11.4a Loosen the axle clamp bolt at the bottom of the right fork leg (it isn't necessary to remove it completely)

11.4b On early models the axle can be unscrewed with a 12 mm Allen wrench

11.5 Support the wheel and pull out the axle

12.3 Remove the cotter pin (split pin) from the axle nut

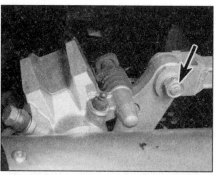

12.4 Loosen the torque link nut (arrow) to allow the brake caliper to swing to the rear as the wheel is removed

12.6 Push the wheel forward and detach the chain from the sprocket

3 Remove the cotter pin (split pin) from the axle nut **(see illustration)** and remove the nut.
4 Loosen the torque link nut **(see illustration)**.
5 Loosen the chain adjusting bolt locknuts (see Chapter 1) and fully loosen both adjusting bolts.
6 Push the rear wheel as far forward as possible. Lift the top of the chain up off the rear sprocket and pull it to the left while rotating the wheel backwards **(see illustration)**. This will disengage the chain from the sprocket.

 Warning: Don't let your fingers slip between the chain and the sprocket.

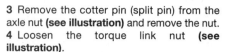

12.8 Check the axle for runout using a dial indicator and V-blocks

7 Support the wheel and slide the axle out. Lower the wheel and remove it from the swingarm, being careful not to lose the spacers on either side of the hub. **Note:** *Do not operate the brake pedal with the wheel removed.*
Caution: Don't lay the wheel down and allow it to rest on the disc or the sprocket - they could become warped. Set the wheel on wood blocks so the disc or the sprocket doesn't support the weight of the wheel.
8 Before installing the wheel, check the axle for straightness. If the axle is corroded, first remove the corrosion with fine emery cloth. Set the axle on V-blocks and check it for runout using a dial indicator **(see illustration)**. If the axle exceeds the maximum allowable runout limit listed in this Chapter's Specifications, it must be replaced.
9 Check the condition of the wheel bearings (see Section 13).

Installation

10 Apply a thin coat of grease to the seal lips, then slide the spacers into their proper positions on the sides of the hub.
11 Slide the wheel into place, making sure the brake disc slides between the brake pads. If it doesn't, spread the pads apart with a piece of wood.
12 Pull the chain up over the sprocket, raise the wheel and install the axle and axle nut.

Don't tighten the axle nut at this time.
13 Adjust the chain slack (see Chapter 1) and tighten the adjuster locknuts.
14 Tighten the axle nut to the torque listed in this Chapter's Specifications. Install a new cotter pin (split pin), tightening the axle nut an additional amount, if necessary, to align the hole in the axle with the castellations on the nut.
15 Tighten the torque link nut to the torque listed in the Chapter 6 Specifications.
16 Check the operation of the brakes carefully before riding the motorcycle.

13 Wheel bearings - inspection and maintenance

1 Set the bike on its centrestand and remove the wheel - see Section 11 (front wheel) or 12 (rear wheel).
2 Set the wheel on blocks so as not to allow the weight of the wheel to rest on the brake disc or sprocket.
3 If you're removing the front wheel bearings, remove the snap-ring securing the speedometer drive **(see illustration)**. Remove the speedometer drive from the hub.
4 If you're removing the rear wheel bearings, lift off the coupling sleeve, rear wheel coupling and coupling collar **(see illustrations)**.

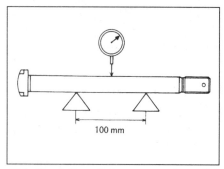

13.3 If you're removing the front wheel bearings, remove the snap-ring that secures the speedometer drive, then remove the drive (arrow)

13.4a If you're removing the rear wheel bearings, lift out the coupling sleeve, . . .

13.4b . . . then lift the coupling from the wheel (don't lose the collar underneath the coupling)

13.5 A screwdriver can be used to pry out the grease seal

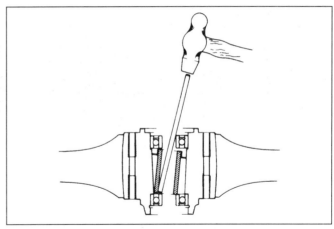

13.7 Once the snap-rings have been removed, drive the bearings from the hub with a metal rod and a hammer

5 Pry out the grease seal **(see illustration)**.

6 Remove the snap-ring from the other side of the hub.

7 Using a metal rod (preferably a brass drift punch) inserted through the centre of the hub bearing, tap evenly around the inner race of the opposite bearing to drive it from the hub **(see illustration)**. The bearing spacer will also come out.

8 Lay the wheel on its other side and remove the remaining bearing using the same technique.

9 Clean the bearings with a high flash-point solvent (one which won't leave any residue) and blow them dry with compressed air (don't let the bearings spin as you dry them). Apply a few drops of oil to the bearing.

 HAYNES HiNT *Refer to Tools and Workshop Tips in the Reference section for information on checking bearings for wear.*

10 If the bearing checks out okay and will be reused, wash it in solvent once again and dry it, then pack the bearing from the open side with high quality bearing grease **(see illustration)**.

11 Thoroughly clean the hub area of the wheel. Install the bearing into the recess in the hub, with the marked or shielded side facing out. Using a bearing driver or a socket large enough to contact the outer race of the bearing, drive it in **(see illustration)** until the snap-ring groove is visible. Install the snap-ring.

12 Turn the wheel over and install the bearing spacer and bearing, driving the bearing into place as described in Step 10, then install the snap-ring. Install the speedometer drive (if you're working on the front wheel) and the snap-ring.

13 Install a new grease seal, using a seal driver, large socket or a flat piece of wood to drive it into place.

14 If you're working on the rear wheel, press a little grease into the bearing in the rear wheel coupling. Install the coupling to the wheel, making sure the coupling collar is located in the inside of the inner race (between the wheel and the coupling).

15 Clean off all grease from the brake disc(s) using acetone or brake system cleaner. Install the wheel.

14 Tyres - general information and fitting

General information

1 The wheels fitted to all models are designed to take tubeless tyres only. Tyre sizes are given in the Specifications at the beginning of this chapter.

2 Refer to the Daily (pre-ride) checks listed at the beginning of this manual for tyre maintenance.

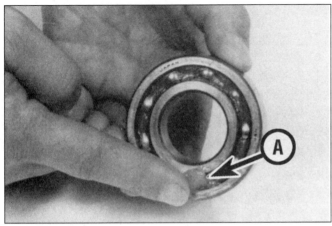

13.10 Press grease into the open side of the bearing until it's full

13.11 A socket of the appropriate diameter can be used to drive the bearing into the hub

Fitting new tyres

3 When selecting new tyres, refer to the tyre information label on the swingarm and the tyre options listed in the owners handbook. Ensure that front and rear tyre types are compatible, the correct size and correct speed rating; if necessary seek advice from a Kawasaki dealer or tyre fitting specialist **(see illustration).**

4 It is recommended that tyres are fitted by a motorcycle tyre specialist rather than attempted in the home workshop. This is particularly relevant in the case of tubeless tyres because the force required to break the seal between the wheel rim and tyre bead is substantial, and is usually beyond the capabilities of an individual working with normal tyre levers. Additionally, the specialist will be able to balance the wheels after tyre fitting.

5 Note that punctured tubeless tyres can in some cases be repaired. Kawasaki recommend that a repaired tyre should not be used at speeds above 60 mph (100 kmh) for the first 24 hrs after the repair, and thereafter not above 110 mph (180 kmh).

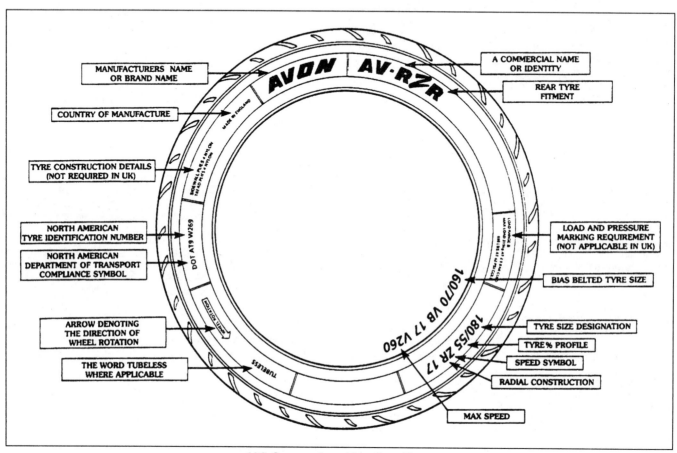

14.3 Common tyre sidewall markings

Chapter 8
Fairing and bodywork

Contents

Degrees of difficulty

Easy, suitable for novice with little experience		**Fairly easy,** suitable for beginner with some experience		**Fairly difficult,** suitable for competent DIY mechanic		**Difficult,** suitable for experienced DIY mechanic		**Very difficult,** suitable for expert DIY or professional	

---

1 General information

This Chapter covers the procedures necessary to remove and install the fairing and other body parts **(see illustrations)**.

Since many service and repair operations on these motorcycles require removal of the fairing and/or other body parts, the procedures are grouped here and referred to from other Chapters.

In the case of damage to the fairing or other body part, it is usually necessary to remove the broken component and replace it with a new (or used) one. The material that the fairing and other body parts is composed of doesn't lend itself to conventional repair techniques. There are, however, some shops that specialize in "plastic welding", so it would be advantageous to check around first before throwing the damaged part away.

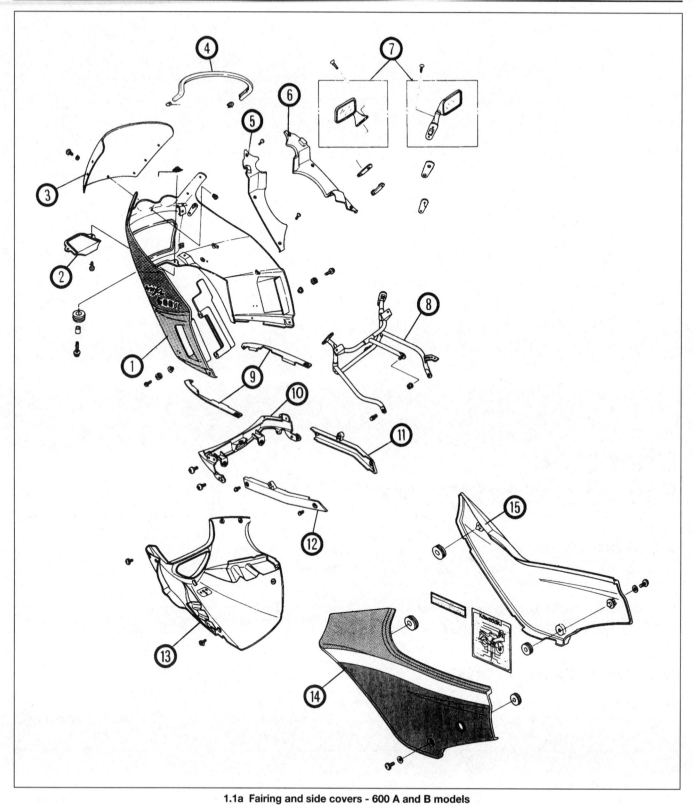

1.1a Fairing and side covers - 600 A and B models

1 Upper fairing
2 Access panel
3 Windshield
4 Trim

5 Left inner panel
6 Right inner panel
7 Rear view mirrors
8 Fairing mount

9 Rubber strips
10 Fairing stay (front)
11 Fairing stay (right - A models only)

12 Fairing stay (left - A models only)
13 Lower fairing
14 Left side cover
15 Right side cover

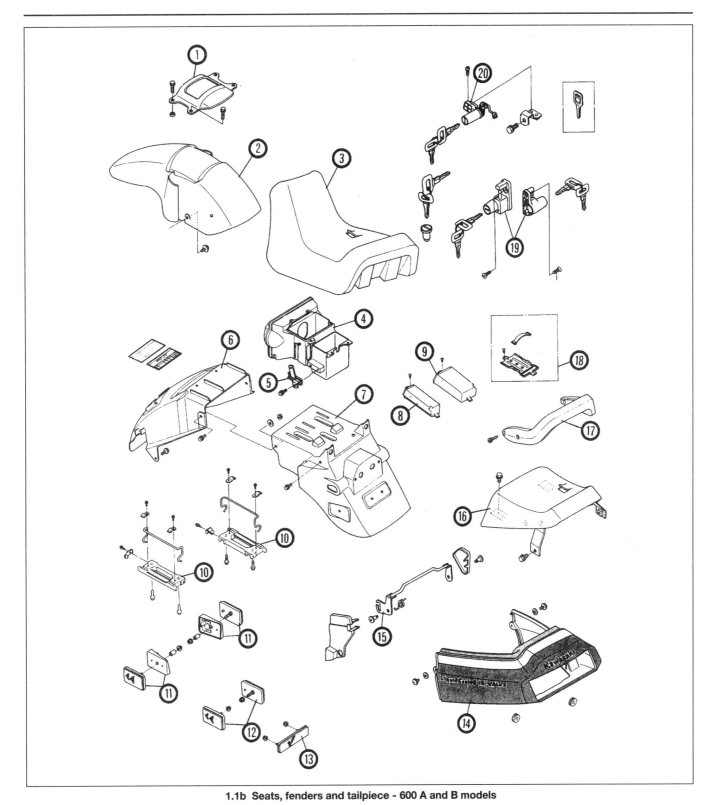

1.1b Seats, fenders and tailpiece - 600 A and B models

1 Fork brace
2 Front fender
3 Rider's seat
4 Air filter housing
5 Voltage regulator/rectifier mounting bracket
6 Rear fender (front section)
7 Rear fender (rear section)
8 Tool kit compartment
9 Document container (or evaporative emission control canister - California models)
10 Tie-down hook and bracket
11 Upper fairing reflectors
12 Rear fender side reflectors
13 Rear reflector
14 Tailpiece
15 Seat latch
16 Passenger seat
17 Grab handle
18 Canister mount (California models only)
19 Helmet locks
20 Seat latch lock

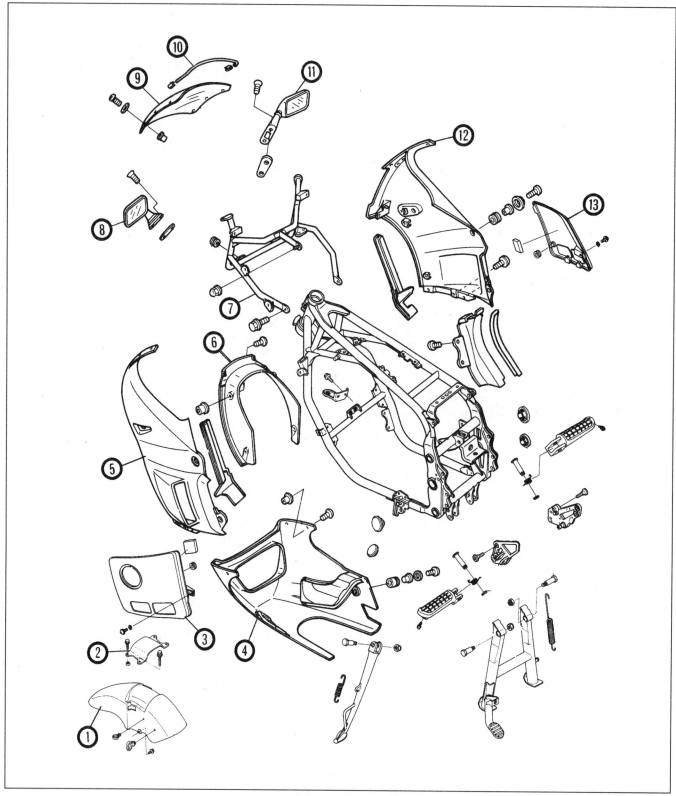

1.1c Fairing - 600 C models

1   Front fender
2   Fender brace
3   Knee grip cover (left side)
4   Lower fairing

5   Upper fairing (left side)
6   Headlight lower cover
7   Fairing mount

8   Rear view mirror (left side)
9   Windshield
10 Trim

11 Rear view mirror (right side)
12 Upper fairing (right side)
13 Knee grip cover (right side)

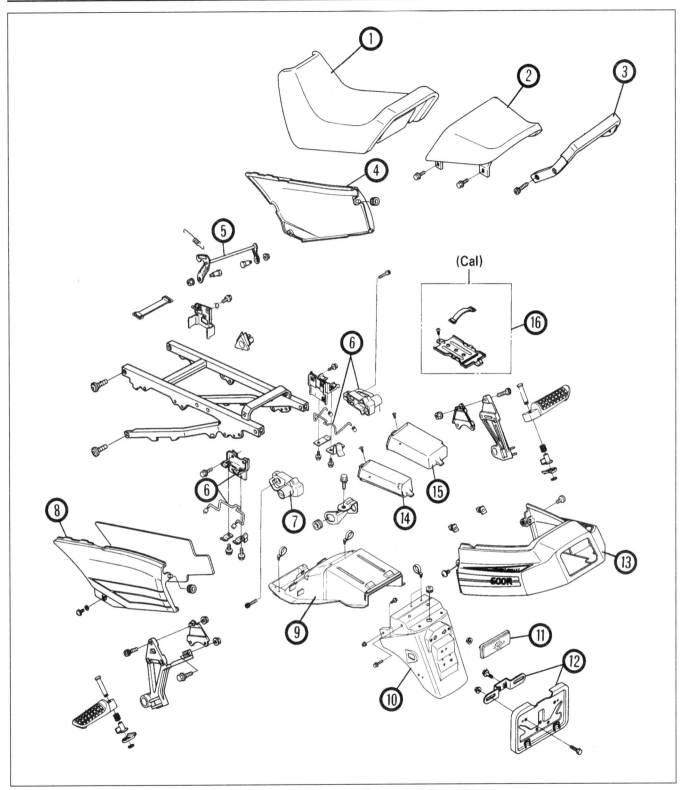

1.1d Seats, rear fender and tailpiece - 600 C models

1 Rider's seat
2 Passenger seat
3 Grab handle
4 Right side cover
5 Seat latch

6 Tie-down hook and bracket
7 Helmet lock
8 Left side cover
9 Rear fender (front section)
10 Rear fender (rear section)

11 Rear reflector
12 License plate bracket and mount
13 Tailpiece
14 Tool kit compartment

15 Document container (or evaporative emission control canister - California models)
16 Canister mount (California models only)

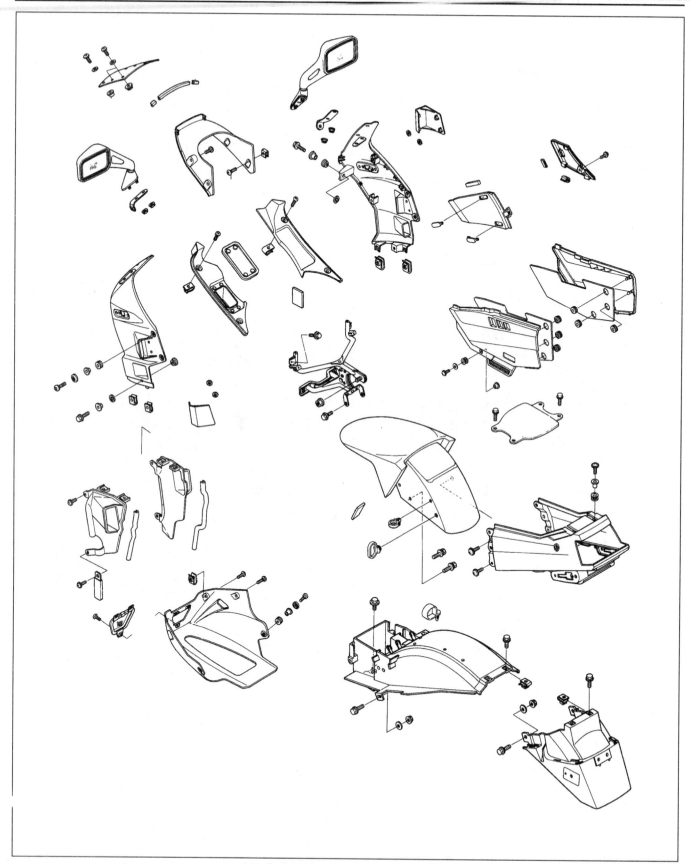

1.1e Bodywork components - 750 models

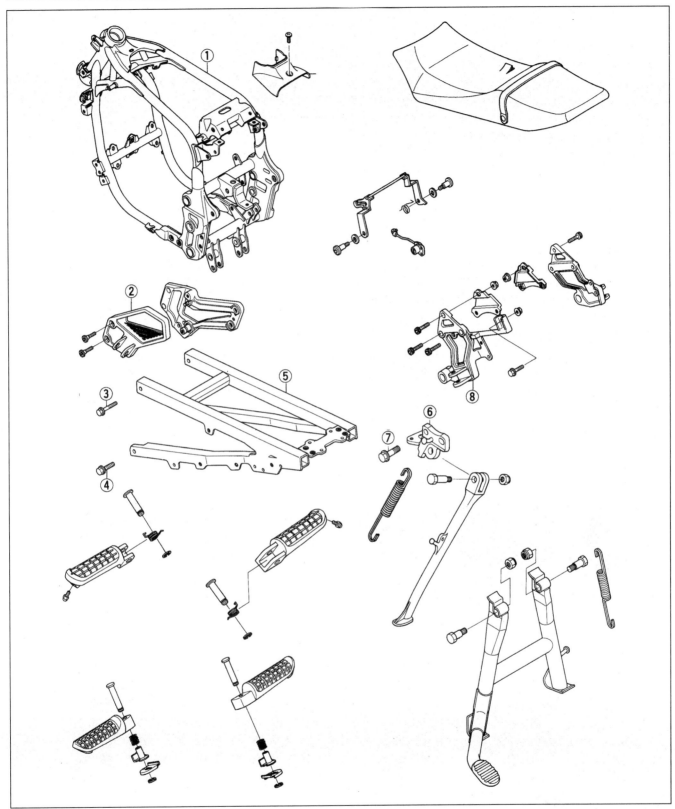

1.1f Frame fittings - 750 models

1 Main frame section
2 Front footpeg brackets
3 Rear frame upper mounting bolts
4 Rear frame lower mounting bolts
5 Rear frame section
6 Sidestand bracket
7 Sidestand bracket bolts
8 Muffler (silencer) brackets

2.2 Remove these screws from each side of the fairing (A model shown, others similar)

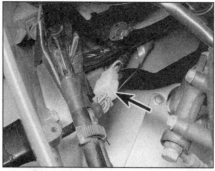

3.4 The electrical connector for the headlight (arrow) can be unplugged after the left inner panel has been removed

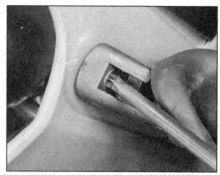

3.5 Each rear view mirror serves as an upper mounting point for the fairing, and is retained by two screws hidden under the mirror cover

2 Lower fairing - removal and installation

600 models

1 Set the bike on its centrestand.
2 Support the lower fairing and remove the mounting screws **(see illustration)**.
3 Carefully maneuver the fairing out from under the bike. On A and B models, mark and disconnect the hoses from the coolant reservoir (see Chapter 3).

3.6a On A and B models, remove the two screws from each side of the fairing (arrows)

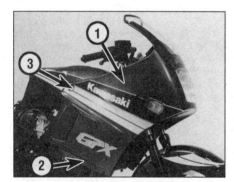

3.6b On C models, remove the screw and the Allen bolt from each side of the fairing

1 Fairing   2 Screw   3 Allen bolt

4 Installation is the reverse of removal. On A and B models, check the coolant level and add some, if necessary (see *Daily (pre-ride) checks*).

750 models

5 Set the bike on its centrestand.
6 Remove two screws and one Allen bolt on each side of the lower fairing.
7 Carefully manoeuvre the fairing out from under the bike.
8 Installation is the reverse of removal.

3 Upper fairing and mount - removal and installation

600 models

Upper fairing

1 Set the bike on its centrestand.
2 If you're working on a C model, remove the seat, the knee grip covers (see Section 6) and the lower fairing (see Section 2).
3 If you're working on an A or B model, remove the fairing inner panels **(see illustration 1.1a)**.
4 On A and B models, unplug the headlight electrical connector **(see illustration)**.

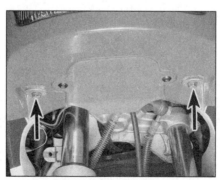

3.7 On A and B models, remove the two front mounting bolts (arrows)

5 Remove the rear view mirrors **(see illustration)**.
6 Remove the side mounting screws/bolts **(see illustrations)**. If you're working on a C model, be sure to support the fairing as this is done.
7 On A and B models, support the fairing and remove the two front mounting bolts **(see illustration)**.
8 Carefully pull the fairing forward and off the bike. It may be necessary to spread the lower sides of the fairing to clear the frame as you do this. If you're working on a C model, unplug the electrical connector for the headlight.
9 If it is necessary to disassemble the fairing (C models only), refer to illustration 1.1c.
10 Installation is the reverse of removal.

Mount

11 Remove the upper fairing, if you haven't already done so.
12 Detach the electrical connectors to the instrument cluster, noting how the harnesses are routed.
13 Detach the upper end of the speedometer cable from the speedometer. On C models, remove the coolant reservoir (see Chapter 3).
14 On A and B models, unbolt the radiator filler neck from the mount, then remove the right mounting bolt **(see illustration)**.

3.14 Unbolt the radiator filler neck from the fairing mount, then remove the right mounting bolt (arrows)

3.15 Location of the fairing mount-to-steering head bolt (arrow)

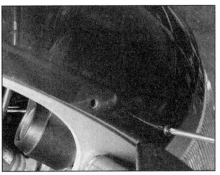

4.1 Remove the windshield screws and carefully detach the windshield from the upper fairing

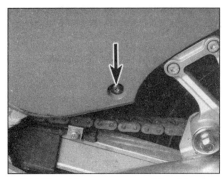

7.2 The side covers are retained by a single screw (arrow) - A model shown, others similar

15 Remove the mount-to-steering head bolt (see illustration).
16 Remove the left side mounting bolt and detach the mount from the frame. If it is necessary to remove the instrument cluster from the mount, refer to Chapter 9.
17 Installation is the reverse of removal.

750 models

18 Remove the lower fairing.
19 Remove two Allen bolts at the rear of the upper fairing on each side.
20 Remove the screws and detach the inner fairing panels on both sides.
21 Remove six screws, washers and nuts that secure the windshield. Carefully separate the windshield from the fairing. If it sticks, don't attempt to pry it off - just keep applying steady pressure with your fingers.
22 Remove the headlight lower cover.
23 Remove two bolts at the lower front of the fairing.
24 On the top side of the fairing, remove one bolt on each side.
25 Lift the fairing off, together with the headlight assembly and coolant reservoir. As you ease the fairing forward, disconnect the coolant reservoir hose and the wiring connectors for the headlight and turn signals.
26 Installation is the reverse of the removal steps.

4 Windshield - removal and installation

1 Remove the screws securing the windshield to the fairing (see illustration).
2 Carefully separate the windshield from the fairing. If it sticks, don't attempt to pry it off - just keep applying steady pressure with your fingers.
3 Installation is the reverse of the removal procedure. Be sure each screw has a plastic washer under its head. Tighten the screws securely, but be careful not to overtighten them, as the windshield might crack.

5 Fairing stays (600 A models only) - removal and installation

Side stays

1 Remove the rear mounting screw from each side of the upper fairing (see illustration 3.6a).
2 Remove the screw from each end of the side stay. Remove the stay and separate the boss for the upper fairing mounting bolt from the rubber strip.
3 Installation is the reverse of removal.

Front stay

4 Refer to Chapter 3, Section 8, Step 10 for the front fairing stay removal procedure.

6 Knee grip covers (600 C and 750 models only) - removal and installation

1 On 600 models, if you're removing the left side knee grip cover, remove the screw from the fuel tap knob and pull the knob off.
2 Remove the seat.
3 Remove the screw from the rear edge of the knee grip cover, then carefully pull the cover off. The resistance that will be felt is the lug on

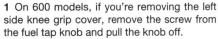

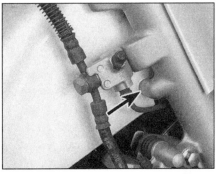

8.4 Remove the bolts from inside the fenders (arrow shows general location), right above the brake calipers (there's one bolt per side)

the front part of the cover pulling out of its rubber grommet.
4 Installation is the reverse of removal.

7 Side covers - removal and installation

1 Remove the seat.
2 Remove the side cover mounting screw (see illustration).
3 Carefully pull the side cover from the bike. There's a lug that fits into a rubber bushing on the fuel tank, so apply a little extra force there (be careful not to break it off, though).
4 Installation is the reverse of the removal procedure.

8 Front fender - removal and installation

1 Set the bike on its centrestand.
2 Disconnect the speedometer cable from the speedometer drive.
3 On 600 C and 750 models, cut the guides that secure the brake hoses and ESCS wires to the fender.
4 Remove the bolts from inside the fender (see illustration).
5 Remove the bolts from the fork brace (see illustration). Remove the brace, being careful

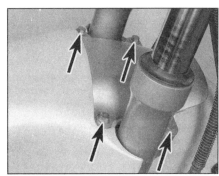

8.5 Remove the four bolts retaining the fork brace (arrows)

9.2 The tailpiece is secured to the frame by a screw on each side

9.4 The passenger seat is retained to the frame by four bolts (arrows)

10.5 Unplug the electrical connectors and remove the bolts (arrows)

not to lose the spacers that sit in the front two holes. Remove the fender toward the front.
6 Installation is the reverse of removal. On 600 C and 750 models, install new guides for the brake hoses and ESCS wires.

9 Tailpiece and passenger seat - removal and installation

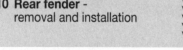

1 On 600 models, remove the rider's seat. On 750 models, remove the seat.
2 Remove the side covers (see Section 7). Remove the two screws securing the tailpiece to the frame **(see illustration)**.
3 Slide the tailpiece to the rear to remove it.
4 To remove the passenger seat on 600 models, remove the four bolts that secure the seat to the frame **(see illustration)**, then lift it off.
5 Installation is the reverse of removal.

10 Rear fender - removal and installation

600 A and B models

Rear section

1 Set the bike on its centrestand.
2 Remove the rider's seat, the tailpiece and the passenger seat (see Section 9).
3 Open the tool compartment and remove the mounting screw. Slide the compartment forward and remove it.
4 Remove the document container the same way. If you're working on a California model, remove the charcoal canister by unclipping the retaining strap.
5 Unplug the electrical connectors for the tail/brake and turn signal lights **(see illustration)**.
6 Remove the four mounting bolts and remove the rear section of the rear fender **(see illustration 10.5)**.
7 Installation is the reverse of the removal procedure.

Front section

8 Remove the rear section (see above).
9 Detach the starter relay from the fender (see Chapter 9).

10 Remove the bolt and detach the rear brake fluid reservoir from the fender.
11 Remove the mounting screws from inside the fender **(see illustration)**.
12 Slide the junction box out of its mount on the fender. Push the fender down toward the wheel, then remove it from the machine.
13 Installation is the reverse of removal. Be sure to route the brake fluid reservoir hose between the fender and the frame tube.

600 C models

Rear section

14 Remove the rider's seat and knee grip covers (see Section 6).
15 Remove the fuel tank (see Chapter 4) and side covers (see Section 7).
16 Remove the tailpiece and passenger seat (see Section 9).
17 Remove the tool kit compartment and document container by removing the screw inside the door of each one. If you're working on a California model, remove the charcoal canister.
18 Unplug the electrical connectors for the tail/brake, turn signal and license plate lights.
19 Remove the mounting bolts and remove the rear fender section **(see illustration)**.
20 Installation is the reverse of removal.

Front section

21 Remove the rear section (see above).
22 Detach the turn signal relay and the starter relay from the left side of the fender (see Chapter 9).
23 Detach the brake fluid reservoir and the air valve mounting bracket from the right side of the fender.

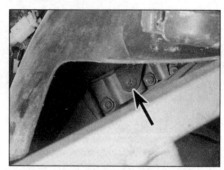

10.11 Location of the mounting screws for the fender front section (arrow)

24 Remove the battery (see Chapter 9).
25 Slide the junction box out of its mount (see Chapter 9).
26 Detach any breather hoses that might interfere with fender removal.
27 Slide the fender to the rear and remove it.
28 Installation is the reverse of the removal procedure.

750 models

29 Remove the seat.
30 Remove the knee grip covers and side covers on both sides of the bike.
31 Remove the cover from the air cleaner housing.
32 Remove the battery (see Chapter 1).
33 Disconnect the electrical connectors on the left side, on the right side, at the rear inside the tailpiece and just forward of the starter solenoid.
34 Unbolt the rear master cylinder (but leave the brake line connected).
35 Support the exhaust system and remove the muffler mounting bolts on both sides.
36 Unbolt the rear frame section and take it off together with the rear fender.
37 Unbolt the rear section of the rear fender from the front section and take it off.
38 Unbolt the front section of the rear fender from the rear frame section.
39 If necessary, detach the rear footpeg brackets from the rear frame section.
40 Installation is the reverse of the removal steps.

10.19 Fender rear section mounting details (C models)

1 Fender rear section   3 Turn signal
2 Mounting bolts   4 Licence plate light

Chapter 9
Electrical system

Contents

Degrees of difficulty

Easy, suitable for novice with little experience		Fairly easy, suitable for beginner with some experience		Fairly difficult, suitable for competent DIY mechanic		Difficult, suitable for experienced DIY mechanic		Very difficult, suitable for expert DIY or professional	

Specifications

Battery

Type	12 volt, 12 Ah (amp hours)
Specific gravity	
Fully charged	1.280 at 20° C (68° F)
Minimum	1.260 at 20° C (68° F)

Charging system - 600 models

Charging voltage	14.5 ± 0.5 volts at 4000 rpm
Stator coil resistance	0.1 to 0.8 ohms

Charging system - 750 models

Charging voltage	13.5 volts at 4000 rpm (minimum)
Stator coil resistance	Less than 1.0 ohm
Rotor coil resistance	Approx. 4 ohms
Slip ring diameter	
Standard	14.4 mm (0.567 in)
Minimum	14.0 mm (0.551 in)
Alternator brush length	
Standard	10.5 mm (0.413 in)
Minimum	4.5 mm (0.177 in)

Starter motor

Brush length
  Standard .... 12 mm (0.472 in)
  Minimum
    600 models .... 6 mm (0.236 in)
    750 models .... 8.5 mm (0.335 in)
Commutator diameter
  Standard .... 28 mm (1.1024 in)
  Minimum .... 27 mm (1.0630 in)

Fuse ratings

Taillight 10A
Accessory 10A
Headlight 10A
Fan (where fitted) 10A
Main fuse 30A

Torque specifications

Alternator rotor bolt - 600 models .... 145 Nm (110 ft-lbs)
Alternator stator bolts - 600 models .... 12 Nm (104 in-lbs)
Alternator cover bolts - 600 models .... 5 Nm (43 in-lbs)
Alternator mounting bolts - 750 models .... 39 Nm (29 ft-lbs)
Alternator belt cover bolts - 750 models .... 5 Nm (43 in-lbs)
Alternator pulley nut - 750 models .... 110 Nm (80 ft-lbs)
Neutral switch - all models .... 15 Nm (11 ft-lbs)

1 General information

The machines covered by this manual are equipped with a 12-volt electrical system. On 600 models, the charging system consists of a crankshaft mounted permanent magnet alternator and a solid state voltage regulator/rectifier unit. On 750 models, the external alternator houses separate regulator and rectifier units and is belt driven off the starter clutch shaft.

The regulator maintains the charging system output within the specified range to prevent overcharging. The rectifier converts the AC output of the alternator to DC current to power the lights and other components and to charge the battery.

The starting system includes the motor, the battery, the solenoid, the starter circuit relay (part of the junction box) and the various wires and switches. If the engine STOP switch and the main key switch are both in the On position, the circuit relay allows the starter motor to operate only if the transmission is in Neutral (Neutral switch on) or the clutch lever is pulled to the handlebar (clutch switch on) and the sidestand is up (sidestand switch on).

Note: *Keep in mind that electrical parts, once purchased, can't be returned. To avoid unnecessary expense, make very sure the faulty component has been positively identified before buying a replacement part.*

2 Electrical troubleshooting

1 A typical electrical circuit consists of an electrical component, the switches, relays, etc. related to that component and the wiring and connectors that hook the component to both the battery and the frame. To aid in locating a problem in any electrical circuit, complete wiring diagrams of each model are included at the end of this Chapter.

2 Before tackling any troublesome electrical circuit, first study the appropriate diagrams thoroughly to get a complete picture of what makes up that individual circuit. Trouble spots, for instance, can often be narrowed down by noting if other components related to that circuit are operating properly or not. If several components or circuits fail at one time, chances are the fault lies in the fuse or ground (earth) connection, as several circuits often are routed through the same fuse and ground (earth) connections.

3 Electrical problems often stem from simple causes, such as loose or corroded connections or a blown fuse. Prior to any electrical troubleshooting, always visually check the condition of the fuse, wires and connections in the problem circuit.

4 If testing instruments are going to be utilized, use the diagrams to plan where you will make the necessary connections in order to accurately pinpoint the trouble spot.

5 The basic tools needed for electrical troubleshooting include a battery and bulb test circuit, a continuity tester, test light and a jumper wire. For more extensive checks, a multimeter capable of measuring ohms, volts and amps will be required. Full details on the use of this test equipment are given in *Fault Finding Equipment* in the Reference section at the end of this manual.

3 Battery - inspection and maintenance

1 Most battery damage is caused by heat, vibration, and/or low electrolyte levels, so keep the battery securely mounted, check the electrolyte level frequently and make sure the charging system is functioning properly.

2 Refer to Chapter 1 for electrolyte level and specific gravity checking procedures.

3 Check around the base inside of the battery for sediment, which is the result of sulfation caused by low electrolyte levels. These deposits will cause internal short circuits, which can quickly discharge the battery. Look for cracks in the case and replace the battery if either of these conditions is found.

4 Check the battery terminals and cable ends for tightness and corrosion. If corrosion is evident, remove the cables from the battery and clean the terminals and cable ends with a wire brush or knife and emery paper. Reconnect the cables and apply a thin coat of petroleum jelly to the connections to slow further corrosion.

5 The battery case should be kept clean to prevent current leakage, which can discharge the battery over a period of time (especially when it sits unused). Wash the outside of the case with a solution of baking soda and water. Do not get any baking soda solution in the battery cells. Rinse the battery thoroughly, then dry it.

6 If acid has been spilled on the frame or battery box, neutralize it with the baking soda and water solution, dry it thoroughly, then touch up any damaged paint. Make sure the battery vent tube is directed away from the frame and is not kinked or pinched.

7 If the motorcycle sits unused for long periods of time, disconnect the cables from the battery terminals. Refer to Section 4 and charge the battery approximately once every month.

4 Battery - charging

1 If the machine sits idle for extended periods or if the charging system malfunctions, the battery can be charged from an external source.

2 To properly charge the battery, you will need a charger of the correct rating, an hydrometer, a clean rag and a syringe for adding distilled water to the battery cells.

3 The maximum charging rate for any battery is 1/10 of the rated amp/hour capacity. As an example, the maximum charging rate for a 12 amp/hour battery would be 1.2 amps. If the battery is charged at a higher rate, it could be damaged.

4 Do not allow the battery to be subjected to a so-called quick charge (high rate of charge over a short period of time) unless you are prepared to buy a new battery.

5 When charging the battery, always remove it from the machine and be sure to check the electrolyte level before hooking up the charger. Add distilled water to any cells that are low.

6 Loosen the cell caps, hook up the battery charger leads (red to positive, black to negative), cover the top of the battery with a clean rag, then, and only then, plug in the battery charger.

⚠️ **Warning: Remember, the gas escaping from a charging battery is explosive, so keep open flames and sparks well away from the area. Also, the electrolyte is extremely corrosive and will damage anything it comes in contact with.**

7 Allow the battery to charge until the specific gravity is as specified (refer to Chapter 1 for specific gravity checking procedures). The charger must be unplugged and disconnected from the battery when making specific gravity checks. If the battery overheats or gases excessively, the charging rate is too high. Either disconnect the charger or lower the charging rate to prevent damage to the battery.

8 If one or more of the cells do not show an increase in specific gravity after a long slow charge, or if the battery as a whole does not seem to want to take a charge, it is time for a new battery.

9 When the battery is fully charged, unplug the charger first, then disconnect the leads from the battery. Install the cell caps and wipe any electrolyte off the outside of the battery case.

5 Fuses - check and replacement

1 The fusebox is located under the seat on 600 models **(see illustration)** and under the left sidepanel on 750 models; the fusebox is part of the junction box. The fuses are protected by a plastic cover, which snaps into place. The main fuse (30A) is located in the fusebox on 600 A and B models. On 600 C models the main fuse is attached to the starter relay, and on 750 models it is located in a separate holder in the battery white/red wire.

2 If you have a test light, the fuses can be checked without removing them. Turn the ignition to the On position, connect one end of the test light to a good ground, then probe each terminal on top of the fuse **(see illustration)**. If the fuse is good, there will be voltage available at both terminals. If the fuse is blown, there will only be voltage present at one of the terminals.

3 The fuses can be removed and checked visually. If you can't pull the fuse out with your fingertips, use a pair of needle-nose pliers. A blown fuse is easily identified by a break in the element **(see illustration)**.

4 If a fuse blows, be sure to check the wiring harnesses very carefully for evidence of a short circuit. Look for bare wires and chafed, melted or burned insulation. If a fuse is replaced before the cause is located, the new fuse will blow immediately.

5 Never, under any circumstances, use a higher rated fuse or bridge the fuse block terminals, as damage to the electrical system could result.

6 Occasionally a fuse will blow or cause an open circuit for no obvious reason. Corrosion of the fuse ends and fuse block terminals may occur and cause poor fuse contact. If this happens, remove the corrosion with a wire brush or emery paper, then spray the fuse end and terminals with electrical contact cleaner.

6 Junction box - check

1 Aside from serving as the fuse block, the junction box also houses three relays - the fan relay, the starter circuit relay (not the starter solenoid) and the headlight relay. None of these relays are replaceable individually. If one of them fails, the junction box must be replaced.

2 In addition to the relay checks, the fuse circuits and diode circuits should be checked also, to rule out the possibility of an open circuit condition or blown diode within the

5.1 Unlatch the right side of the fuse cover and lift it up for access to the fuses

5.2 The fuses can be checked with a test light - with the ignition On, each fuse should have voltage available on both sides

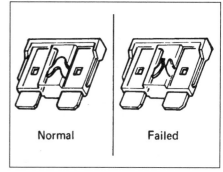

5.3 A blown fuse can be identified by a broken element - be sure to replace a blown fuse with one of the same amperage rating

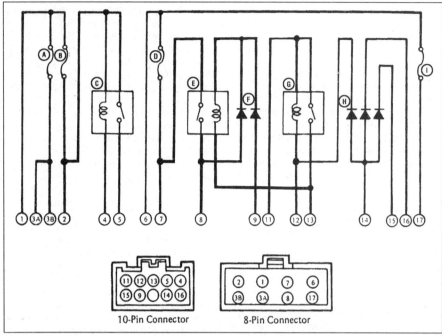

10-Pin Connector   8-Pin Connector

6.2a Junction box circuitry and terminal identification - 600 C models

A   Accessory fuse
B   Fan fuse
C   Fan relay - models up to frame number
    JKAZX4C1*KB505600 (US) or ZX600C-
    011126 (UK)
D   Headlight fuse

E   Headlight relay - US/Canada models
F   Diodes - US/Canada models
G   Starter circuit relay
H   Diodes
I   Taillight fuse

9-Pin Connector   8-Pin Connector

6.2b Junction box circuitry and terminal identification - 750 models

A   Accessory fuse
B   Fan fuse
C   Fan relay
D   Headlight fuse
E   Headlight relay - US/Canada models

F   Diodes - US/Canada models
G   Starter circuit relay
H   Diodes
I   Taillight fuse

junction block as the cause of an electrical problem **(see illustrations)**.

Fuse circuit check

3 Remove the junction box by sliding it out of its holder. Unplug the electrical connectors from the box.
4 If the terminals are dirty or bent, clean and straighten them. Using the accompanying table as a guide, check the continuity across the indicated terminals with an ohmmeter - some should have no resistance and others should have infinite resistance **(see illustration)**.
5 If the resistance values are not as specified, replace the junction box.

Diode circuit check

6 Remove the junction box by sliding it out of its holder. Unplug the electrical connectors from the box.
7 Using an ohmmeter, check the resistance across the following pairs of terminals, then write down the readings. Here are the terminal pairs to be checked:

13 and 8 (US and Canadian models only)
13 and 9 (US and Canadian models only)
12 and 14
15 and 14
16 and 14

8 Now, reverse the ohmmeter leads and check the resistances again, writing down the readings. The resistances should be low in one direction and more than ten times as much in the other direction. If the readings for any pair of terminals are low or high in both directions, a diode is defective and the junction box must be replaced.

Relay checks

9 Remove the junction box by sliding it out of its holder. Unplug the electrical connectors from the box.
10 Using an ohmmeter, check the conductivity across the terminals indicated in

Fuse Circuit Inspection	
Meter Connection	Meter Reading (Ω)
1 — 2	0
1 — 3A	0
6 — 7	0
6 — 17	0
1 — 7	∞
3A — 8	∞
8 — 17	∞

6.4 Using an ohmmeter, check the conductivity across the indicated pairs of terminals

	Meter Connection	Meter Reading (Ω)
Fan Relay	2 – 5	∞
	4 – 5	∞
Headlight Relay	*7 – 8	∞
	*7 – 13	∞
Starter Relay	11 – 13	∞
	12 – 13	∞

6.10a With the junction box unplugged, there should be infinite resistance between the indicated terminals

*US and Canadian models only

	Meter Connection	Battery Connection + –	Meter Reading (Ω)
Fan	2 – 5	2 – 4	0
Headlight	*7 – 8	9 – 13	0
Starter	11 – 13	11 – 12	0

6.10b Use jumper wires to connect battery voltage to the indicated terminals, then check the conductivity across the corresponding terminals - the resistance should be zero

*US and Canadian models only

the accompanying table (see illustration). Then, energize each relay by applying battery voltage across the indicated terminals and check the conductivity across the corresponding terminals shown on the table (see illustration).

11 If the junction box fails any of these tests, it must be replaced.

7 Lighting system - check

1 The battery provides power for operation of the headlight, taillight, brake light, license plate light and instrument cluster lights. If none of the lights operate, always check battery voltage before proceeding. Low battery voltage indicates either a faulty battery, low battery electrolyte level or a defective charging system. Refer to Chapter 1 for battery checks and Section 29 and 30 for charging system tests. Also, check the condition of the fuses and replace any blown fuses with new ones.

Headlight

2 If the headlight is out with the engine running (US and Canadian models) or with the lighting switch in the On position (UK models), check the fuse first with the key On (see Section 5), then unplug the electrical connector for the headlight (see illustration)

and use jumper wires to connect the bulb directly to the battery terminals. If the light comes on, the problem lies in the wiring or one of the switches in the circuit. Refer to Sections 20 and 21 for the switch testing procedures, and also the wiring diagrams at the end of this Chapter.

3 US and Canadian models (except the 600 C7, C8, C9 and C10) use an additional relay in the system, called the reserve lighting device. On these models, the headlight doesn't come on when the ignition switch is first turned on, but comes on when the starter button is pressed and stays on until the ignition is turned off. The light will go out whenever the starter is operated after the engine has stalled (this prevents excessive strain on the battery). The reserve lighting device is located behind the steering head, mounted to a bracket that is bolted to the frame (see illustration).

Taillight/licence plate light

4 If the taillight fails to work, check the bulbs and the bulb terminals first, then check for battery voltage at the red wire in the taillight. If voltage is present, check the ground (earth) circuit for an open or poor connection.

5 If no voltage is indicated, check the wiring between the taillight and the main (key) switch, then check the switch.

Brake light

6 See Section 14 for the brake light circuit checking procedure.

Neutral indicator light

7 If the neutral light fails to operate when the transmission is in Neutral, check the fuses and the bulb (see Section 18 for bulb removal procedures). If the bulb and fuses are in good condition, check for battery voltage at the green wire attached to the neutral switch on the left side of the engine. If battery voltage is present, refer to Section 23 for the neutral switch check and replacement procedures.

8 If no voltage is indicated, check the brown wire between the junction box and the bulb, and the light green wire between the junction box and the switch and between the switch and the bulb for open circuits and poor connections.

Oil pressure warning light

9 See Section 19 for the oil pressure warning light circuit check.

8 Headlight bulb - replacement

1 On 600 A and B models, remove the access cover from the underside of the upper fairing (see illustration). On 600 C and 750 models, remove the headlight lower cover screws and remove the cover (the piece that joins the two halves of the upper fairing together).

7.2 Unplug the headlight electrical connector and apply battery voltage to the red/yellow wire and ground (earth) the black/yellow wire - the light should come on

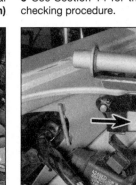

7.3 The reserve lighting device is attached to a bracket behind the steering head (arrow)

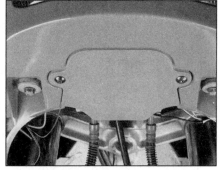

8.1 Remove the screws and detach the panel for access to the headlight bulb (600 A and B models)

8.2 The dust cover can be removed after the electrical connector is unplugged (fairing removed for clarity)

8.3 Unsnap the headlight retaining clip, then pull the bulb out of the headlight housing (fairing removed for clarity)

9.2 Remove the screws (arrows) and separate the headlight assembly from the fairing

2 Unplug the electrical connector from the headlight, then remove the dust cover **(see illustration)**.

3 Lift up the retaining clip and swing it out of the way **(see illustration)**. Remove the bulb.

4 When installing the new bulb, reverse the removal procedure.

Caution: Be sure not to touch the bulb with your fingers - oil from your skin will cause the bulb to overheat and fail prematurely. If you do touch the bulb, wipe it off with a clean rag dampened with rubbing alcohol.

5 The parking (or city) light on UK models is positioned in the base of the headlight unit.

10.4 The horizontal adjuster is located on the upper right corner of the headlight assembly - use a Phillips screwdriver (arrow), inserted from below

10.5 The vertical adjuster is located on the lower left corner of the headlight housing

Peel back the rubber dust cover and pull the bulbholder out of the grommet in the headlight. Twist the bulb counterclockwise to release it.

9 Headlight assembly - removal and installation

1 Remove the upper fairing (see Chapter 8).
2 Remove the screws holding the headlight assembly to the fairing **(see illustration)**. Separate the headlight assembly from the fairing.
3 Installation is the reverse of removal. Be sure to adjust the headlight aim (see Section 10).

10 Headlight aim - check and adjustment

1 An improperly adjusted headlight may cause problems for oncoming traffic or provide poor, unsafe illumination of the road ahead. Before adjusting the headlight, be sure to consult with local traffic laws and regulations. UK owners should refer to *MOT Test Checks* in the Reference section.
2 The headlight beam can be adjusted both vertically and horizontally. Before performing the adjustment, make sure the fuel tank has at

least a half tank of fuel, and have an assistant sit on the seat.

3 If you're working on a 600 A or B model, remove the right side inner panel from the upper fairing (see Chapter 8). Also, unbolting the instrument cluster makes it easier to see what you're doing (see Section 15).
4 Insert a Phillips screwdriver into the horizontal adjuster guide **(see illustration)**, then turn the adjuster as necessary to centre the beam.
5 To adjust the vertical position of the beam, insert the screwdriver into the vertical adjuster guide **(see illustration)** and turn the adjuster as necessary to raise or lower the beam.
6 Install the right side inner panel into the upper fairing, if you're working on a 600 A or B model. Install the instrument cluster if it was removed.

11 Turn signal and taillight bulbs - replacement

Turn signal bulbs

1 Bulb replacement for the turn signals is the same for the front and rear.
2 Remove the screw that holds the lens/reflector assembly to the turn signal housing **(see illustration)**. Pull out the lens/reflector assembly.
3 Remove the screws securing the lens to the reflector **(see illustration)**.

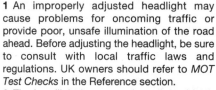
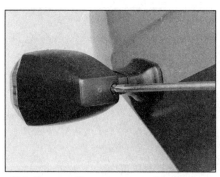

11.2 Remove the screw securing the turn signal lens/reflector assembly to the turn signal housing, then pull the lens/reflector assembly out

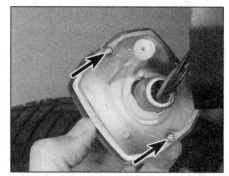

11.3 Remove these screws (arrows) to detach the lens from the reflector

11.8 Turn the bulb holder counterclockwise until it stops, then pull it out of the taillight housing

4 Push the bulb in and turn it counterclockwise to remove it. Check the socket terminals for corrosion and clean them if necessary. Line up the pins on the new bulb with the slots in the socket, push in and turn the bulb clockwise until it locks in place. **Note:** *The pins on the bulb are offset so it can only be installed one way. It is a good idea to use a paper towel or dry cloth when handling the new bulb to prevent injury if the bulb should break and to increase bulb life.*
5 Position the lens on the reflector and install the screws. Be careful not to overtighten them.
6 Place the lens/reflector assembly into the housing and install the screw, tightening it securely.

Taillight bulbs

7 To remove the taillight bulbs, remove the tail section and the passenger seat (see Chapter 8).
8 Turn the bulb holders counterclockwise **(see illustration)** until they stop, then pull straight out to remove them from the taillight housing. The bulbs can be removed from the holders by turning them counterclockwise and pulling straight out.
9 Check the socket terminals for corrosion and clean them if necessary. Line up the pins on the new bulb with the slots in the socket, push in and turn the bulb clockwise until it locks in place. **Note:** *The pins on the bulb are offset so it can only be installed one way. It is*

a good idea to use a paper towel or dry cloth when handling the new bulb to prevent injury if the bulb should break and to increase bulb life.
10 Make sure the rubber gaskets are in place and in good condition, then line up the tabs on the holder with the slots in the housing and push the holder into the mounting hole. Turn it clockwise until it stops to lock it in place. **Note:** *The tabs and slots are two different sizes so the holders can only be installed one way.*
11 Reinstall the seat and the tail section.

12 Turn signal assemblies - removal and installation

1 The turn signal assemblies can be removed individually in the event of damage or failure.
2 To remove a turn signal assembly, first follow the wiring harness from the turn signal to its electrical connectors. Mark the wires with pieces of numbered tape then unplug the electrical connectors.
3 Unscrew the nut that secures the turn signal to the fairing or rear fender **(see illustrations)**. If you're removing a rear turn signal, don't lose the spring plate.
4 Detach the turn signal from the fairing or fender. If you're installing a new turn signal, separate the stalk trim from the old stalk and transfer it to the new one.
5 Installation is the reverse of the removal procedure.

13 Turn signal circuit - check

1 The battery provides power for operation of the signal lights, so if they do not operate, always check the battery voltage and specific gravity first. Low battery voltage indicates either a faulty battery, low electrolyte level or a defective charging system. Refer to Chapter 1 for battery checks and Sections 29 and 30 for

charging system tests. Also, check the fuses (see Section 5).
2 Most turn signal problems are the result of a burned out bulb or corroded socket. This is especially true when the turn signals function properly in one direction, but fail to flash in the other direction. Check the bulbs and the sockets (see Section 11).
3 If the bulbs and sockets check out okay, check for power at the brown wire to the turn signal flasher **(see illustration)** with the ignition On. If there's no power at the brown wire, check the junction box (see Section 6).
4 If the junction box is okay, check the wiring between the turn signal flasher and the turn signal lights (see the wiring diagrams at the end of this Chapter).
5 If the wiring checks out okay, replace the turn signal flasher.

14 Brake light switches - check and replacement

Circuit check

1 Before checking any electrical circuit, check the fuses (see Section 5).
2 Using a test light connected to a good ground, check for voltage to the brown wire at the brake light switch. If there's no voltage present, check the brown wire between the switch and the junction box (see the wiring diagrams at the end of this Chapter).
3 If voltage is available, touch the probe of the test light to the other terminal of the switch, then pull the brake lever or depress the brake pedal - if the test light doesn't light up, replace the switch.
4 If the test light does light, check the wiring between the switch and the brake lights (see the wiring diagrams at the end of this Chapter).

Switch replacement

Brake lever switch

5 Unplug the electrical connectors from the switch.

12.3a Front turn signal nut (arrow)

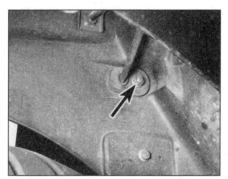
12.3b Rear turn signal nut (arrow)

13.3 Location of the turn signal flasher (arrow)

14.6 The brake light switch mounted on the brake lever is retained by a screw

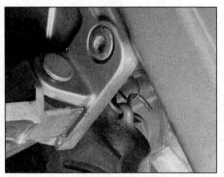

14.9 Unhook the spring from the brake pedal spring . . .

14.10 . . . then loosen the adjuster nut and unscrew the switch

6 Remove the mounting screw **(see illustration)** and detach the switch from the brake lever bracket/front master cylinder.
7 Installation is the reverse of the removal procedure. The brake lever switch isn't adjustable.

Brake pedal switch

8 Unplug the electrical connector in the switch harness.
9 Disconnect the spring from the brake pedal spring **(see illustration)**.
10 Loosen the adjuster nut **(see illustration)** and unscrew the switch.
11 Install the switch by reversing the removal procedure, then adjust by following the procedure described in Chapter 1, Section 6.

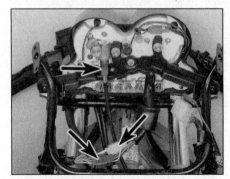

15.2 Unplug the electrical connectors in the cluster harness and unscrew the speedometer cable end from the speedometer (arrows) (fairing removed for clarity)

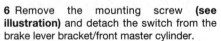

15 Instrument cluster -
removal and installation

1 If you're working on a 600 C model, remove the lower and upper fairings (see Chapter 8). This will also make removal of the instrument cluster easier on 600 A and B models, although it isn't absolutely necessary. On 750 models, remove the windshield and the upper fairing inner panels (see Chapter 8).
2 Detach the speedometer cable from the speedometer and unplug the electrical connectors from the cluster harness **(see illustration)**.
3 Remove the instrument cluster mounting bolts **(see illustration)** and detach the cluster from the upper fairing mount.
4 Installation is the reverse of the removal procedure.

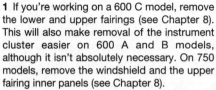

16 Meters and gauges -
check and replacement

Fuel gauge

1 To check the operation of the fuel gauge, unplug the electrical connector from the fuel level sending unit (see Chapter 4, Fuel tank - removal and installation).

2 Turn the ignition switch to the On position. The gauge should read Empty. Connect a jumper wire between the two pins in the gauge side of the connector (not the wires that lead back to the fuel tank). If the fuel level gauge is working properly it should read Full. Turn the ignition switch Off and reconnect the sending unit connector.
Caution: Don't leave the wire grounded (earthed) longer than necessary to perform this check. If you do, the gauge could be damaged. With the wire disconnected, the needle should fall to the empty mark.
3 If the gauge doesn't respond as described, either the wiring is defective or the gauge itself is malfunctioning. If the gauge does pass the above test, the fuel level sensor is defective (see Section 17).
4 If it's necessary to replace the gauge, remove the instrument cluster (see Section 15). Remove the fasteners that secure the cluster mounting bracket to the cluster **(see illustration)** and detach the bracket.
5 Remove the three screws that secure the instrument cluster cover **(see illustration)**. Detach the cover.
Caution: Always store the cluster with the gauges facing up or in a horizontal position - otherwise, the unit could be damaged.
6 Mark the positions of the wires and remove

15.3 The instrument cluster is retained by two bolts (arrows)

16.4 The instrument cluster mounting bracket is secured by two nuts and a screw (arrows) on A and B models (on C models it's retained by three nuts)

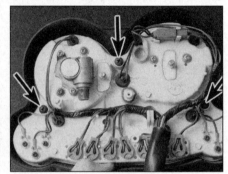

16.5 To remove the cover from the instrument cluster, remove these three screws and carefully lift the cover off

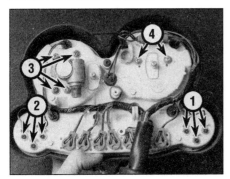

16.6 Instrument cluster details

1 *Fuel gauge screws*
2 *Temperature gauge screws*
3 *Speedometer screws*
4 *Tachometer screws*

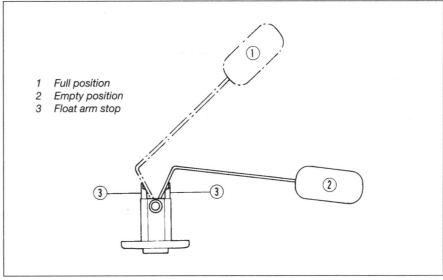

1 *Full position*
2 *Empty position*
3 *Float arm stop*

17.3 Fuel level sensor details

the small screws that secure the gauge to the cluster housing **(see illustration)**.

7 Detach the gauge from the housing, being careful not to disturb the other components.

8 Installation is the reverse of the removal procedure.

Temperature gauge

9 Refer to Chapter 3 for the temperature gauge checking procedure.

10 To replace the gauge, follow steps 4 through 8.

Tachometer and speedometer

11 Special instruments are required to properly check the operation of these meters. Take the instrument cluster to a Kawasaki dealer service department or other qualified repair shop for diagnosis.

12 The replacement procedure for either of these meters is essentially the same as the gauge replacement procedure. Follow Steps 4 through 8.

17 Fuel level sensor - check and replacement

Warning: Gasoline (petrol) is extremely flammable, so take extra precautions when you work on any part of the fuel system. Don't smoke or allow open flames or bare light bulbs near the work area, and don't work in a garage where a natural gas-type appliance (such as a water heater or clothes dryer) is present. If you spill any fuel on your skin, rinse it off immediately with soap and water. When you perform any kind of work on the fuel system, wear safety glasses and have a class B type fire extinguisher on hand.

1 Remove the fuel tank (see Chapter 4). Drain the fuel into an approved fuel container.

2 Unscrew the sending unit fasteners and remove the sending unit from the tank.

3 Using an ohmmeter, measure the resistance across the terminals of the sensor

electrical connector. With the float in the full position **(see illustration)**, the resistance should be 4 to 10 ohms (600 A and B models and 750 models) and 2 to 8 ohms (600 C models). With the float in the empty position, the reading should be 90 to 100 ohms.

4 If the sensor fails either test, replace it.

18 Instrument and warning light bulbs - replacement

1 Remove the instrument cluster (see Section 15).

2 To replace a bulb, pull the appropriate rubber socket out of the back of the instrument cluster housing **(see illustration)**, then pull the bulb out of the socket. If the socket contacts are dirty or corroded, they should be scraped clean and sprayed with electrical contact cleaner before new bulbs are installed.

3 Carefully push the new bulb into position, then push the socket into the cluster housing.

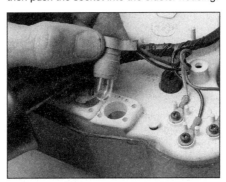

18.2 The bulb sockets are just pressed into the cluster housing - sometimes they're tough to remove and have to be pried out with a small screwdriver

19 Oil pressure sending unit - check and replacement

1 If the oil pressure warning light fails to operate properly, check the oil level and make sure it is correct.

2 If the oil level is correct, disconnect the wire from the oil pressure sending unit, which is located on the bottom of the engine. Turn the main switch On and ground (earth) the end of the wire **(see illustration)**. If the light comes on, the oil pressure sending unit is defective and must be replaced with a new one (only after draining the engine oil).

3 If the light does not come on, check the oil pressure warning light bulb, the wiring between the oil pressure sending unit and the light, and the light and the junction box (see the wiring diagrams at the end of this Chapter).

4 To replace the sending unit, drain the engine oil (see Chapter 1) and unscrew the sending unit from the case. Wrap the threads

19.2 When the ignition is On and the wire to the oil pressure sending unit is grounded (earthed), the warning light should come on

	BR	W	Y	BL	R	W/BK	O/G
OFF, LOCK							
ON	●━━━━●━━━━●			●━━━━●		●━━━━●	
P (Park)		●━━━━━━━━━━━●				●━━━━●	
						US, Canada	

20.2 Check the continuity of the ignition switch in the different switch positions across the indicated terminals

of the new sending unit with Teflon tape or apply a thin coat of sealant on them, then screw the unit into its hole, tightening it securely.

5 Fill the crankcase with the recommended type and amount of oil (see *Daily (pre-ride) checks*) and check for leaks.

20 Ignition main (key) switch - check and replacement

Check

1 Remove the upper fairing (see Chapter 8). This isn't absolutely necessary, but it sure makes access to the switch electrical connector easier.

2 Using an ohmmeter, check the continuity of the terminal pairs indicated in the accompanying table **(see illustration)**. Continuity should exist between the terminals connected by a solid line when the switch is in the indicated position.

3 If the switch fails any of the tests, replace it.

Replacement

4 Remove the upper fairing, if you haven't already done so (see Chapter 8).

5 Remove the instrument cluster (see Section 15).

20.7 The shear bolts (arrows) must be carefully drilled and removed with a screw extractor or knocked in a counterclockwise direction using a hammer and punch

6 Unplug the switch electrical connector.

7 The switch is held to the upper clamp with two shear-head bolts **(see illustration)**. Using a hammer and a sharp punch, knock the shear-head bolts in a counterclockwise direction to unscrew them. If they're too tight and won't turn, carefully drill holes through the centres of the bolts and unscrew them using a screw extractor. If necessary, remove the fairing mount for better access to the bolts. Detach the switch from the upper clamp.

8 Hold the new switch in position and install the new shear-head bolts. Tighten the bolts until the heads break off.

9 The remainder of installation is the reverse of the removal procedure.

> **HAYNES HINT** *Refer to Tools and Workshop Tips in the Reference section for details of how to use a screw extractor.*

21 Handlebar switches - check

1 Generally speaking, the switches are reliable and trouble-free. Most troubles, when they do occur, are caused by dirty or corroded contacts, but wear and breakage of internal parts is a possibility that should not be overlooked. If breakage does occur, the entire switch and related wiring harness will have to be replaced with a new one, since individual parts are not usually available.

2 The switches can be checked for continuity with an ohmmeter or a continuity test light. Always disconnect the battery ground (earth) cable, which will prevent the possibility of a short circuit, before making the checks.

3 Trace the wiring harness of the switch in question and unplug the electrical connectors.

4 Using the ohmmeter or test light, check for continuity between the terminals of the switch

harness with the switch in the various positions **(see illustration on opposite page)**. Continuity should exist between the terminals connected by a solid line when the switch is in the indicated position.

5 If the continuity check indicates a problem exists, refer to Section 22, disassemble the switch and spray the switch contacts with electrical contact cleaner. If they are accessible, the contacts can be scraped clean with a knife or polished with crocus cloth. If switch components are damaged or broken, it will be obvious when the switch is disassembled.

22 Handlebar switches - removal and installation

1 The handlebar switches are composed of two halves that clamp around the bars. They are easily removed for cleaning or inspection by taking out the clamp screws and pulling the switch halves away from the handlebars.

2 To completely remove the switches, the electrical connectors in the wiring harness should be unplugged. The right side switch must be separated from the throttle cables, also.

3 When installing the switches, make sure the wiring harnesses are properly routed to avoid pinching or stretching the wires.

23 Neutral switch - check and replacement

Check

1 Mark the position of the shift lever to the shift lever shaft **(see illustration)**. Remove the shift lever pinch bolt and slide the lever off the shaft.

23.1 Mark the relationship of the shift lever to the shift shaft so it can be installed in the same position

Starter Lockout Switch Connections

	BK/Y	Y/G (BK)	LG (BK/R)
When clutch lever is pulled in	●━━━━●		
When clutch lever is released		●━━━━━━●	

( ): Late Model

Engine Stop Switch Connections

	R	Y/R
OFF		
RUN	●━━●	

Dimmer Switch Connections (US, Canada)

	BL/Y	BL/O	R/Y	R/BK
HI	●━━━━━━━━━━━━━━━━●			
		●━━━━●		
LO	●━━━━●			
		●━━━━━━●		

Starter Button Connections

	BK (BK/R)	BK (BK/R)
Free		
Push on	●━━━━●	

( ): Late Model

Headlight switch connections - 600 UK models

	R/W	R/BL	BL	BL/Y
OFF				
■	●━━━●			
ON	●━━━●		●━━━●	

Dimmer Switch Connections (Other than US, Canada)

	R/BK	BL/Y	R/Y
HI	●━━━━●		
LO		●━━━━●	

Headlight switch connections - 750 UK models

	R/W	R/BL	BL	BL/Y
OFF				
■	●━━━●			
ON	●━━○━━○━━━━○			

Turn Signal Switch Connections

	GY	O	G
R	●━━━●		
N			
L		●━━━●	

Hazard Switch Connections

	GY	O	G
Off ▄			
On ▄	●━━○━━━●		

Front Brake Light Switch Connections

	BR (BK)	BL or BL/R (BK)
When brake lever is pulled in	●━━━━●	

( ): Late Model

Passing Button Connections (Other than US, Canada)

	BR	R/BK
Free		
Push on	●━━━━●	

Horn Button Connections

		BK/W	BK/Y
Free			
Push on		●━━━━●	

Continuity tables for the handlebar switches

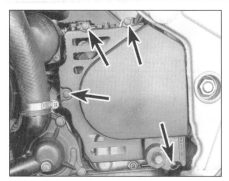

23.2 Remove the engine sprocket cover bolts and slide the cover off

2 Remove the bolts securing the engine sprocket cover to the engine case (see illustration). Slide the sprocket cover off.
3 Disconnect the wire from the neutral switch. Connect one lead of an ohmmeter to a good ground (earth) and the other lead to the post on the switch (see illustration).
4 When the transmission is in neutral, the ohmmeter should read 0 ohms - in any other gear, the meter should indicate infinite resistance.
5 If the switch doesn't check out as described, replace it.

Replacement

6 Unscrew the neutral switch from the case.
7 Wrap the threads of the new switch with Teflon tape or apply a thin coat of RTV sealant to them. Install the switch in the case and tighten it to the torque listed in this Chapter's Specifications.

24 Sidestand switch - check and replacement

Check

1 Following Steps 1 and 2 of Section 23, remove the engine sprocket cover.
2 Follow the wiring harness from the switch to the connector, then unplug the connector. Connect the leads of an ohmmeter to the brown and black/yellow wire terminals. With the sidestand in the UP position, there should be continuity through the switch (0 ohms).
3 Connect the leads of the ohmmeter to the black/yellow and green/white wires. With the sidestand in the DOWN position, the meter should indicate continuity.
4 If the switch fails either of these tests, replace it.

Replacement

5 Remove the engine sprocket cover, if you haven't already done so.
6 Unscrew the two countersunk Phillips head screws and remove the switch. Disconnect the switch electrical connector.
7 Installation is the reverse of the removal procedure.

23.3 If the switch is working properly, the meter should read 0 ohms when the transmission is in neutral

25 Horn - check, replacement and adjustment

Check

1 Remove the lower fairing.
2 Unplug the electrical connectors from the horn. Using two jumper wires, apply battery voltage directly to the terminals on the horn. If the horn sounds, check the switch (see Section 21) and the wiring between the switch and the horn (see the Wiring diagrams at the end of this Chapter).
3 If the horn doesn't sound, replace it. If it makes noise, but sounds "sick", try adjusting the tone as described below.

Replacement

4 Remove the lower fairing, if you haven't already done so.
5 Unbolt the horn bracket from the frame (see illustration) and detach the electrical connectors.
6 Unbolt the horn from the bracket and transfer the bracket to the new horn.
7 Installation is the reverse of removal.

Adjustment

8 Loosen the locknut on the adjustment screw (see illustration 25.5). Have an

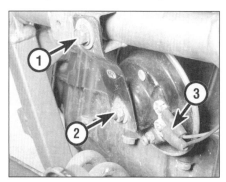

25.5 Horn installation details (radiator and exhaust system removed for clarity)

1 Horn bracket-to-frame bolt
2 Horn-to-bracket nut
3 Electrical connectors

assistant operate the horn. Turn the adjustment screw in or out until the tone is satisfactory. Tighten the locknut.

26 Starter solenoid - check and replacement

Check

1 On 600 models the solenoid is located under the left side cover. On 750 models the solenoid is situated behind the battery; remove the seat for access.
2 Disconnect the battery positive cable and the starter wire from the terminals on the starter solenoid (see illustration).
Caution: Don't let the battery positive cable make contact with anything, as it would be a direct short to ground.
3 Connect the leads of an ohmmeter to the terminals of the starter solenoid. Press the starter button - the solenoid should click and the ohmmeter should indicate 0 ohms.
4 If the solenoid clicks but the ohmmeter doesn't indicate zero ohms, replace the solenoid.
5 If the solenoid doesn't click, it may be defective or there may be a problem in the starter circuit. To determine which, disconnect the electrical connector from the solenoid and connect a voltmeter or 12-volt test lamp between the terminals of the black/yellow and yellow/red wires in the wiring harness. Press the starter button again - the voltmeter should indicate approximately 12 volts or the test lamp should light.
a) If the voltmeter indicates 12 volts or the test lamp lights, the circuit is good. Replace the solenoid.
b) If the voltmeter indicates no voltage or the test lamp stays out, check all wiring connections in the starter circuit (refer to the Wiring diagrams at the end of this Chapter). Also test the starter circuit relay in the junction box, the starter lockout switch, starter switch (button), engine stop switch and ignition switch.

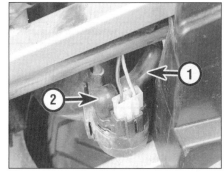

26.2 Starter solenoid mounting details

1 Battery positive cable
2 Cable to starter motor

27.5 Remove the nut that retains the starter wire (arrow) and detach the wire

27.6 Remove the starter mounting bolts (arrows)

27.7 Lift the left end of the starter up slightly, then pull the starter out

Replacement

6 Disconnect the cable from the negative terminal of the battery.

7 Detach the battery positive cable, the starter cable and two wire electrical connector from the solenoid **(see illustration 26.2)**.

8 Slide the solenoid off its mounting tabs.

9 Installation is the reverse of removal. Reconnect the negative battery cable after all the other electrical connections are made.

27 Starter motor -
removal and installation

Removal

1 Remove the lower fairing (see Chapter 8).

2 Disconnect the cable from the negative terminal of the battery.

3 Drain the engine oil and coolant (see Chapter 1).

4 Remove the coolant hose from the water pump to the lower coolant pipe. Remove the lower coolant pipe (see Chapter 3).

5 Remove the nut retaining the starter wire to the starter **(see illustration)**.

6 Remove the starter mounting bolts **(see illustration)**.

7 Lift the outer end of the starter up a little bit and slide the starter out of the engine case **(see illustration)**.

8 Check the condition of the O-ring on the end of the starter and replace it if necessary.

Installation

9 Remove any corrosion or dirt from the mounting lugs on the starter and the mounting points on the crankcase.

10 Apply a little engine oil to the O-ring and install the starter by reversing the removal procedure. Refill the cooling system and the engine crankcase following the procedures outlined in Chapter 1.

28 Starter motor - disassembly, inspection and reassembly

1 Remove the starter motor (see Section 27).

Disassembly

600 models

2 Mark the position of the housing to each end cover. Remove the two long screws and detach both end covers **(see illustration)**.

3 Pull the armature out of the housing (toward the pinion gear side).

4 Remove the brush plate from the housing **(see illustration)**.

5 Remove the nut and push the terminal bolt through the housing. Remove the two brushes

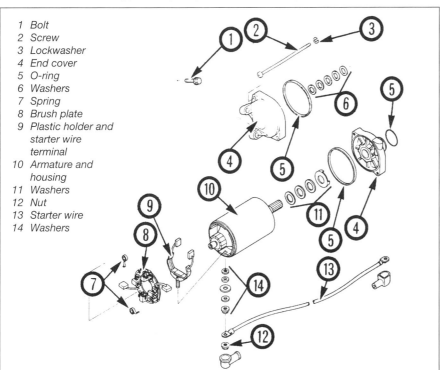

1 *Bolt*
2 *Screw*
3 *Lockwasher*
4 *End cover*
5 *O-ring*
6 *Washers*
7 *Spring*
8 *Brush plate*
9 *Plastic holder and starter wire terminal*
10 *Armature and housing*
11 *Washers*
12 *Nut*
13 *Starter wire*
14 *Washers*

28.2 Exploded view of the starter motor - 600 models

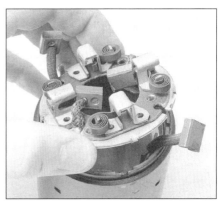

28.4 Remove the brush plate from the housing (four-brush starter shown; two-brush similar)

28.5 Push the terminal bolt through the housing and remove the plastic brush holder

with the plastic holder from the housing **(see illustration)**.

750 models

6 Mark the position of the housing to each end cover. Remove the two long screws and detach both end covers **(see illustration)**.
7 Pull the armature out of the housing.
8 Remove the brush plate from the housing. Disengage the brushes from the plate and detach the terminal bolt with its brush from the housing.

Inspection

9 The parts of the starter motor that most likely will require attention are the brushes.

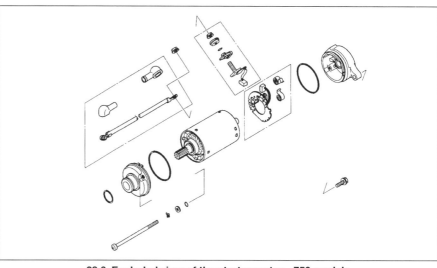

28.6 Exploded view of the starter motor - 750 models

Measure the length of the brushes and compare the results to the brush length listed in this Chapter's Specifications **(see illustration)**. If any of the brushes are worn beyond the specified limits, replace the brush holder assembly with a new one. If the brushes are not worn excessively, cracked, chipped, or otherwise damaged, they may be reused.
10 Inspect the commutator **(see illustration)**

for scoring, scratches and discoloration. The commutator can be cleaned and polished with crocus cloth, but do not use sandpaper or emery paper. After cleaning, wipe away any residue with a cloth soaked in an electrical system cleaner or denatured alcohol. Measure the commutator diameter and compare it to the diameter listed in this Chapter's Specifications. If it is less than the service limit, the motor must be replaced with a new one.
11 Using an ohmmeter or a continuity test light, check for continuity between the commutator bars **(see illustration)**. Continuity should exist between each bar and all of the others. Also, check for continuity between the commutator bars and the armature shaft **(see illustration)**. There should be no continuity between the commutator and the shaft. If the checks indicate otherwise, the armature is defective.
12 Check for continuity between the brush plate and the brushes **(see illustration)**. The meter should read close to 0 ohms. If it doesn't, the brush plate has an open and must be replaced.
13 Using the highest range on the ohmmeter, measure the resistance between the brush

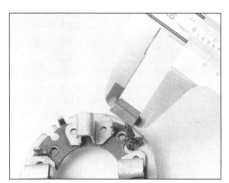

28.9 Measure the length of the brushes and compare the length of the shortest brush with the length listed in this Chapter's Specifications

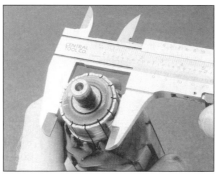

28.10 Check the commutator for cracks and discoloring, then measure the diameter and compare it with the minimum diameter listed in this Chapter's Specifications

28.11a Continuity should exist between the commutator bars

28.11b There should be no continuity between the commutator bars and the armature shaft

28.12 There should be almost no resistance (0 ohms) between the brushes and the brush plate

28.13 There should be no continuity between the brush plate and the brush holders (the resistance reading should be infinite)

holders and the brush plate **(see illustration)**. The reading should be infinite. If there is any reading at all, replace the brush plate.

14 Check the starter pinion gear for worn, cracked, chipped and broken teeth. If the gear is damaged or worn, replace the starter motor.

Reassembly

600 models

15 Install the plastic brush holder into the housing. Make sure the terminal bolt and washers are assembled correctly **(see illustration)**. Tighten the terminal nut securely.
16 Detach the brush springs from the brush plate (this will make armature installation much easier). Install the brush plate into the housing, routing the brush leads into the notches in the plate **(see illustration)**. Make sure the tongue on the brush plate fits into the notch in the housing.
17 Install the brushes into their holders and slide the armature into place. Install the brush springs **(see illustrations)**.
18 Install any washers that were present on the end of the armature shaft. Install the end covers, aligning the previously applied matchmarks. Install the two long screws and tighten them securely.

750 models

19 Assembly is the reverse of the disassembly steps, with the following additions:

28.17a Install each brush spring on the post in this position . . .

28.15 Install the washers on the starter terminal as shown

a) Be sure to install a large O-ring at each end of the starter housing and a small O-ring on the end that fits into the crankcase.
b) Align the mark on the starter housing with the notches in the brush cover and plate **(see illustration)**.
c) Align the cover screw hole with the marks on the starter housing.

29 Charging system testing - general information and precautions

1 If the performance of the charging system is suspect, the system as a whole should be checked first, followed by testing of the individual components (the alternator and the voltage regulator/rectifier). **Note:** Before beginning the checks, make sure the battery is fully charged and that all system connections are clean and tight.
2 Checking the output of the charging system and the performance of the various components within the charging system requires the use of special electrical test equipment. A voltmeter and ammeter or a multimeter are the absolute minimum tools required. In addition, an ohmmeter is generally required for checking the remainder of the system.

28.17b . . . then pull the end of the spring 1/2 turn clockwise and seat the end of it in the groove in the end of the brush

28.16 When installing the brush plate, make sure the brush leads fit into the notches in the plate (arrow) - also, make sure the tongue on the plate fits into the notch in the housing (arrows)

3 When making the checks, follow the procedures carefully to prevent incorrect connections or short circuits, as irreparable damage to electrical system components may result if short circuits occur. Because of the special tools and expertise required, it is recommended that the job of checking the charging system be left to a dealer service department or a reputable motorcycle repair shop.

30 Charging system - output test

Caution: Never disconnect the battery cables from the battery while the engine is running. If the battery is disconnected, the altemator and regulator/rectifier will be damaged.

All models

1 To check the charging system output, you will need a voltmeter or a multimeter with a voltmeter function.
2 The battery must be fully charged (charge it from an external source if necessary) and the engine must be at normal operating temperature to obtain an accurate reading.

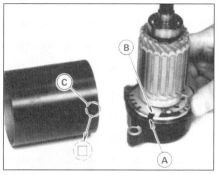

28.19 Align the notches in the cover and brush plate with the housing mark

A   Cover notch   C   Housing mark
B   Brush plate notch

3 Attach the positive (red) voltmeter lead to the positive (+) battery terminal and the negative (black) lead to the battery negative (-) terminal. The voltmeter selector switch (if so equipped) must be in a DC volt range greater than 15 volts.

4 Start the engine.

5 The charging system output should be within the specified range (see Specifications).

600 models

6 If the output is as specified, the alternator is functioning properly. If the charging system as a whole is not performing as it should, refer to Section 33 and check the voltage regulator/rectifier.

7 Low voltage output may be the result of damaged windings in the alternator stator coils, loss of magnetism in the alternator rotor or wiring problems. Make sure all electrical connections are clean and tight, then refer to Section 31 and check the alternator stator coil windings and leads for continuity.

750 models

8 If the output is as specified, the alternator is functioning properly. If the charging system as a whole is not performing as it should, refer to Step 34 and check the alternator brushes.

9 Low voltage output may be the result of damaged windings in the alternator stator coils or wiring problems. Make sure all electrical connections are clean and tight, then refer to Section 34 and inspect the brushes.

10 High voltage output (above the specified range) indicates a defective voltage regulator (see Section 34).

31 Alternator stator coil - continuity test (600 models)

1 If charging system output is low or non-existent, the alternator stator coil windings and leads should be checked for proper continuity. The test can be made with the stator in place on the machine.

2 To gain access to the stator coil wiring harness connector, remove engine sprocket cover (see Section 23).

3 Locate the stator coil electrical connector and unplug it (the connector contains three yellow wires on the chassis side of the harness and three black wires on the alternator side of the harness).

4 Using an ohmmeter (preferred) or a continuity test light, check for continuity between each of the wires coming from the alternator stator. Continuity should exist between any one wire and each of the others (Kawasaki actually specifies a resistance of 0.1 to 0.8 ohms).

5 Check for continuity between each of the wires and the engine. No continuity should exist between any of the wires and the case.

6 If there is no continuity between any two of the wires, or if there is continuity between the

wires and an engine ground, an open circuit or a short exists within the stator coils. Since repair of the stator is not feasible, it must be replaced with a new one.

32 Alternator - removal and installation (600 models)

Removal

1 Disconnect the cable from the negative terminal of the battery.

2 Remove the lower fairing.

3 Remove the alternator cover **(see illustration)**. Remove the engine sprocket cover (see Section 23, Steps 1 and 2).

4 Prevent the alternator rotor from turning by holding it with Kawasaki tool no. 57001-308 or a pin spanner wrench. Remove the rotor bolt **(see illustration)**.

5 Hold the rotor from turning again, and using tool no. 57001-1216 or equivalent, remove the rotor from the crankshaft **(see illustration)**.

6 To remove the stator coil, follow the wiring harness back from the stator and disconnect the electrical connector. Remove the three Allen bolts and detach the stator from the engine **(see illustration)**.

Installation

7 Position the stator coil on the engine and install the bolts, tightening them to the torque listed in this Chapter's Specifications. Install the wiring harness grommet in the slot in the case. Route the wiring harness into position, making sure it's secured by the clamp behind the water pump, then plug in the electrical connector.

8 Clean the end of the crankshaft, the rotor bolt, the threads in the crankshaft and the tapered portion of the rotor with an oil-less cleaning solvent (brake system cleaner works well, as it leaves no residue) **(see illustration)**.

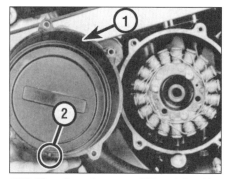

32.3 The alternator cover (1) is retained by four bolts - when installing it, make sure the notch (2) is at the bottom

32.4 A pin spanner wrench can be used to hold the rotor stationary while the bolt is loosened

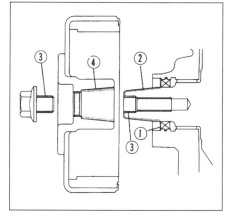

32.8 Clean the areas indicated before installing the alternator rotor - 600 models

1 *The surface of the crankshaft oil seal*
2 *The tapered portion of the crankshaft*
3 *The alternator rotor bolt and the threads in the crankshaft*
4 *The tapered portion of the rotor*

32.5 A special tool (A) is needed to pull the alternator rotor off - the rotor must be held stationary while doing this

32.6 The alternator stator is retained by three Allen-head bolts (arrows)

33.2 The voltage regulator/rectifier is held to its bracket by two bolts - 600 models

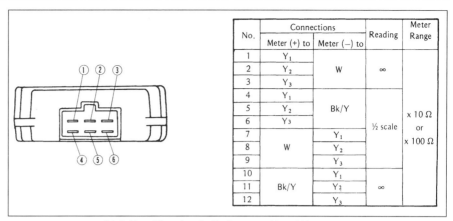

No.	Connections		Reading	Meter Range
	Meter (+) to	Meter (−) to		
1	Y₁			
2	Y₂	W	∞	
3	Y₃			
4	Y₁			x 10 Ω
5	Y₂	Bk/Y		or
6	Y₃		½ scale	x 100 Ω
7		Y₁		
8	W	Y₂		
9		Y₃		
10		Y₁		
11	Bk/Y	Y₂	∞	
12		Y₃		

33.3 Rectifier circuit diode test table - 600 models

9 Install the rotor and the bolt. Prevent the rotor from turning using the method described in Step 4, and tighten the rotor bolt to the torque listed in this Chapter's Specifications.

10 Install the alternator cover, making sure the notch in the edge of the cover is at the bottom (it serves as a drain for moisture).

11 Install the engine sprocket cover.

12 Install the lower fairing and connect the cable to the negative terminal of the battery.

33 Voltage regulator/rectifier - check and replacement (600 models)

1 Remove the rider's seat and the left side cover.

2 Remove the two bolts securing the regulator/rectifier to its bracket, then detach the electrical connector **(see illustration)**.

3 Using an ohmmeter, check the resistance across the terminals indicated in the accompanying table **(see illustration)**. If the meter readings are not as specified, replace the regulator.

4 This check, combined with the charging system output test described in Section 30 and the alternator stator coil test outlined in Section 31, should diagnose most charging system problems. If the voltage regulator/rectifier passes the test described in Step 3, and the stator coil passes the test in Section 31, take the regulator/rectifier to a dealer service department or other repair shop for further checks, or substitute a known good unit and recheck the charging system output.

34 Alternator/regulator/rectifier - removal, inspection and installation (750 models)

Removal

1 Disconnect the cable from the negative terminal of the battery.

2 Remove the left side cover and clutch slave cylinder (see Chapter 2B).

3 Make a written note of which mark on the alternator scale aligns with the pointer **(see illustration 25.2 in Chapter 1)**, then remove the alternator bracket and engine sprocket cover.

4 Unplug the alternator electrical connector.

5 Remove the alternator mounting bolts and lift the alternator off the engine. Disengage the alternator pulley from the drive belt and take the alternator off **(see illustration)**.

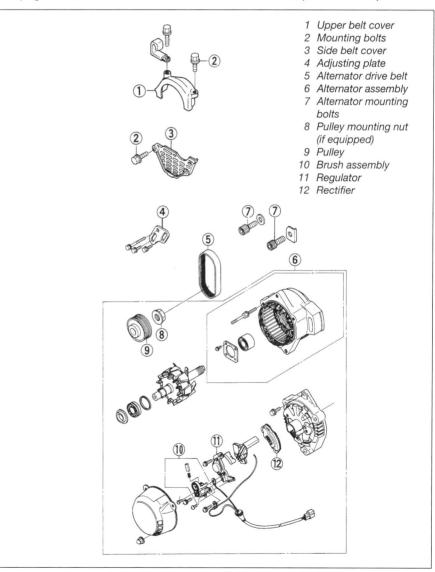

1 Upper belt cover
2 Mounting bolts
3 Side belt cover
4 Adjusting plate
5 Alternator drive belt
6 Alternator assembly
7 Alternator mounting bolts
8 Pulley mounting nut (if equipped)
9 Pulley
10 Brush assembly
11 Regulator
12 Rectifier

34.5 Alternator exploded view - 750 models

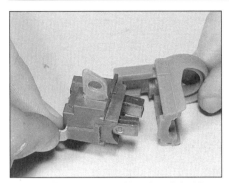

34.7a Remove the rubber brush holder cover . . .

34.7b . . . and measure projected length of each brush

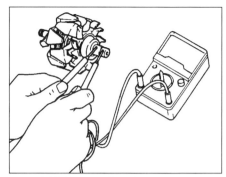

34.9 Rotor coil resistance test

34.10a Regulator is retained by two screws (arrows)

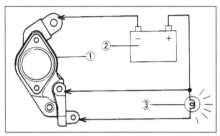

34.10b Regulator test with single battery

1 Regulator   3 Illuminated bulb
2 Battery

34.11 Regulator test with two batteries

1 Regulator   3 Extinguished bulb
2 Batteries

Inspection

Brushes and slip rings

6 Unscrew the three nuts securing the alternator end cover and remove the cover.
7 Remove the three screws securing the brush holder to the regulator and rectifier and remove the holder, noting the position of the wire **(see illustration)**. Inspect the holder for any signs of damage. Measure the brush lengths and compare the measurements with the figures given in the Specifications at the beginning of the Chapter **(see illustration)**.
8 Clean the slip rings with a rag moistened with some solvent. Slip ring diameter can be measured and compared with the figures given in the Specifications at the beginning of the Chapter, although access is difficult with the rotor installed in the alternator body.
9 Check the rotor coil resistance by measuring between the slip rings **(see illustration)**. There should be continuity (low resistance - about 4 ohms). If there is no continuity (infinite resistance), replace the rotor.

Regulator

10 Remove the two screws securing the regulator to the alternator and remove the regulator **(see illustration)**. Testing the regulator requires two fully-charged 12 volt batteries, a 12 V 3.4 W bulb and connecting wires. Identify the regulator terminals and set up the circuit as shown **(see illustration)**, taking great care not to allow either wire to

contact the regulator's metal case. The bulb should illuminate.
11 Replace the one battery with two batteries joined together in series as shown **(see illustration)** and repeat the test. In this case, the bulb should not light up.
12 If either circuit fails to produce the required results, the regulator is defective and must be renewed.

Rectifier

13 To test the rectifier it is necessary to unsolder the rectifier connections and remove it from the alternator **(see illustration)**. *Caution: The rectifier wires must be unsoldered quickly, otherwise the heat from the soldering iron will damage the rectifier diodes.*
14 With the rectifier removed from the alternator, use a multimeter set to the ohms x 1 scale to check its six diodes **(see**

34.13 Unsolder terminal wires quickly to avoid damaging the rectifier

illustration). Each diode is checked in both directions by reversing the meter probes. Continuity should only exist in one direction only; if no continuity is shown in both directions, or continuity is shown if both directions, the diode is fault. When installing the rectifier, solder its connections quickly to avoid damage (see **Caution**).

B to P1	P1 to B
B to P2	P2 to B
B to P3	P3 to B
E to P1	P1 to E
E to P2	P2 to E
E to P3	P3 to E

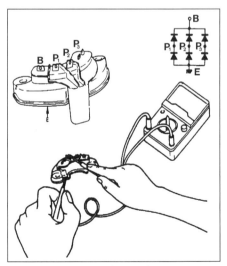

34.14 Rectifier diode test

Alternator bearings and further dismantling

15 Any further testing or dismantling of the alternator assembly must be carried out by an electrical specialist or Kawasaki dealer.

Installation

16 Installation is the reverse of the removal steps, noting that the previously noted alignment marks must be matched and the belt tension set (see Chapter 1). Tighten the alternator mounting bolts to the torque listed in this Chapter's Specifications.

35 Wiring diagrams

1 Prior to troubleshooting a circuit, check the fuses to make sure they're in good condition. Make sure the battery is fully charged and check the cable connections.

2 When checking a circuit, make sure all connectors are clean, with no broken or loose terminals or wires. When unplugging a connector, don't pull on the wires - pull only on the connector housings themselves.

Wiring diagrams commence overleaf

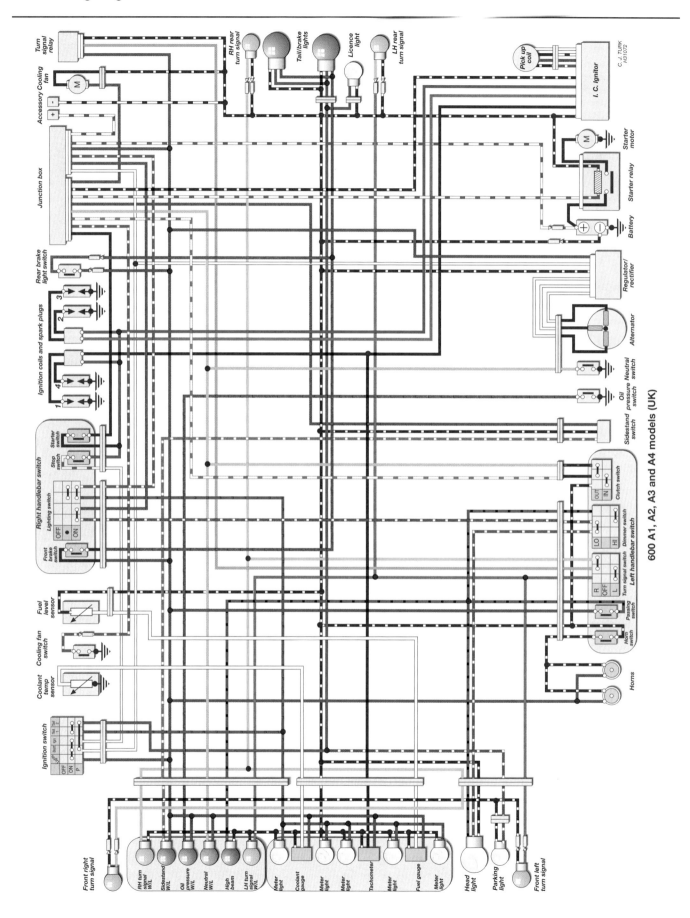

600 A1, A2, A3 and A4 models (UK)

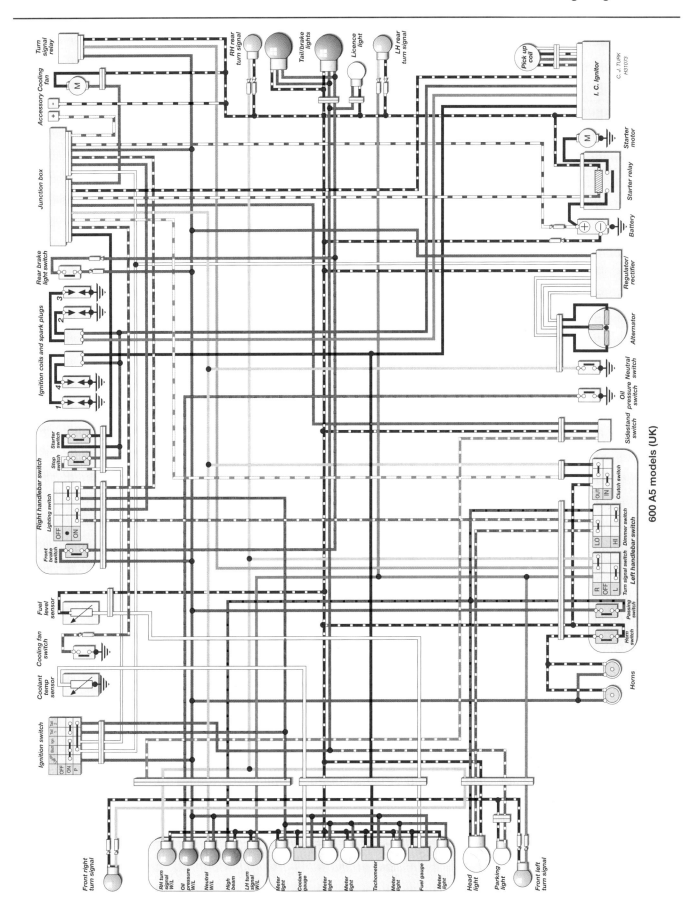

600 A5 models (UK)

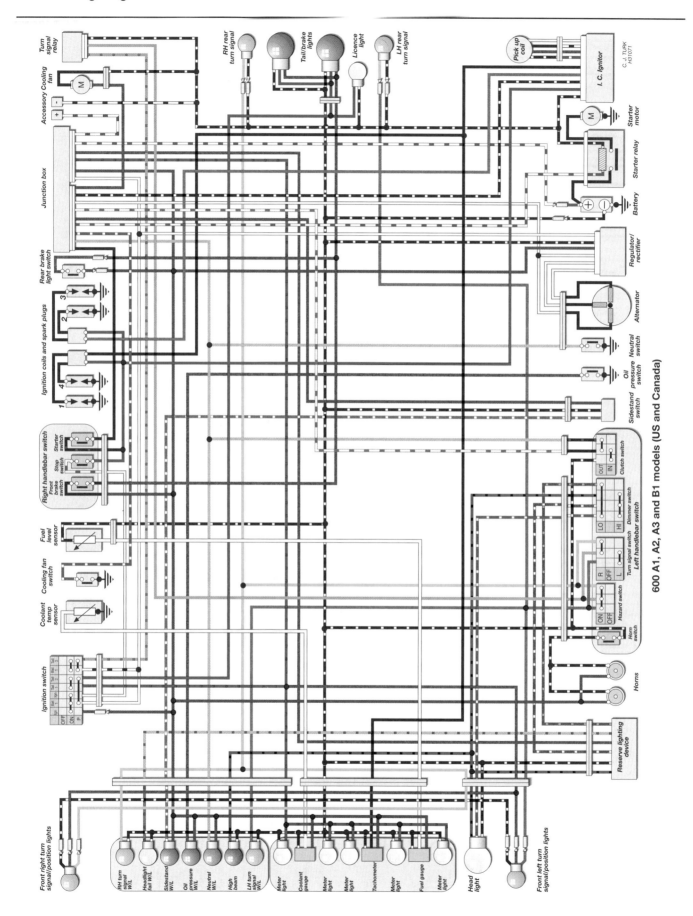

600 A1, A2, A3 and B1 models (US and Canada)

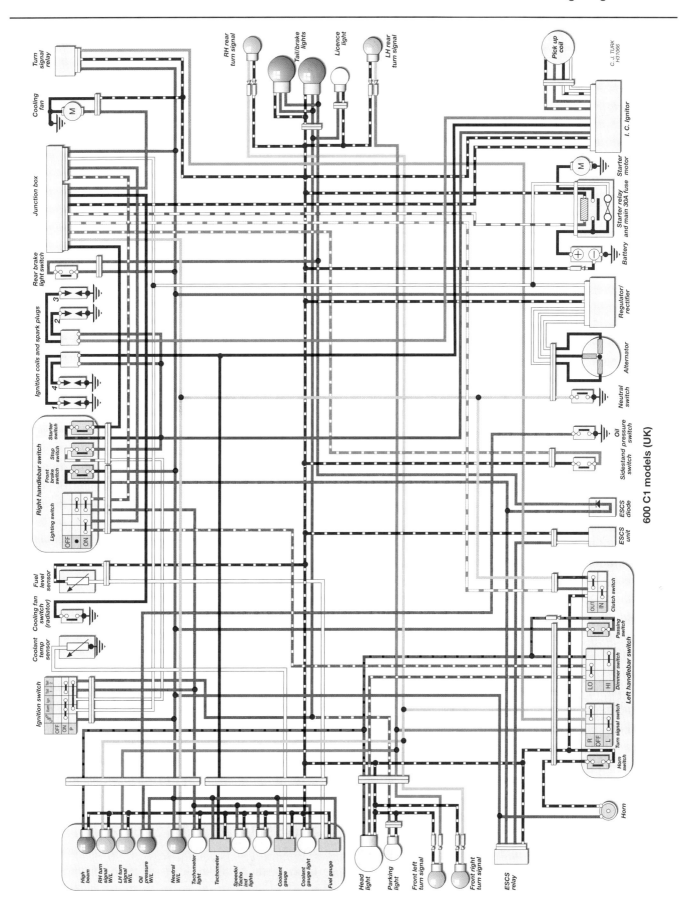

Turn signal relay

Cooling fan

Junction box

Rear brake light switch

Ignition coils and spark plugs

3
2
4
1

Right handlebar switch

Starter switch

Stop switch

Front brake switch

Lighting switch

OFF   ON

Fuel level sensor

Cooling fan switch (radiator)

Coolant temp sensor

Ignition switch

Off   Back   Ign   Tail 1   Tail 2   Dim   1   2

OFF
ON
P

High beam

RH turn signal W/L

LH turn signal W/L

Oil pressure W/L

Neutral W/L

Tachometer light

Tachometer

Speedo/ Tacho ind lights

Coolant gauge

Coolant gauge light

Fuel gauge

RH rear turn signal

Tail/brake lights

Licence light

LH rear turn signal

Head light

Parking light

Front left turn signal

Front right turn signal

ESCS relay

Pick up coil

I.C. Ignitor

C. J. TURK
H31066

Starter motor

Starter relay and main 30A fuse

Battery

Regulator/ rectifier

Alternator

Neutral switch

Oil pressure switch

Sidestand switch

ESCS diode

ESCS unit

Clutch switch

OUT   IN

Passing switch

Dimmer switch

LO   HI

Left handlebar switch

Turn signal switch

R   OFF   L

Horn switch

Horn

600 C1 models (UK)

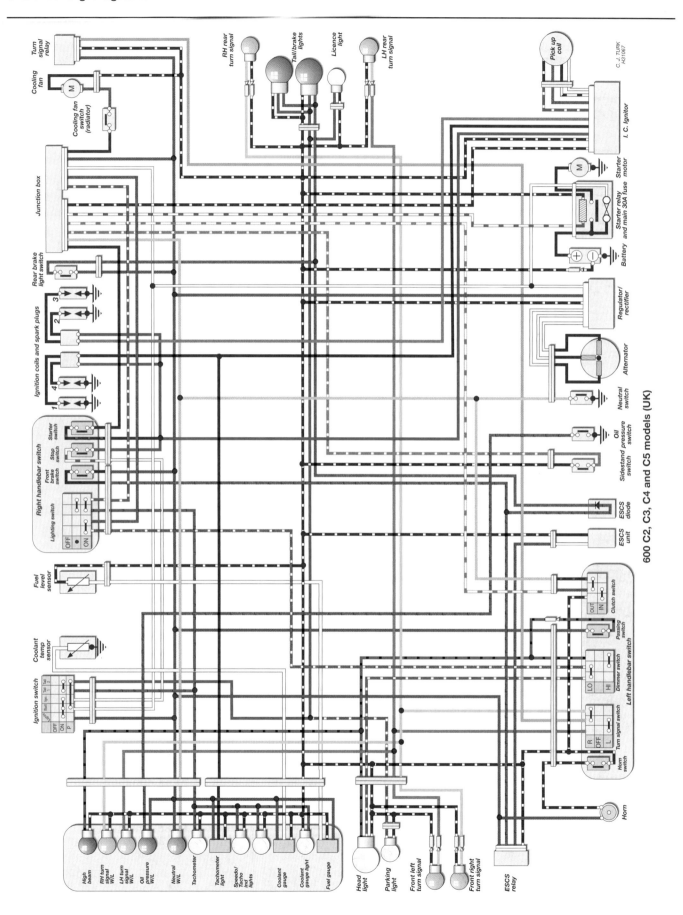

600 C2, C3, C4 and C5 models (UK)

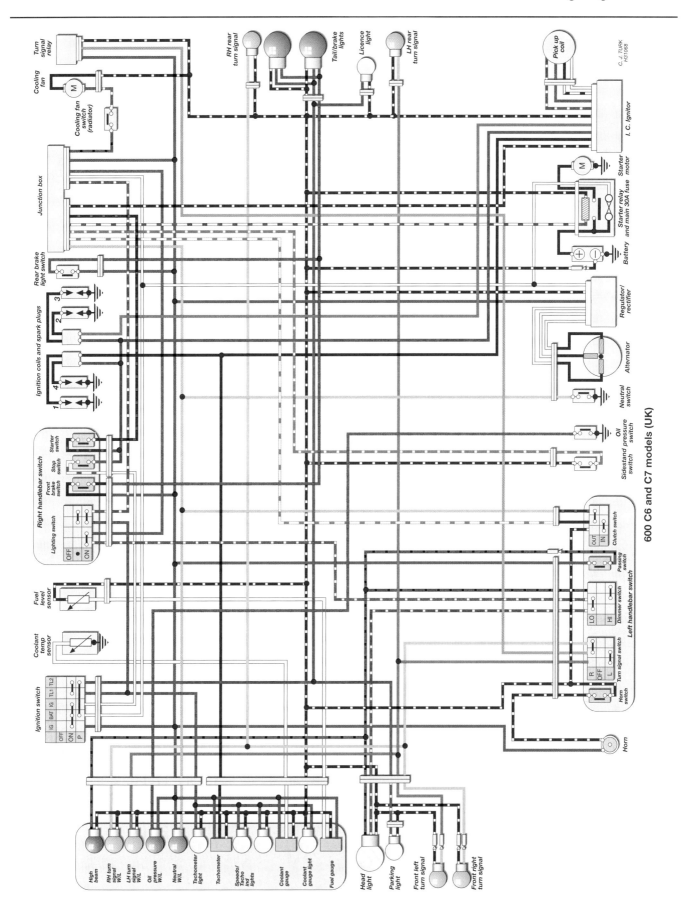

600 C6 and C7 models (UK)

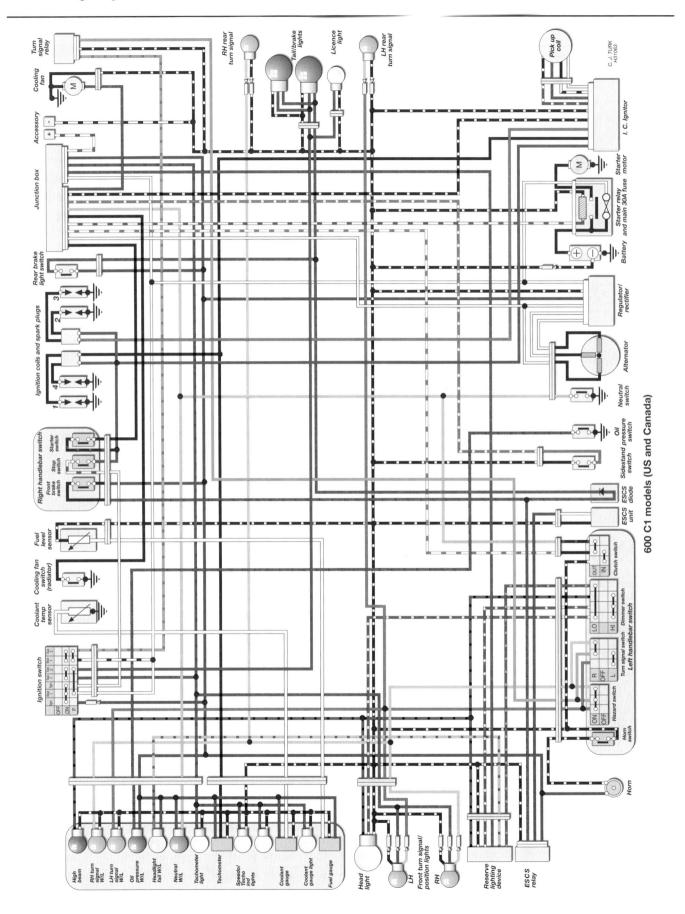

600 C1 models (US and Canada)

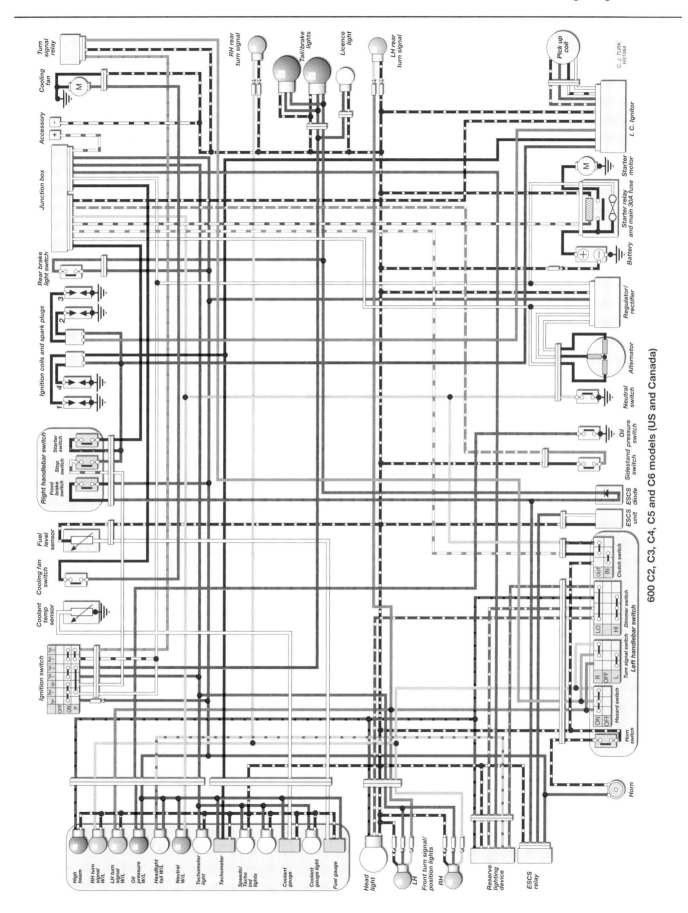

600 C2, C3, C4, C5 and C6 models (US and Canada)

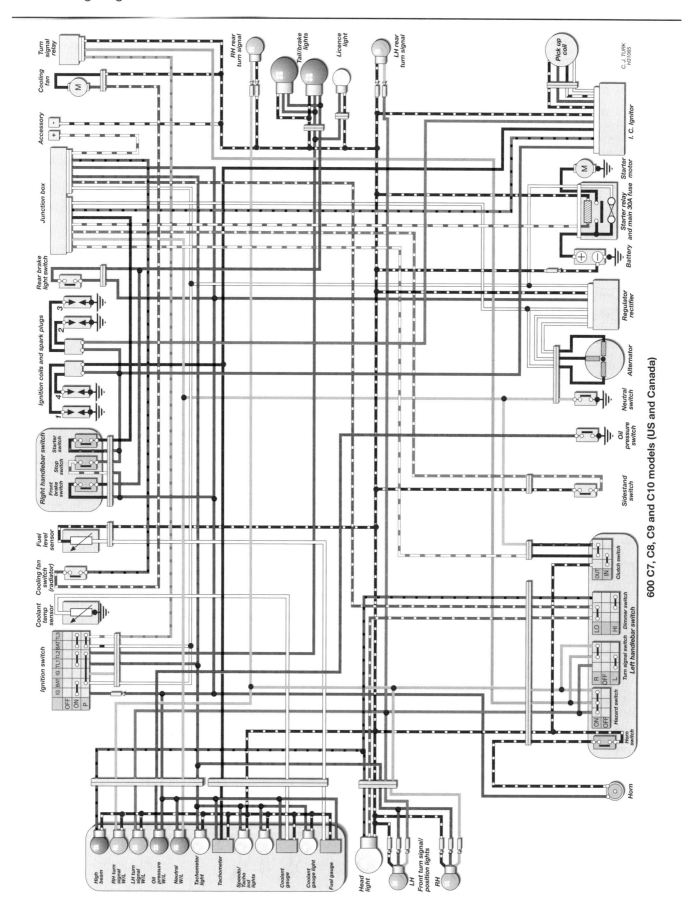

600 C7, C8, C9 and C10 models (US and Canada)

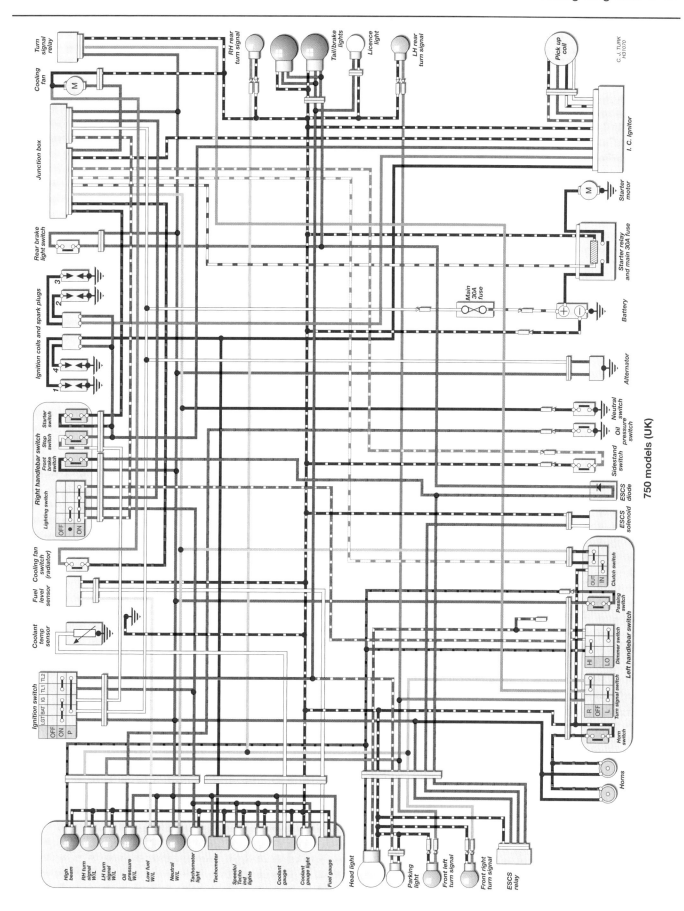

750 models (UK)

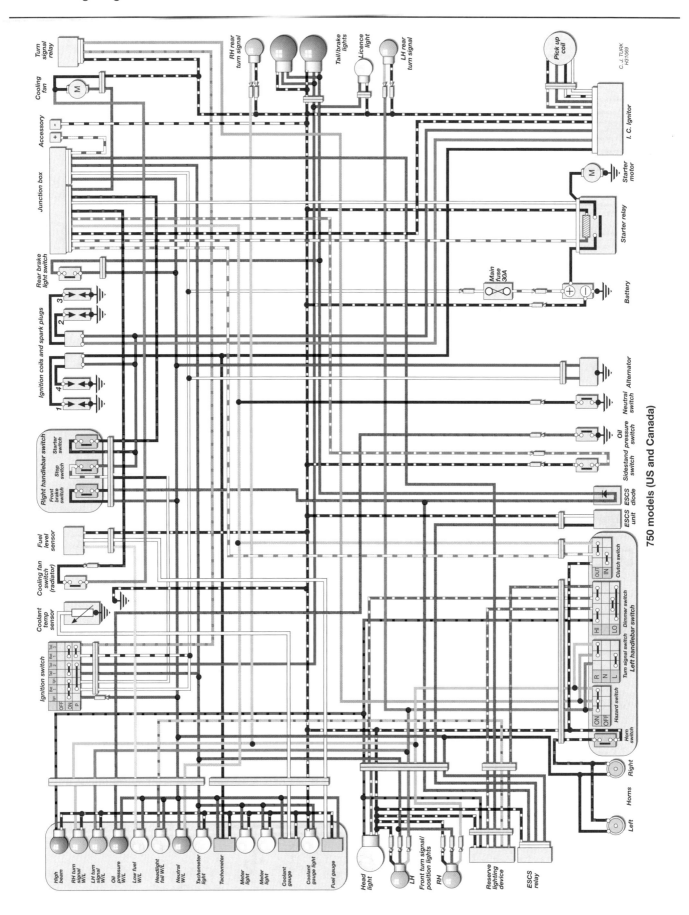

750 models (US and Canada)

Buying tools

A toolkit is a fundamental requirement for servicing and repairing a motorcycle. Although there will be an initial expense in building up enough tools for servicing, this will soon be offset by the savings made by doing the job yourself. As experience and confidence grow, additional tools can be added to enable the repair and overhaul of the motorcycle. Many of the specialist tools are expensive and not often used so it may be preferable to hire them, or for a group of friends or motorcycle club to join in the purchase.

As a rule, it is better to buy more expensive, good quality tools. Cheaper tools are likely to wear out faster and need to be renewed more often, nullifying the original saving.

 Warning: To avoid the risk of a poor quality tool breaking in use, causing injury or damage to the component being worked on, always aim to purchase tools which meet the relevant national safety standards.

The following lists of tools do not represent the manufacturer's service tools, but serve as a guide to help the owner decide which tools are needed for this level of work. In addition, items such as an electric drill, hacksaw, files, hammers, soldering iron and a workbench equipped with a vice, may be needed. Although not classed as tools, a selection of bolts, screws, nuts, washers and pieces of tubing always come in useful.

For more information about tools, refer to the Haynes *Motorcycle Workshop Practice Manual* (Bk. No. 1454).

Manufacturer's service tools

Inevitably certain tasks require the use of a service tool. Where possible an alternative tool or method of approach is recommended, but sometimes there is no option if personal injury or damage to the component is to be avoided. Where required, service tools are referred to in the relevant procedure.

Service tools can usually only be purchased from a motorcycle dealer and are identified by a part number. Some of the commonly-used tools, such as rotor pullers, are available in aftermarket form from mail-order motorcycle tool and accessory suppliers.

Maintenance and minor repair tools

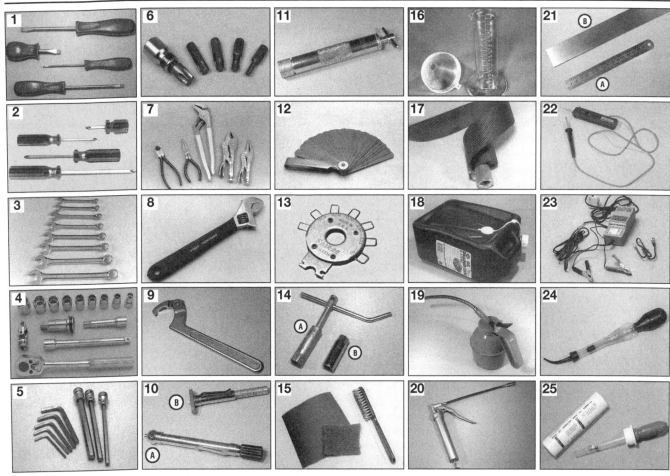

1 Set of flat-bladed screwdrivers	6 Set of Torx keys or bits	11 Cable pressure oiler	16 Funnel and measuring vessel	21 Steel rule (A) and straight-edge (B)
2 Set of Phillips head screwdrivers	7 Pliers and self-locking grips (Mole grips)	12 Feeler gauges	17 Strap wrench, chain wrench or oil filter	22 Continuity tester
3 Combination open-end & ring spanners	8 Adjustable spanner	13 Spark plug gap measuring and adjusting tool	removal tool	23 Battery charger
4 Socket set (3/8 inch or 1/2 inch drive)	9 C-spanner (ideally adjustable type)	14 Spark plug spanner (A) or deep plug socket (B)	18 Oil drainer can or tray	24 Hydrometer (for battery specific gravity check)
5 Set of Allen keys or bits	10 Tyre pressure gauge (A) & tread depth gauge (B)	15 Wire brush and emery paper	19 Pump type oil can	25 Anti-freeze tester (for liquid-cooled engines)
			20 Grease gun	

Dimensions and Weights

Wheelbase
600 A and B models . . . . .1430 mm (56.3 in)
600 C models . . . . .1425 mm (56.1 in)
750 models . . . . .1460 mm (57.5 in)

Overall length
600 A and B models . . . . .2140 mm (84.3 in)
600 C models . . . . .2095 mm (82.5 in)
750 models
   US . . . . .2115 mm (83.3 in)
   UK . . . . .2170 mm (85.4 in)

Overall width
600 A and B models . . . . .670 mm (26.3 in)
600 C models . . . . .690 mm (27.1 in)
750 models . . . . .715 mm (28.2 in)

Overall height
600 A and B models and 750 models . . . . .1185 mm (46.6 in)
600 C models . . . . .1150 mm (45.3 in)

Seat height
600 A and B models . . . . .770 mm (30.3 in)
600 C models . . . . .755 mm (29.7 in)
750 models . . . . .775 mm (30.5 in)

Dry weight
600 A models and 750 models . . . . .195 kg (430 lbs)
600 B model . . . . .191 kg (420 lbs)
600 C models . . . . .180 kg (397 lbs)

Repair and overhaul tools

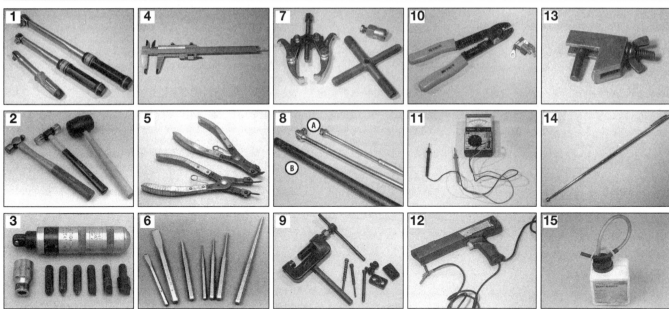

1 Torque wrench
  (small and mid-ranges)
2 Conventional, plastic or
  soft-faced hammers
3 Impact driver set

4 Vernier gauge
5 Circlip pliers (internal and
  external, or combination)
6 Set of punches
  and cold chisels

7 Selection of pullers
8 Breaker bars (A)
  and length of tubing (B)
9 Chain breaking/
  riveting tool

10 Wire crimper tool
11 Multimeter (measures
   amps, volts and ohms)
12 Stroboscope (for
   dynamic timing checks)

13 Hose clamp
   (wingnut type shown)
14 Magnetic arm
   (telescopic type shown)
15 One-man brake/clutch
   bleeder kit

Specialist tools

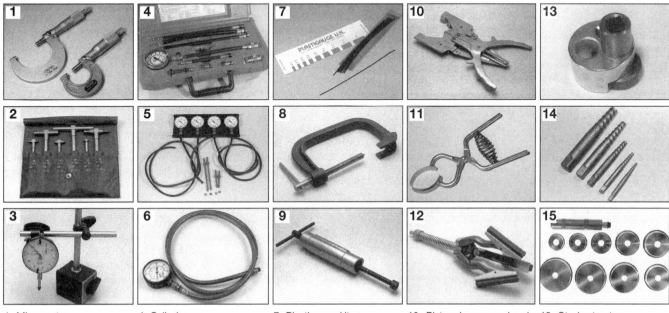

1 Micrometer
  (external type)
2 Telescoping gauges or
  small-hole gauges
3 Dial gauge

4 Cylinder
  compression gauge
5 Vacuum gauges (shown)
  or manometer
6 Oil pressure gauge

7 Plastigauge kit
8 Valve spring compressor
  (4-stroke engines)
9 Piston pin drawbolt tool

10 Piston ring removal and
   installation tool
11 Piston ring clamp
12 Cylinder bore hone
   (stone type shown)

13 Stud extractor
14 Screw extractor set
15 Bearing driver set

1 Workshop equipment and facilities

The workbench

● Work is made much easier by raising the bike up on a ramp - components are much more accessible if raised to waist level. The hydraulic or pneumatic types seen in the dealer's workshop are a sound investment if you undertake a lot of repairs or overhauls (see illustration 1.1).

1.1 Hydraulic motorcycle ramp

● If raised off ground level, the bike must be supported on the ramp to avoid it falling. Most ramps incorporate a front wheel locating clamp which can be adjusted to suit different diameter wheels. When tightening the clamp, take care not to mark the wheel rim or damage the tyre - use wood blocks on each side to prevent this.
● Secure the bike to the ramp using tie-downs (see illustration 1.2). If the bike has only a sidestand, and hence leans at a dangerous angle when raised, support the bike on an auxiliary stand.

1.2 Tie-downs are used around the passenger footrests to secure the bike

● Auxiliary (paddock) stands are widely available from mail order companies or motorcycle dealers and attach either to the wheel axle or swingarm pivot (see illustration 1.3). If the motorcycle has a centrestand, you can support it under the crankcase to prevent it toppling whilst either wheel is removed (see illustration 1.4).

1.3 This auxiliary stand attaches to the swingarm pivot

1.4 Always use a block of wood between the engine and jack head when supporting the engine in this way

Fumes and fire

● Refer to the Safety first! page at the beginning of the manual for full details. Make sure your workshop is equipped with a fire extinguisher suitable for fuel-related fires (Class B fire - flammable liquids) - it is not sufficient to have a water-filled extinguisher.
● Always ensure adequate ventilation is available. Unless an exhaust gas extraction system is available for use, ensure that the engine is run outside of the workshop.
● If working on the fuel system, make sure the workshop is ventilated to avoid a build-up of fumes. This applies equally to fume build-up when charging a battery. Do not smoke or allow anyone else to smoke in the workshop.

Fluids

● If you need to drain fuel from the tank, store it in an approved container marked as suitable for the storage of petrol (gasoline) (see illustration 1.5). Do not store fuel in glass jars or bottles.

1.5 Use an approved can only for storing petrol (gasoline)

● Use proprietary engine degreasers or solvents which have a high flash-point, such as paraffin (kerosene), for cleaning off oil, grease and dirt - never use petrol (gasoline) for cleaning. Wear rubber gloves when handling solvent and engine degreaser. The fumes from certain solvents can be dangerous - always work in a well-ventilated area.

Dust, eye and hand protection

● Protect your lungs from inhalation of dust particles by wearing a filtering mask over the nose and mouth. Many frictional materials still contain asbestos which is dangerous to your health. Protect your eyes from spouts of liquid and sprung components by wearing a pair of protective goggles (see illustration 1.6).

1.6 A fire extinguisher, goggles, mask and protective gloves should be at hand in the workshop

● Protect your hands from contact with solvents, fuel and oils by wearing rubber gloves. Alternatively apply a barrier cream to your hands before starting work. If handling hot components or fluids, wear suitable gloves to protect your hands from scalding and burns.

What to do with old fluids

● Old cleaning solvent, fuel, coolant and oils should not be poured down domestic drains or onto the ground. Package the fluid up in old oil containers, label it accordingly, and take it to a garage or disposal facility. Contact your local authority for location of such sites or ring the oil care hotline.

OIL CARE

FOLLOW THE CODE

OIL BANK LINE
0800 66 33 66

Note: It is antisocial and illegal to dump oil down the drain. To find the location of your local oil recycling bank, call this number free.

In the USA, note that any oil supplier must accept used oil for recycling.

2 Fasteners -
screws, bolts and nuts

Fastener types and applications

Bolts and screws

● Fastener head types are either of hexagonal, Torx or splined design, with internal and external versions of each type **(see illustrations 2.1 and 2.2)**; splined head fasteners are not in common use on motorcycles. The conventional slotted or Phillips head design is used for certain screws. Bolt or screw length is always measured from the underside of the head to the end of the item **(see illustration 2.11)**.

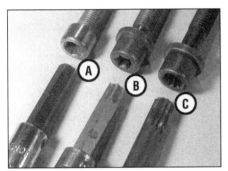

2.1 Internal hexagon/Allen (A), Torx (B) and splined (C) fasteners, with corresponding bits

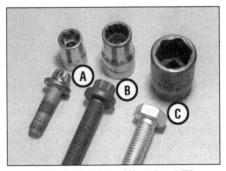

2.2 External Torx (A), splined (B) and hexagon (C) fasteners, with corresponding sockets

● Certain fasteners on the motorcycle have a tensile marking on their heads, the higher the marking the stronger the fastener. High tensile fasteners generally carry a 10 or higher marking. Never replace a high tensile fastener with one of a lower tensile strength.

Washers (see illustration 2.3)

● Plain washers are used between a fastener head and a component to prevent damage to the component or to spread the load when torque is applied. Plain washers can also be used as spacers or shims in certain assemblies. Copper or aluminium plain washers are often used as sealing washers on drain plugs.

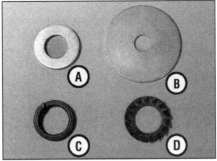

2.3 Plain washer (A), penny washer (B), spring washer (C) and serrated washer (D)

● The split-ring spring washer works by applying axial tension between the fastener head and component. If flattened, it is fatigued and must be renewed. If a plain (flat) washer is used on the fastener, position the spring washer between the fastener and the plain washer.
● Serrated star type washers dig into the fastener and component faces, preventing loosening. They are often used on electrical earth (ground) connections to the frame.
● Cone type washers (sometimes called Belleville) are conical and when tightened apply axial tension between the fastener head and component. They must be installed with the dished side against the component and often carry an OUTSIDE marking on their outer face. If flattened, they are fatigued and must be renewed.
● Tab washers are used to lock plain nuts or bolts on a shaft. A portion of the tab washer is bent up hard against one flat of the nut or bolt to prevent it loosening. Due to the tab washer being deformed in use, a new tab washer should be used every time it is disturbed.
● Wave washers are used to take up endfloat on a shaft. They provide light springing and prevent excessive side-to-side play of a component. Can be found on rocker arm shafts.

Nuts and split pins

● Conventional plain nuts are usually six-sided **(see illustration 2.4)**. They are sized by thread diameter and pitch. High tensile nuts carry a number on one end to denote their tensile strength.

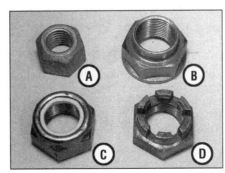

2.4 Plain nut (A), shouldered locknut (B), nylon insert nut (C) and castellated nut (D)

● Self-locking nuts either have a nylon insert, or two spring metal tabs, or a shoulder which is staked into a groove in the shaft - their advantage over conventional plain nuts is a resistance to loosening due to vibration. The nylon insert type can be used a number of times, but must be renewed when the friction of the nylon insert is reduced, ie when the nut spins freely on the shaft. The spring tab type can be reused unless the tabs are damaged. The shouldered type must be renewed every time it is disturbed.
● Split pins (cotter pins) are used to lock a castellated nut to a shaft or to prevent slackening of a plain nut. Common applications are wheel axles and brake torque arms. Because the split pin arms are deformed to lock around the nut a new split pin must always be used on installation - always fit the correct size split pin which will fit snugly in the shaft hole. Make sure the split pin arms are correctly located around the nut **(see illustrations 2.5 and 2.6)**.

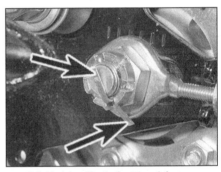

2.5 Bend split pin (cotter pin) arms as shown (arrows) to secure a castellated nut

2.6 Bend split pin (cotter pin) arms as shown to secure a plain nut

Caution: If the castellated nut slots do not align with the shaft hole after tightening to the torque setting, tighten the nut until the next slot aligns with the hole - never slacken the nut to align its slot.

● R-pins (shaped like the letter R), or slip pins as they are sometimes called, are sprung and can be reused if they are otherwise in good condition. Always install R-pins with their closed end facing forwards **(see illustration 2.7)**.

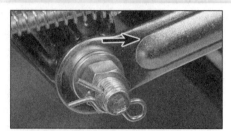

2.7 Correct fitting of R-pin. Arrow indicates forward direction

Circlips (see illustration 2.8)

● Circlips (sometimes called snap-rings) are used to retain components on a shaft or in a housing and have corresponding external or internal ears to permit removal. Parallel-sided (machined) circlips can be installed either way round in their groove, whereas stamped circlips (which have a chamfered edge on one face) must be installed with the chamfer facing away from the direction of thrust load **(see illustration 2.9)**.

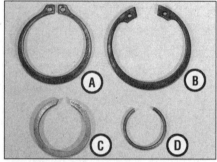

2.8 External stamped circlip (A), internal stamped circlip (B), machined circlip (C) and wire circlip (D)

● Always use circlip pliers to remove and install circlips; expand or compress them just enough to remove them. After installation, rotate the circlip in its groove to ensure it is securely seated. If installing a circlip on a splined shaft, always align its opening with a shaft channel to ensure the circlip ends are well supported and unlikely to catch **(see illustration 2.10)**.

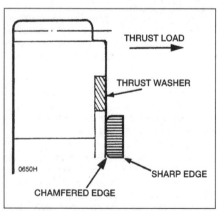

2.9 Correct fitting of a stamped circlip

THRUST LOAD

THRUST WASHER

SHARP EDGE

CHAMFERED EDGE

0650H

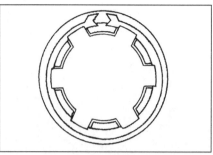

2.10 Align circlip opening with shaft channel

● Circlips can wear due to the thrust of components and become loose in their grooves, with the subsequent danger of becoming dislodged in operation. For this reason, renewal is advised every time a circlip is disturbed.

● Wire circlips are commonly used as piston pin retaining clips. If a removal tang is provided, long-nosed pliers can be used to dislodge them, otherwise careful use of a small flat-bladed screwdriver is necessary. Wire circlips should be renewed every time they are disturbed.

Thread diameter and pitch

● Diameter of a male thread (screw, bolt or stud) is the outside diameter of the threaded portion **(see illustration 2.11)**. Most motorcycle manufacturers use the ISO (International Standards Organisation) metric system expressed in millimetres, eg M6 refers to a 6 mm diameter thread. Sizing is the same for nuts, except that the thread diameter is measured across the valleys of the nut.

● Pitch is the distance between the peaks of the thread **(see illustration 2.11)**. It is expressed in millimetres, thus a common bolt size may be expressed as 6.0 x 1.0 mm (6 mm thread diameter and 1 mm pitch). Generally pitch increases in proportion to thread diameter, although there are always exceptions.

● Thread diameter and pitch are related for conventional fastener applications and the following table can be used as a guide. Additionally, the AF (Across Flats), spanner or socket size dimension of the bolt or nut **(see illustration 2.11)** is linked to thread and pitch specification. Thread pitch can be measured with a thread gauge **(see illustration 2.12)**.

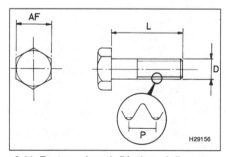

2.11 Fastener length (L), thread diameter (D), thread pitch (P) and head size (AF)

AF

L

D

P

H29156

2.12 Using a thread gauge to measure pitch

AF size	Thread diameter x pitch (mm)
8 mm	M5 x 0.8
8 mm	M6 x 1.0
10 mm	M6 x 1.0
12 mm	M8 x 1.25
14 mm	M10 x 1.25
17 mm	M12 x 1.25

● The threads of most fasteners are of the right-hand type, ie they are turned clockwise to tighten and anti-clockwise to loosen. The reverse situation applies to left-hand thread fasteners, which are turned anti-clockwise to tighten and clockwise to loosen. Left-hand threads are used where rotation of a component might loosen a conventional right-hand thread fastener.

Seized fasteners

● Corrosion of external fasteners due to water or reaction between two dissimilar metals can occur over a period of time. It will build up sooner in wet conditions or in countries where salt is used on the roads during the winter. If a fastener is severely corroded it is likely that normal methods of removal will fail and result in its head being ruined. When you attempt removal, the fastener thread should be heard to crack free and unscrew easily - if it doesn't, stop there before damaging something.

● A smart tap on the head of the fastener will often succeed in breaking free corrosion which has occurred in the threads **(see illustration 2.13)**.

● An aerosol penetrating fluid (such as WD-40) applied the night beforehand may work its way down into the thread and ease removal. Depending on the location, you may be able to make up a Plasticine well around the fastener head and fill it with penetrating fluid.

2.13 A sharp tap on the head of a fastener will often break free a corroded thread

● If you are working on an engine internal component, corrosion will most likely not be a problem due to the well lubricated environment. However, components can be very tight and an impact driver is a useful tool in freeing them **(see illustration 2.14)**.

2.14 Using an impact driver to free a fastener

● Where corrosion has occurred between dissimilar metals (eg steel and aluminium alloy), the application of heat to the fastener head will create a disproportionate expansion rate between the two metals and break the seizure caused by the corrosion. Whether heat can be applied depends on the location of the fastener - any surrounding components likely to be damaged must first be removed **(see illustration 2.15)**. Heat can be applied using a paint stripper heat gun or clothes iron, or by immersing the component in boiling water - wear protective gloves to prevent scalding or burns to the hands.

2.15 Using heat to free a seized fastener

● As a last resort, it is possible to use a hammer and cold chisel to work the fastener head unscrewed **(see illustration 2.16)**. This will damage the fastener, but more importantly extreme care must be taken not to damage the surrounding component.

Caution: Remember that the component being secured is generally of more value than the bolt, nut or screw - when the fastener is freed, do not unscrew it with force, instead work the fastener back and forth when resistance is felt to prevent thread damage.

2.16 Using a hammer and chisel to free a seized fastener

Broken fasteners and damaged heads

● If the shank of a broken bolt or screw is accessible you can grip it with self-locking grips. The knurled wheel type stud extractor tool or self-gripping stud puller tool is particularly useful for removing the long studs which screw into the cylinder mouth surface of the crankcase or bolts and screws from which the head has broken off **(see illustration 2.17)**. Studs can also be removed by locking two nuts together on the threaded end of the stud and using a spanner on the lower nut **(see illustration 2.18)**.

2.17 Using a stud extractor tool to remove a broken crankcase stud

2.18 Two nuts can be locked together to unscrew a stud from a component

● A bolt or screw which has broken off below or level with the casing must be extracted using a screw extractor set. Centre punch the fastener to centralise the drill bit, then drill a hole in the fastener **(see illustration 2.19)**. Select a drill bit which is

2.19 When using a screw extractor, first drill a hole in the fastener . . .

approximately half to three-quarters the diameter of the fastener and drill to a depth which will accommodate the extractor. Use the largest size extractor possible, but avoid leaving too small a wall thickness otherwise the extractor will merely force the fastener walls outwards wedging it in the casing thread.

● If a spiral type extractor is used, thread it anti-clockwise into the fastener. As it is screwed in, it will grip the fastener and unscrew it from the casing **(see illustration 2.20)**.

2.20 . . . then thread the extractor anti-clockwise into the fastener

● If a taper type extractor is used, tap it into the fastener so that it is firmly wedged in place. Unscrew the extractor (anti-clockwise) to draw the fastener out.

 Warning: Stud extractors are very hard and may break off in the fastener if care is not taken - ask an engineer about spark erosion if this happens.

● Alternatively, the broken bolt/screw can be drilled out and the hole retapped for an oversize bolt/screw or a diamond-section thread insert. It is essential that the drilling is carried out squarely and to the correct depth, otherwise the casing may be ruined - if in doubt, entrust the work to an engineer.

● Bolts and nuts with rounded corners cause the correct size spanner or socket to slip when force is applied. Of the types of spanner/socket available always use a six-point type rather than an eight or twelve-point type - better grip

2.21 Comparison of surface drive ring spanner (left) with 12-point type (right)

is obtained. Surface drive spanners grip the middle of the hex flats, rather than the corners, and are thus good in cases of damaged heads **(see illustration 2.21)**.

● Slotted-head or Phillips-head screws are often damaged by the use of the wrong size screwdriver. Allen-head and Torx-head screws are much less likely to sustain damage. If enough of the screw head is exposed you can use a hacksaw to cut a slot in its head and then use a conventional flat-bladed screwdriver to remove it. Alternatively use a hammer and cold chisel to tap the head of the fastener round to slacken it. Always replace damaged fasteners with new ones, preferably Torx or Allen-head type.

HAYNES HiNT

A dab of valve grinding compound between the screw head and screw-driver tip will often give a good grip.

Thread repair

● Threads (particularly those in aluminium alloy components) can be damaged by overtightening, being assembled with dirt in the threads, or from a component working loose and vibrating. Eventually the thread will fail completely, and it will be impossible to tighten the fastener.

● If a thread is damaged or clogged with old locking compound it can be renovated with a thread repair tool (thread chaser) **(see illustrations 2.22 and 2.23)**; special thread

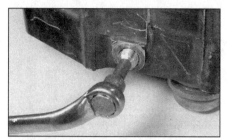

2.22 A thread repair tool being used to correct an internal thread

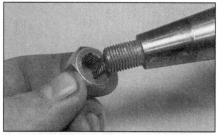

2.23 A thread repair tool being used to correct an external thread

chasers are available for spark plug hole threads. The tool will not cut a new thread, but clean and true the original thread. Make sure that you use the correct diameter and pitch tool. Similarly, external threads can be cleaned up with a die or a thread restorer file **(see illustration 2.24)**.

2.24 Using a thread restorer file

● It is possible to drill out the old thread and retap the component to the next thread size. This will work where there is enough surrounding material and a new bolt or screw can be obtained. Sometimes, however, this is not possible - such as where the bolt/screw passes through another component which must also be suitably modified, also in cases where a spark plug or oil drain plug cannot be obtained in a larger diameter thread size.

● The diamond-section thread insert (often known by its popular trade name of Heli-Coil) is a simple and effective method of renewing the thread and retaining the original size. A kit can be purchased which contains the tap, insert and installing tool **(see illustration 2.25)**. Drill out the damaged thread with the size drill specified **(see illustration 2.26)**. Carefully retap the thread **(see illustration 2.27)**. Install the

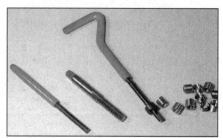

2.25 Obtain a thread insert kit to suit the thread diameter and pitch required

2.26 To install a thread insert, first drill out the original thread . . .

2.27 . . . tap a new thread . . .

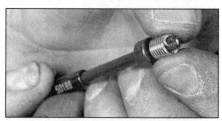

2.28 . . . fit insert on the installing tool . . .

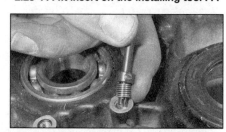

2.29 . . . and thread into the component . . .

2.30 . . . break off the tang when complete

insert on the installing tool and thread it slowly into place using a light downward pressure **(see illustrations 2.28 and 2.29)**. When positioned between a 1/4 and 1/2 turn below the surface withdraw the installing tool and use the break-off tool to press down on the tang, breaking it off **(see illustration 2.30)**.

● There are epoxy thread repair kits on the market which can rebuild stripped internal threads, although this repair should not be used on high load-bearing components.

Thread locking and sealing compounds

● Locking compounds are used in locations where the fastener is prone to loosening due to vibration or on important safety-related items which might cause loss of control of the motorcycle if they fail. It is also used where important fasteners cannot be secured by other means such as lockwashers or split pins.

● Before applying locking compound, make sure that the threads (internal and external) are clean and dry with all old compound removed. Select a compound to suit the component being secured - a non-permanent general locking and sealing type is suitable for most applications, but a high strength type is needed for permanent fixing of studs in castings. Apply a drop or two of the compound to the first few threads of the fastener, then thread it into place and tighten to the specified torque. Do not apply excessive thread locking compound otherwise the thread may be damaged on subsequent removal.

● Certain fasteners are impregnated with a dry film type coating of locking compound on their threads. Always renew this type of fastener if disturbed.

● Anti-seize compounds, such as copper-based greases, can be applied to protect threads from seizure due to extreme heat and corrosion. A common instance is spark plug threads and exhaust system fasteners.

3 Measuring tools and gauges

Feeler gauges

● Feeler gauges (or blades) are used for measuring small gaps and clearances (see illustration 3.1). They can also be used to measure endfloat (sideplay) of a component on a shaft where access is not possible with a dial gauge.

● Feeler gauge sets should be treated with care and not bent or damaged. They are etched with their size on one face. Keep them clean and very lightly oiled to prevent corrosion build-up.

3.1 Feeler gauges are used for measuring small gaps and clearances - thickness is marked on one face of gauge

● When measuring a clearance, select a gauge which is a light sliding fit between the two components. You may need to use two gauges together to measure the clearance accurately.

Micrometers

● A micrometer is a precision tool capable of measuring to 0.01 or 0.001 of a millimetre. It should always be stored in its case and not in the general toolbox. It must be kept clean and never dropped, otherwise its frame or measuring anvils could be distorted resulting in inaccurate readings.

● External micrometers are used for measuring outside diameters of components and have many more applications than internal micrometers. Micrometers are available in different size ranges, eg 0 to 25 mm, 25 to 50 mm, and upwards in 25 mm steps; some large micrometers have interchangeable anvils to allow a range of measurements to be taken. Generally the largest precision measurement you are likely to take on a motorcycle is the piston diameter.

● Internal micrometers (or bore micrometers) are used for measuring inside diameters, such as valve guides and cylinder bores. Telescoping gauges and small hole gauges are used in conjunction with an external micrometer, whereas the more expensive internal micrometers have their own measuring device.

External micrometer

Note: The conventional analogue type instrument is described. Although much easier to read, digital micrometers are considerably more expensive.

● Always check the calibration of the micrometer before use. With the anvils closed (0 to 25 mm type) or set over a test gauge (for

3.2 Check micrometer calibration before use

the larger types) the scale should read zero (see illustration 3.2); make sure that the anvils (and test piece) are clean first. Any discrepancy can be adjusted by referring to the instructions supplied with the tool. Remember that the micrometer is a precision measuring tool - don't force the anvils closed, use the ratchet (4) on the end of the micrometer to close it. In this way, a measured force is always applied.

● To use, first make sure that the item being measured is clean. Place the anvil (1) against the item and use the thimble (2) to bring the spindle (3) lightly into contact with the other side of the item (see illustration 3.3). Don't tighten the thimble down because this will damage the micrometer - instead use the ratchet (4) on the end of the micrometer. The ratchet mechanism applies a measured force preventing damage to the instrument.

● The micrometer is read by referring to the linear scale on the sleeve and the annular scale on the thimble. Read off the sleeve first to obtain the base measurement, then add the fine measurement from the thimble to obtain the overall reading. The linear scale on the sleeve represents the measuring range of the micrometer (eg 0 to 25 mm). The annular scale

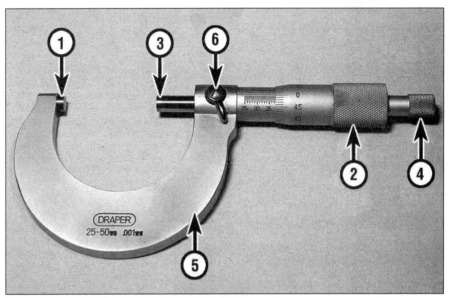

3.3 Micrometer component parts

| 1 Anvil | 3 Spindle | 5 Frame |
| 2 Thimble | 4 Ratchet | 6 Locking lever |

on the thimble will be in graduations of 0.01 mm (or as marked on the frame) - one full revolution of the thimble will move 0.5 mm on the linear scale. Take the reading where the datum line on the sleeve intersects the thimble's scale. Always position the eye directly above the scale otherwise an inaccurate reading will result.

In the example shown the item measures 2.95 mm (see illustration 3.4):

Linear scale	2.00 mm
Linear scale	0.50 mm
Annular scale	0.45 mm
Total figure	**2.95 mm**

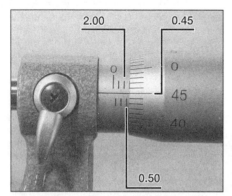

3.4 Micrometer reading of 2.95 mm

Most micrometers have a locking lever (6) on the frame to hold the setting in place, allowing the item to be removed from the micrometer.
● Some micrometers have a vernier scale on their sleeve, providing an even finer measurement to be taken, in 0.001 increments of a millimetre. Take the sleeve and thimble measurement as described above, then check which graduation on the vernier scale aligns with that of the annular scale on the thimble **Note:** *The eye must be perpendicular to the scale when taking the vernier reading - if necessary rotate the body of the micrometer to ensure this.* Multiply the vernier scale figure by 0.001 and add it to the base and fine measurement figures.

In the example shown the item measures 46.994 mm (see illustrations 3.5 and 3.6):

Linear scale (base)	46.000 mm
Linear scale (base)	00.500 mm
Annular scale (fine)	00.490 mm
Vernier scale	00.004 mm
Total figure	**46.994 mm**

Internal micrometer

● Internal micrometers are available for measuring bore diameters, but are expensive and unlikely to be available for home use. It is suggested that a set of telescoping gauges and small hole gauges, both of which must be used with an external micrometer, will suffice for taking internal measurements on a motorcycle.

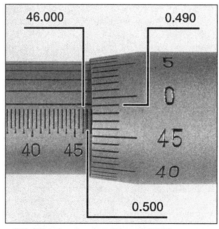

3.5 Micrometer reading of 46.99 mm on linear and annular scales . . .

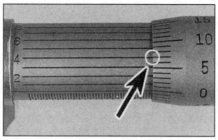

3.6 . . . and 0.004 mm on vernier scale

● Telescoping gauges can be used to measure internal diameters of components. Select a gauge with the correct size range, make sure its ends are clean and insert it into the bore. Expand the gauge, then lock its position and withdraw it from the bore (see illustration 3.7). Measure across the gauge ends with a micrometer (see illustration 3.8).

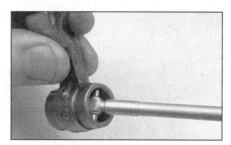

3.7 Expand the telescoping gauge in the bore, lock its position . . .

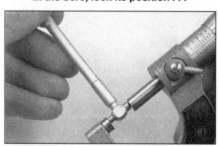

3.8 . . . then measure the gauge with a micrometer

3.9 Expand the small hole gauge in the bore, lock its position . . .

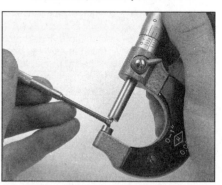

3.10 . . . then measure the gauge with a micrometer

● Very small diameter bores (such as valve guides) are measured with a small hole gauge. Once adjusted to a slip-fit inside the component, its position is locked and the gauge withdrawn for measurement with a micrometer (see illustrations 3.9 and 3.10).

Vernier caliper

Note: *The conventional linear and dial gauge type instruments are described. Digital types are easier to read, but are far more expensive.*
● The vernier caliper does not provide the precision of a micrometer, but is versatile in being able to measure internal and external diameters. Some types also incorporate a depth gauge. It is ideal for measuring clutch plate friction material and spring free lengths.
● To use the conventional linear scale vernier, slacken off the vernier clamp screws (1) and set its jaws over (2), or inside (3), the item to be measured (see illustration 3.11). Slide the jaw into contact, using the thumb-wheel (4) for fine movement of the sliding scale (5) then tighten the clamp screws (1). Read off the main scale (6) where the zero on the sliding scale (5) intersects it, taking the whole number to the left of the zero; this provides the base measurement. View along the sliding scale and select the division which lines up exactly with any of the divisions on the main scale, noting that the divisions usually represents 0.02 of a millimetre. Add this fine measurement to the base measurement to obtain the total reading.

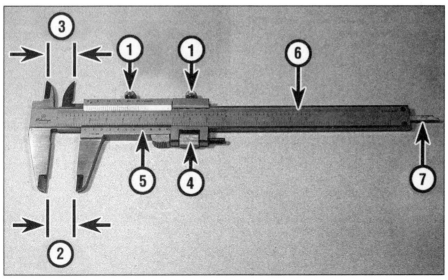

3.11 Vernier component parts (linear gauge)

1	Clamp screws	3	Internal jaws	5	Sliding scale	7	Depth gauge
2	External jaws	4	Thumbwheel	6	Main scale		

In the example shown the item measures 55.92 mm **(see illustration 3.12)**:

Base measurement	55.00 mm
Fine measurement	00.92 mm
Total figure	**55.92 mm**

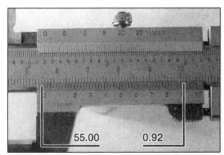

3.12 Vernier gauge reading of 55.92 mm

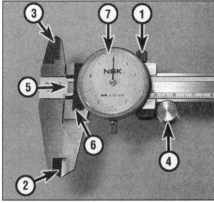

3.13 Vernier component parts (dial gauge)

1	Clamp screw	5	Main scale
2	External jaws	6	Sliding scale
3	Internal jaws	7	Dial gauge
4	Thumbwheel		

● Some vernier calipers are equipped with a dial gauge for fine measurement. Before use, check that the jaws are clean, then close them fully and check that the dial gauge reads zero. If necessary adjust the gauge ring accordingly. Slacken the vernier clamp screw (1) and set its jaws over (2), or inside (3), the item to be measured **(see illustration 3.13)**. Slide the jaws into contact, using the thumbwheel (4) for fine movement. Read off the main scale (5) where the edge of the sliding scale (6) intersects it, taking the whole number to the left of the zero; this provides the base measurement. Read off the needle position on the dial gauge (7) scale to provide the fine measurement; each division represents 0.05 of a millimetre. Add this fine measurement to the base measurement to obtain the total reading.

In the example shown the item measures 55.95 mm **(see illustration 3.14)**:

Base measurement	55.00 mm
Fine measurement	00.95 mm
Total figure	**55.95 mm**

3.14 Vernier gauge reading of 55.95 mm

Plastigauge

● Plastigauge is a plastic material which can be compressed between two surfaces to measure the oil clearance between them. The width of the compressed Plastigauge is measured against a calibrated scale to determine the clearance.

● Common uses of Plastigauge are for measuring the clearance between crankshaft journal and main bearing inserts, between crankshaft journal and big-end bearing inserts, and between camshaft and bearing surfaces. The following example describes big-end oil clearance measurement.

● Handle the Plastigauge material carefully to prevent distortion. Using a sharp knife, cut a length which corresponds with the width of the bearing being measured and place it carefully across the journal so that it is parallel with the shaft **(see illustration 3.15)**. Carefully install both bearing shells and the connecting rod. Without rotating the rod on the journal tighten its bolts or nuts (as applicable) to the specified torque. The connecting rod and bearings are then disassembled and the crushed Plastigauge examined.

3.15 Plastigauge placed across shaft journal

● Using the scale provided in the Plastigauge kit, measure the width of the material to determine the oil clearance **(see illustration 3.16)**. Always remove all traces of Plastigauge after use using your fingernails.

Caution: Arriving at the correct clearance demands that the assembly is torqued correctly, according to the settings and sequence (where applicable) provided by the motorcycle manufacturer.

3.16 Measuring the width of the crushed Plastigauge

Dial gauge or DTI (Dial Test Indicator)

● A dial gauge can be used to accurately measure small amounts of movement. Typical uses are measuring shaft runout or shaft endfloat (sideplay) and setting piston position for ignition timing on two-strokes. A dial gauge set usually comes with a range of different probes and adapters and mounting equipment.

● The gauge needle must point to zero when at rest. Rotate the ring around its periphery to zero the gauge.

● Check that the gauge is capable of reading the extent of movement in the work. Most gauges have a small dial set in the face which records whole millimetres of movement as well as the fine scale around the face periphery which is calibrated in 0.01 mm divisions. Read off the small dial first to obtain the base measurement, then add the measurement from the fine scale to obtain the total reading.

In the example shown the gauge reads 1.48 mm (see illustration 3.17):

Base measurement	1.00 mm
Fine measurement	0.48 mm
Total figure	**1.48 mm**

3.17 Dial gauge reading of 1.48 mm

● If measuring shaft runout, the shaft must be supported in vee-blocks and the gauge mounted on a stand perpendicular to the shaft. Rest the tip of the gauge against the centre of the shaft and rotate the shaft slowly whilst watching the gauge reading (see illustration 3.18). Take several measurements along the length of the shaft and record the

3.18 Using a dial gauge to measure shaft runout

maximum gauge reading as the amount of runout in the shaft. **Note:** *The reading obtained will be total runout at that point - some manufacturers specify that the runout figure is halved to compare with their specified runout limit.*

● Endfloat (sideplay) measurement requires that the gauge is mounted securely to the surrounding component with its probe touching the end of the shaft. Using hand pressure, push and pull on the shaft noting the maximum endfloat recorded on the gauge (see illustration 3.19).

3.19 Using a dial gauge to measure shaft endfloat

● A dial gauge with suitable adapters can be used to determine piston position BTDC on two-stroke engines for the purposes of ignition timing. The gauge, adapter and suitable length probe are installed in the place of the spark plug and the gauge zeroed at TDC. If the piston position is specified as 1.14 mm BTDC, rotate the engine back to 2.00 mm BTDC, then slowly forwards to 1.14 mm BTDC.

Cylinder compression gauges

● A compression gauge is used for measuring cylinder compression. Either the rubber-cone type or the threaded adapter type can be used. The latter is preferred to ensure a perfect seal against the cylinder head. A 0 to 300 psi (0 to 20 Bar) type gauge (for petrol/gasoline engines) will be suitable for motorcycles.

● The spark plug is removed and the gauge either held hard against the cylinder head (cone type) or the gauge adapter screwed into the cylinder head (threaded type) (see illustration 3.20). Cylinder compression is measured with the engine turning over, but not running - carry out the compression test as described in

3.20 Using a rubber-cone type cylinder compression gauge

Fault Finding Equipment. The gauge will hold the reading until manually released.

Oil pressure gauge

● An oil pressure gauge is used for measuring engine oil pressure. Most gauges come with a set of adapters to fit the thread of the take-off point (see illustration 3.21). If the take-off point specified by the motorcycle manufacturer is an external oil pipe union, make sure that the specified replacement union is used to prevent oil starvation.

3.21 Oil pressure gauge and take-off point adapter (arrow)

● Oil pressure is measured with the engine running (at a specific rpm) and often the manufacturer will specify pressure limits for a cold and hot engine.

Straight-edge and surface plate

● If checking the gasket face of a component for warpage, place a steel rule or precision straight-edge across the gasket face and measure any gap between the straight-edge and component with feeler gauges (see illustration 3.22). Check diagonally across the component and between mounting holes (see illustration 3.23).

3.22 Use a straight-edge and feeler gauges to check for warpage

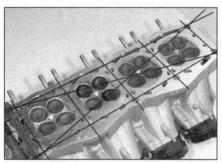

3.23 Check for warpage in these directions

● Checking individual components for warpage, such as clutch plain (metal) plates, requires a perfectly flat plate or piece or plate glass and feeler gauges.

4 Torque and leverage

What is torque?

● Torque describes the twisting force about a shaft. The amount of torque applied is determined by the distance from the centre of the shaft to the end of the lever and the amount of force being applied to the end of the lever; distance multiplied by force equals torque.

● The manufacturer applies a measured torque to a bolt or nut to ensure that it will not slacken in use and to hold two components securely together without movement in the joint. The actual torque setting depends on the thread size, bolt or nut material and the composition of the components being held.

● Too little torque may cause the fastener to loosen due to vibration, whereas too much torque will distort the joint faces of the component or cause the fastener to shear off. Always stick to the specified torque setting.

Using a torque wrench

● Check the calibration of the torque wrench and make sure it has a suitable range for the job. Torque wrenches are available in Nm (Newton-metres), kgf m (kilograms-force metre), lbf ft (pounds-feet), lbf in (inch-pounds). Do not confuse lbf ft with lbf in.

● Adjust the tool to the desired torque on the scale (see illustration 4.1). If your torque wrench is not calibrated in the units specified, carefully convert the figure (see Conversion Factors). A manufacturer sometimes gives a torque setting as a range (8 to 10 Nm) rather than a single figure - in this case set the tool midway between the two settings. The same torque may be expressed as 9 Nm ± 1 Nm. Some torque wrenches have a method of locking the setting so that it isn't inadvertently altered during use.

4.1 Set the torque wrench index mark to the setting required, in this case 12 Nm

● Install the bolts/nuts in their correct location and secure them lightly. Their threads must be clean and free of any old locking compound. Unless specified the threads and flange should be dry - oiled threads are necessary in certain circumstances and the manufacturer will take this into account in the specified torque figure. Similarly, the manufacturer may also specify the application of thread-locking compound.

● Tighten the fasteners in the specified sequence until the torque wrench clicks, indicating that the torque setting has been reached. Apply the torque again to double-check the setting. Where different thread diameter fasteners secure the component, as a rule tighten the larger diameter ones first.

● When the torque wrench has been finished with, release the lock (where applicable) and fully back off its setting to zero - do not leave the torque wrench tensioned. Also, do not use a torque wrench for slackening a fastener.

Angle-tightening

● Manufacturers often specify a figure in degrees for final tightening of a fastener. This usually follows tightening to a specific torque setting.

● A degree disc can be set and attached to the socket (see illustration 4.2) or a protractor can be used to mark the angle of movement on the bolt/nut head and the surrounding casting (see illustration 4.3).

4.2 Angle tightening can be accomplished with a torque-angle gauge . . .

4.3 . . . or by marking the angle on the surrounding component

Loosening sequences

● Where more than one bolt/nut secures a component, loosen each fastener evenly a little at a time. In this way, not all the stress of the joint is held by one fastener and the components are not likely to distort.

● If a tightening sequence is provided, work in the REVERSE of this, but if not, work from the outside in, in a criss-cross sequence (see illustration 4.4).

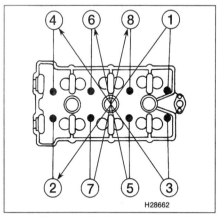

4.4 When slackening, work from the outside inwards

Tightening sequences

● If a component is held by more than one fastener it is important that the retaining bolts/nuts are tightened evenly to prevent uneven stress build-up and distortion of sealing faces. This is especially important on high-compression joints such as the cylinder head.

● A sequence is usually provided by the manufacturer, either in a diagram or actually marked in the casting. If not, always start in the centre and work outwards in a criss-cross pattern (see illustration 4.5). Start off by securing all bolts/nuts finger-tight, then set the torque wrench and tighten each fastener by a small amount in sequence until the final torque is reached. By following this practice,

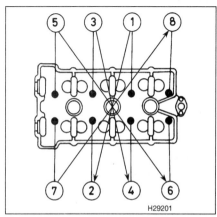

4.5 When tightening, work from the inside outwards

the joint will be held evenly and will not be distorted. Important joints, such as the cylinder head and big-end fasteners often have two- or three-stage torque settings.

Applying leverage

● Use tools at the correct angle. Position a socket wrench or spanner on the bolt/nut so that you pull it towards you when loosening. If this can't be done, push the spanner without curling your fingers around it **(see illustration 4.6)** - the spanner may slip or the fastener loosen suddenly, resulting in your fingers being crushed against a component.

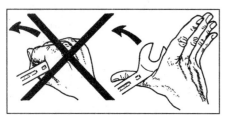

4.6 If you can't pull on the spanner to loosen a fastener, push with your hand open

● Additional leverage is gained by extending the length of the lever. The best way to do this is to use a breaker bar instead of the regular length tool, or to slip a length of tubing over the end of the spanner or socket wrench.
● If additional leverage will not work, the fastener head is either damaged or firmly corroded in place (see *Fasteners*).

5 Bearings

Bearing removal and installation

Drivers and sockets

● Before removing a bearing, always inspect the casing to see which way it must be driven out - some casings will have retaining plates or a cast step. Also check for any identifying markings on the bearing and if installed to a certain depth, measure this at this stage. Some roller bearings are sealed on one side - take note of the original fitted position.
● Bearings can be driven out of a casing using a bearing driver tool (with the correct size head) or a socket of the correct diameter. Select the driver head or socket so that it contacts the outer race of the bearing, not the balls/rollers or inner race. Always support the casing around the bearing housing with wood blocks, otherwise there is a risk of fracture. The bearing is driven out with a few blows on the driver or socket from a heavy mallet. Unless access is severely restricted (as with wheel bearings), a pin-punch is not recommended unless it is moved around the bearing to keep it square in its housing.

● The same equipment can be used to install bearings. Make sure the bearing housing is supported on wood blocks and line up the bearing in its housing. Fit the bearing as noted on removal - generally they are installed with their marked side facing outwards. Tap the bearing squarely into its housing using a driver or socket which bears only on the bearing's outer race - contact with the bearing balls/rollers or inner race will destroy it **(see illustrations 5.1 and 5.2)**.
● Check that the bearing inner race and balls/rollers rotate freely.

5.1 Using a bearing driver against the bearing's outer race

5.2 Using a large socket against the bearing's outer race

Pullers and slide-hammers

● Where a bearing is pressed on a shaft a puller will be required to extract it **(see illustration 5.3)**. Make sure that the puller clamp or legs fit securely behind the bearing and are unlikely to slip out. If pulling a bearing

5.3 This bearing puller clamps behind the bearing and pressure is applied to the shaft end to draw the bearing off

off a gear shaft for example, you may have to locate the puller behind a gear pinion if there is no access to the race and draw the gear pinion off the shaft as well **(see illustration 5.4)**. *Caution: Ensure that the puller's centre bolt locates securely against the end of the shaft and will not slip when pressure is applied. Also ensure that puller does not damage the shaft end.*

5.4 Where no access is available to the rear of the bearing, it is sometimes possible to draw off the adjacent component

● Operate the puller so that its centre bolt exerts pressure on the shaft end and draws the bearing off the shaft.
● When installing the bearing on the shaft, tap only on the bearing's inner race - contact with the balls/rollers or outer race with destroy the bearing. Use a socket or length of tubing as a drift which fits over the shaft end **(see illustration 5.5)**.

5.5 When installing a bearing on a shaft use a piece of tubing which bears only on the bearing's inner race

● Where a bearing locates in a blind hole in a casing, it cannot be driven or pulled out as described above. A slide-hammer with knife-edged bearing puller attachment will be required. The puller attachment passes through the bearing and when tightened expands to fit firmly behind the bearing **(see illustration 5.6)**. By operating the slide-hammer part of the tool the bearing is jarred out of its housing **(see illustration 5.7)**.
● It is possible, if the bearing is of reasonable weight, for it to drop out of its housing if the casing is heated as described below. If this

5.6 Expand the bearing puller so that it locks behind the bearing . . .

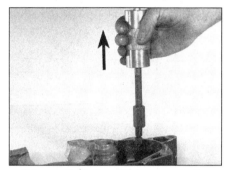

5.7 . . . attach the slide hammer to the bearing puller

method is attempted, first prepare a work surface which will enable the casing to be tapped face down to help dislodge the bearing - a wood surface is ideal since it will not damage the casing's gasket surface. Wearing protective gloves, tap the heated casing several times against the work surface to dislodge the bearing under its own weight **(see illustration 5.8)**.

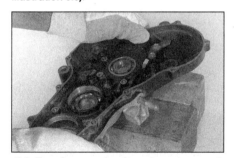

5.8 Tapping a casing face down on wood blocks can often dislodge a bearing

● Bearings can be installed in blind holes using the driver or socket method described above.

Drawbolts

● Where a bearing or bush is set in the eye of a component, such as a suspension linkage arm or connecting rod small-end, removal by drift may damage the component. Furthermore, a rubber bushing in a shock absorber eye cannot successfully be driven out of position. If access is available to a engineering press, the task is straightforward. If not, a drawbolt can be fabricated to extract the bearing or bush.

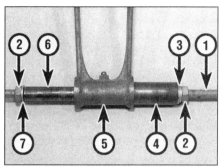

5.9 Drawbolt component parts assembled on a suspension arm

1 *Bolt or length of threaded bar*
2 *Nuts*
3 *Washer (external diameter greater than tubing internal diameter)*
4 *Tubing (internal diameter sufficient to accommodate bearing)*
5 *Suspension arm with bearing*
6 *Tubing (external diameter slightly smaller than bearing)*
7 *Washer (external diameter slightly smaller than bearing)*

5.10 Drawing the bearing out of the suspension arm

● To extract the bearing/bush you will need a long bolt with nut (or piece of threaded bar with two nuts), a piece of tubing which has an internal diameter larger than the bearing/bush, another piece of tubing which has an external diameter slightly smaller than the bearing/ bush, and a selection of washers **(see illustrations 5.9 and 5.10)**. Note that the pieces of tubing must be of the same length, or longer, than the bearing/bush.

● The same kit (without the pieces of tubing) can be used to draw the new bearing/bush back into place **(see illustration 5.11)**.

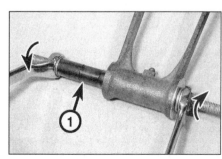

5.11 Installing a new bearing (1) in the suspension arm

Temperature change

● If the bearing's outer race is a tight fit in the casing, the aluminium casing can be heated to release its grip on the bearing. Aluminium will expand at a greater rate than the steel bearing outer race. There are several ways to do this, but avoid any localised extreme heat (such as a blow torch) - aluminium alloy has a low melting point.

● Approved methods of heating a casing are using a domestic oven (heated to 100°C) or immersing the casing in boiling water **(see illustration 5.12)**. Low temperature range localised heat sources such as a paint stripper heat gun or clothes iron can also be used **(see illustration 5.13)**. Alternatively, soak a rag in boiling water, wring it out and wrap it around the bearing housing.

⚠️ *Warning: All of these methods require care in use to prevent scalding and burns to the hands. Wear protective gloves when handling hot components.*

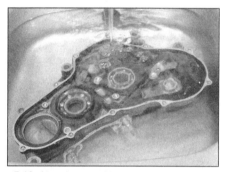

5.12 A casing can be immersed in a sink of boiling water to aid bearing removal

5.13 Using a localised heat source to aid bearing removal

● If heating the whole casing note that plastic components, such as the neutral switch, may suffer - remove them beforehand.

● After heating, remove the bearing as described above. You may find that the expansion is sufficient for the bearing to fall out of the casing under its own weight or with a light tap on the driver or socket.

● If necessary, the casing can be heated to aid bearing installation, and this is sometimes the recommended procedure if the motorcycle manufacturer has designed the housing and bearing fit with this intention.

● Installation of bearings can be eased by placing them in a freezer the night before installation. The steel bearing will contract slightly, allowing easy insertion in its housing. This is often useful when installing steering head outer races in the frame.

Bearing types and markings

● Plain shell bearings, ball bearings, needle roller bearings and tapered roller bearings will all be found on motorcycles (see illustrations **5.14** and **5.15**). The ball and roller types are usually caged between an inner and outer race, but uncaged variations may be found.

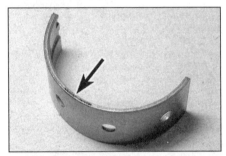

5.14 Shell bearings are either plain or grooved. They are usually identified by colour code (arrow)

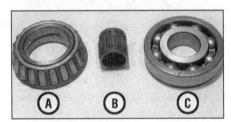

5.15 Tapered roller bearing (A), needle roller bearing (B) and ball journal bearing (C)

● Shell bearings (often called inserts) are usually found at the crankshaft main and connecting rod big-end where they are good at coping with high loads. They are made of a phosphor-bronze material and are impregnated with self-lubricating properties.

● Ball bearings and needle roller bearings consist of a steel inner and outer race with the balls or rollers between the races. They require constant lubrication by oil or grease and are good at coping with axial loads. Taper roller bearings consist of rollers set in a tapered cage set on the inner race; the outer race is separate. They are good at coping with axial loads and prevent movement along the shaft - a typical application is in the steering head.

● Bearing manufacturers produce bearings to ISO size standards and stamp one face of the bearing to indicate its internal and external diameter, load capacity and type (see illustration **5.16**).

● Metal bushes are usually of phosphor-bronze material. Rubber bushes are used in suspension mounting eyes. Fibre bushes have also been used in suspension pivots.

5.16 Typical bearing marking

Bearing fault finding

● If a bearing outer race has spun in its housing, the housing material will be damaged. You can use a bearing locking compound to bond the outer race in place if damage is not too severe.

● Shell bearings will fail due to damage of their working surface, as a result of lack of lubrication, corrosion or abrasive particles in the oil (see illustration **5.17**). Small particles of dirt in the oil may embed in the bearing material whereas larger particles will score the bearing and shaft journal. If a number of short journeys are made, insufficient heat will be generated to drive off condensation which has built up on the bearings.

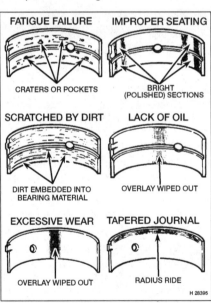

5.17 Typical bearing failures

● Ball and roller bearings will fail due to lack of lubrication or damage to the balls or rollers. Tapered-roller bearings can be damaged by overloading them. Unless the bearing is sealed on both sides, wash it in paraffin (kerosene) to remove all old grease then allow it to dry. Make a visual inspection looking to dented balls or rollers, damaged cages and worn or pitted races (see illustration **5.18**).

● A ball bearing can be checked for wear by listening to it when spun. Apply a film of light oil to the bearing and hold it close to the ear - hold the outer race with one hand and spin the inner

5.18 Example of ball journal bearing with damaged balls and cages

5.19 Hold outer race and listen to inner race when spun

race with the other hand (see illustration **5.19**). The bearing should be almost silent when spun; if it grates or rattles it is worn.

6 Oil seals

Oil seal removal and installation

● Oil seals should be renewed every time a component is dismantled. This is because the seal lips will become set to the sealing surface and will not necessarily reseal.

● Oil seals can be prised out of position using a large flat-bladed screwdriver (see illustration **6.1**). In the case of crankcase seals, check first that the seal is not lipped on the inside, preventing its removal with the crankcases joined.

6.1 Prise out oil seals with a large flat-bladed screwdriver

● New seals are usually installed with their marked face (containing the seal reference code) outwards and the spring side towards the fluid being retained. In certain cases, such as a two-stroke engine crankshaft seal, a double lipped seal may be used due to there being fluid or gas on each side of the joint.

● Use a bearing driver or socket which bears only on the outer hard edge of the seal to install it in the casing - tapping on the inner edge will damage the sealing lip.

Oil seal types and markings

● Oil seals are usually of the single-lipped type. Double-lipped seals are found where a liquid or gas is on both sides of the joint.
● Oil seals can harden and lose their sealing ability if the motorcycle has been in storage for a long period - renewal is the only solution.
● Oil seal manufacturers also conform to the ISO markings for seal size - these are moulded into the outer face of the seal **(see illustration 6.2)**.

6.2 These oil seal markings indicate inside diameter, outside diameter and seal thickness

7 Gaskets and sealants

Types of gasket and sealant

● Gaskets are used to seal the mating surfaces between components and keep lubricants, fluids, vacuum or pressure contained within the assembly. Aluminium gaskets are sometimes found at the cylinder joints, but most gaskets are paper-based. If the mating surfaces of the components being joined are undamaged the gasket can be installed dry, although a dab of sealant or grease will be useful to hold it in place during assembly.
● RTV (Room Temperature Vulcanising) silicone rubber sealants cure when exposed to moisture in the atmosphere. These sealants are good at filling pits or irregular gasket faces, but will tend to be forced out of the joint under very high torque. They can be used to replace a paper gasket, but first make sure that the width of the paper gasket is not essential to the shimming of internal components. RTV sealants should not be used on components containing petrol (gasoline).
● Non-hardening, semi-hardening and hard setting liquid gasket compounds can be used with a gasket or between a metal-to-metal joint. Select the sealant to suit the application: universal non-hardening sealant can be used on virtually all joints; semi-hardening on joint faces which are rough or damaged; hard setting sealant on joints which require a permanent bond and are subjected to high temperature and pressure. **Note:** *Check first if*

the paper gasket has a bead of sealant impregnated in its surface before applying additional sealant.
● When choosing a sealant, make sure it is suitable for the application, particularly if being applied in a high-temperature area or in the vicinity of fuel. Certain manufacturers produce sealants in either clear, silver or black colours to match the finish of the engine. This has a particular application on motorcycles where much of the engine is exposed.
● Do not over-apply sealant. That which is squeezed out on the outside of the joint can be wiped off, whereas an excess of sealant on the inside can break off and clog oilways.

Breaking a sealed joint

● Age, heat, pressure and the use of hard setting sealant can cause two components to stick together so tightly that they are difficult to separate using finger pressure alone. Do not resort to using levers unless there is a pry point provided for this purpose **(see illustration 7.1)** or else the gasket surfaces will be damaged.
● Use a soft-faced hammer **(see illustration 7.2)** or a wood block and conventional hammer to strike the component near the mating surface. Avoid hammering against cast extremities since they may break off. If this method fails, try using a wood wedge between the two components.
Caution: If the joint will not separate, double-check that you have removed all the fasteners.

7.1 If a pry point is provided, apply gently pressure with a flat-bladed screwdriver

7.2 Tap around the joint with a soft-faced mallet if necessary - don't strike cooling fins

Removal of old gasket and sealant

● Paper gaskets will most likely come away complete, leaving only a few traces stuck on

Most components have one or two hollow locating dowels between the two gasket faces. If a dowel cannot be removed, do not resort to gripping it with pliers - it will almost certainly be distorted. Install a close-fitting socket or Phillips screwdriver into the dowel and then grip the outer edge of the dowel to free it.

the sealing faces of the components. It is imperative that all traces are removed to ensure correct sealing of the new gasket.
● Very carefully scrape all traces of gasket away making sure that the sealing surfaces are not gouged or scored by the scraper **(see illustrations 7.3, 7.4 and 7.5)**. Stubborn deposits can be removed by spraying with an aerosol gasket remover. Final preparation of

7.3 Paper gaskets can be scraped off with a gasket scraper tool . . .

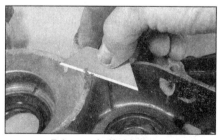

7.4 . . . a knife blade . . .

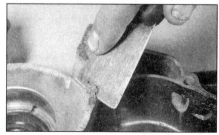

7.5 . . . or a household scraper

7.6 Fine abrasive paper is wrapped around a flat file to clean up the gasket face

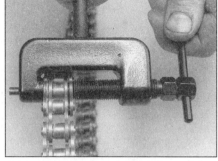

8.1 Tighten the chain breaker to push the pin out of the link . . .

8.4 Insert the new soft link, with O-rings, through the chain ends . . .

7.7 A kitchen scourer can be used on stubborn deposits

8.2 . . . withdraw the pin, remove the tool . . .

8.5 . . . install the O-rings over the pin ends . . .

the gasket surface can be made with very fine abrasive paper or a plastic kitchen scourer **(see illustrations 7.6 and 7.7).**
● Old sealant can be scraped or peeled off components, depending on the type originally used. Note that gasket removal compounds are available to avoid scraping the components clean; make sure the gasket remover suits the type of sealant used.

8.3 . . . and separate the chain link

8.6 . . . followed by the sideplate

ends **(see illustration 8.4).** Install a new O-ring over the end of each pin, followed by the sideplate (with the chain manufacturer's marking facing outwards) **(see illustrations 8.5 and 8.6).** On an unsealed chain, insert the link through the two chain ends, then install the sideplate with the chain manufacturer's marking facing outwards.
● Note that it may not be possible to install the sideplate using finger pressure alone. If using a joining tool, assemble it so that the plates of the tool clamp the link and press the sideplate over the pins **(see illustration 8.7).** Otherwise, use two small sockets placed over

8 Chains

Breaking and joining final drive chains

● Drive chains for all but small bikes are continuous and do not have a clip-type connecting link. The chain must be broken using a chain breaker tool and the new chain securely riveted together using a new soft rivet-type link. Never use a clip-type connecting link instead of a rivet-type link, except in an emergency. Various chain breaking and riveting tools are available, either as separate tools or combined as illustrated in the accompanying photographs - read the instructions supplied with the tool carefully.

⚠️ **Warning: The need to rivet the new link pins correctly cannot be overstressed - loss of control of the motorcycle is very likely to result if the chain breaks in use.**

● Rotate the chain and look for the soft link. The soft link pins look like they have been deeply centre-punched instead of peened over

like all the other pins **(see illustration 8.9)** and its sideplate may be a different colour. Position the soft link midway between the sprockets and assemble the chain breaker tool over one of the soft link pins **(see illustration 8.1).** Operate the tool to push the pin out through the chain **(see illustration 8.2).** On an O-ring chain, remove the O-rings **(see illustration 8.3).** Carry out the same procedure on the other soft link pin.
Caution: Certain soft link pins (particularly on the larger chains) may require their ends to be filed or ground off before they can be pressed out using the tool.
● Check that you have the correct size and strength (standard or heavy duty) new soft link - do not reuse the old link. Look for the size marking on the chain sideplates **(see illustration 8.10).**
● Position the chain ends so that they are engaged over the rear sprocket. On an O-ring chain, install a new O-ring over each pin of the link and insert the link through the two chain

8.7 Push the sideplate into position using a clamp

8.8 Assemble the chain riveting tool over one pin at a time and tighten it fully

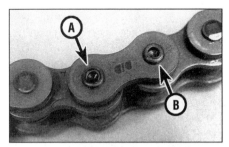

8.9 Pin end correctly riveted (A), pin end unriveted (B)

the rivet ends and two pieces of the wood between a G-clamp. Operate the clamp to press the sideplate over the pins.

● Assemble the joining tool over one pin (following the maker's instructions) and tighten the tool down to spread the pin end securely **(see illustrations 8.8 and 8.9)**. Do the same on the other pin.

 Warning: Check that the pin ends are secure and that there is no danger of the sideplate coming loose. If the pin ends are cracked the soft link must be renewed.

Final drive chain sizing

● Chains are sized using a three digit number, followed by a suffix to denote the chain type **(see illustration 8.10)**. Chain type is either standard or heavy duty (thicker sideplates), and also unsealed or O-ring/ X-ring type.

● The first digit of the number relates to the pitch of the chain, ie the distance from the centre of one pin to the centre of the next pin **(see illustration 8.11)**. Pitch is expressed in eighths of an inch, as follows:

8.10 Typical chain size and type marking

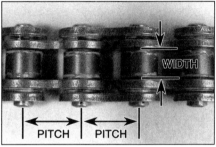

8.11 Chain dimensions

Sizes commencing with a 4 (eg 428) have a pitch of 1/2 inch (12.7 mm)

Sizes commencing with a 5 (eg 520) have a pitch of 5/8 inch (15.9 mm)

Sizes commencing with a 6 (eg 630) have a pitch of 3/4 inch (19.1 mm)

● The second and third digits of the chain size relate to the width of the rollers, again in imperial units, eg the 525 shown has 5/16 inch (7.94 mm) rollers **(see illustration 8.11)**.

9 Hoses

Clamping to prevent flow

● Small-bore flexible hoses can be clamped to prevent fluid flow whilst a component is worked on. Whichever method is used, ensure that the hose material is not permanently distorted or damaged by the clamp.

a) A brake hose clamp available from auto accessory shops **(see illustration 9.1)**.
b) A wingnut type hose clamp **(see illustration 9.2)**.

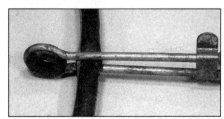

9.1 Hoses can be clamped with an automotive brake hose clamp . . .

9.2 . . . a wingnut type hose clamp . . .

c) Two sockets placed each side of the hose and held with straight-jawed self-locking grips **(see illustration 9.3)**.
d) Thick card each side of the hose held between straight-jawed self-locking grips **(see illustration 9.4)**.

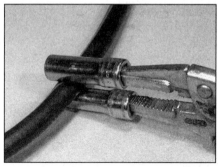

9.3 . . . two sockets and a pair of self-locking grips . . .

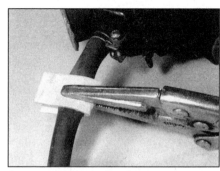

9.4 . . . or thick card and self-locking grips

Freeing and fitting hoses

● Always make sure the hose clamp is moved well clear of the hose end. Grip the hose with your hand and rotate it whilst pulling it off the union. If the hose has hardened due to age and will not move, slit it with a sharp knife and peel its ends off the union **(see illustration 9.5)**.

● Resist the temptation to use grease or soap on the unions to aid installation; although it helps the hose slip over the union it will equally aid the escape of fluid from the joint. It is preferable to soften the hose ends in hot water and wet the inside surface of the hose with water or a fluid which will evaporate.

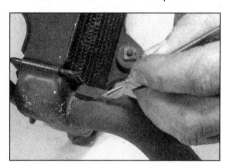

9.5 Cutting a coolant hose free with a sharp knife

Conversion Factors

Length (distance)

Inches (in)	x 25.4	= Millimetres (mm)	x 0.0394	= Inches (in)
Feet (ft)	x 0.305	= Metres (m)	x 3.281	= Feet (ft)
Miles	x 1.609	= Kilometres (km)	x 0.621	= Miles

Volume (capacity)

Cubic inches (cu in; in³)	x 16.387	= Cubic centimetres (cc; cm³)	x 0.061	= Cubic inches (cu in; in³)
Imperial pints (Imp pt)	x 0.568	= Litres (l)	x 1.76	= Imperial pints (Imp pt)
Imperial quarts (Imp qt)	x 1.137	= Litres (l)	x 0.88	= Imperial quarts (Imp qt)
Imperial quarts (Imp qt)	x 1.201	= US quarts (US qt)	x 0.833	= Imperial quarts (Imp qt)
US quarts (US qt)	x 0.946	= Litres (l)	x 1.057	= US quarts (US qt)
Imperial gallons (Imp gal)	x 4.546	= Litres (l)	x 0.22	= Imperial gallons (Imp gal)
Imperial gallons (Imp gal)	x 1.201	= US gallons (US gal)	x 0.833	= Imperial gallons (Imp gal)
US gallons (US gal)	x 3.785	= Litres (l)	x 0.264	= US gallons (US gal)

Mass (weight)

Ounces (oz)	x 28.35	= Grams (g)	x 0.035	= Ounces (oz)
Pounds (lb)	x 0.454	= Kilograms (kg)	x 2.205	= Pounds (lb)

Force

Ounces-force (ozf; oz)	x 0.278	= Newtons (N)	x 3.6	= Ounces-force (ozf; oz)
Pounds-force (lbf; lb)	x 4.448	= Newtons (N)	x 0.225	= Pounds-force (lbf; lb)
Newtons (N)	x 0.1	= Kilograms-force (kgf; kg)	x 9.81	= Newtons (N)

Pressure

Pounds-force per square inch (psi; lbf/in²; lb/in²)	x 0.070	= Kilograms-force per square centimetre (kgf/cm²; kg/cm²)	x 14.223	= Pounds-force per square inch (psi; lbf/in²; lb/in²)
Pounds-force per square inch (psi; lbf/in²; lb/in²)	x 0.068	= Atmospheres (atm)	x 14.696	= Pounds-force per square inch (psi; lbf/in²; lb/in²)
Pounds-force per square inch (psi; lbf/in²; lb/in²)	x 0.069	= Bars	x 14.5	= Pounds-force per square inch (psi; lbf/in²; lb/in²)
Pounds-force per square inch (psi; lbf/in²; lb/in²)	x 6.895	= Kilopascals (kPa)	x 0.145	= Pounds-force per square inch (psi; lbf/in²; lb/in²)
Kilopascals (kPa)	x 0.01	= Kilograms-force per square centimetre (kgf/cm²; kg/cm²)	x 98.1	= Kilopascals (kPa)
Millibar (mbar)	x 100	= Pascals (Pa)	x 0.01	= Millibar (mbar)
Millibar (mbar)	x 0.0145	= Pounds-force per square inch (psi; lbf/in²; lb/in²)	x 68.947	= Millibar (mbar)
Millibar (mbar)	x 0.75	= Millimetres of mercury (mmHg)	x 1.333	= Millibar (mbar)
Millibar (mbar)	x 0.401	= Inches of water (inH₂O)	x 2.491	= Millibar (mbar)
Millimetres of mercury (mmHg)	x 0.535	= Inches of water (inH₂O)	x 1.868	= Millimetres of mercury (mmHg)
Inches of water (inH₂O)	x 0.036	= Pounds-force per square inch (psi; lbf/in²; lb/in²)	x 27.68	= Inches of water (inH₂O)

Torque (moment of force)

Pounds-force inches (lbf in; lb in)	x 1.152	= Kilograms-force centimetre (kgf cm; kg cm)	x 0.868	= Pounds-force inches (lbf in; lb in)
Pounds-force inches (lbf in; lb in)	x 0.113	= Newton metres (Nm)	x 8.85	= Pounds-force inches (lbf in; lb in)
Pounds-force inches (lbf in; lb in)	x 0.083	= Pounds-force feet (lbf ft; lb ft)	x 12	= Pounds-force inches (lbf in; lb in)
Pounds-force feet (lbf ft; lb ft)	x 0.138	= Kilograms-force metres (kgf m; kg m)	x 7.233	= Pounds-force feet (lbf ft; lb ft)
Pounds-force feet (lbf ft; lb ft)	x 1.356	= Newton metres (Nm)	x 0.738	= Pounds-force feet (lbf ft; lb ft)
Newton metres (Nm)	x 0.102	= Kilograms-force metres (kgf m; kg m)	x 9.804	= Newton metres (Nm)

Power

Horsepower (hp)	x 745.7	= Watts (W)	x 0.0013	= Horsepower (hp)

Velocity (speed)

Miles per hour (miles/hr; mph)	x 1.609	= Kilometres per hour (km/hr; kph)	x 0.621	= Miles per hour (miles/hr; mph)

Fuel consumption*

Miles per gallon (mpg)	x 0.354	= Kilometres per litre (km/l)	x 2.825	= Miles per gallon (mpg)

Temperature

Degrees Fahrenheit = (°C x 1.8) + 32   Degrees Celsius (Degrees Centigrade; °C) = (°F - 32) x 0.56

*It is common practice to convert from miles per gallon (mpg) to litres/100 kilometres (l/100km), where mpg x l/100 km = 282

A number of chemicals and lubricants are available for use in motorcycle maintenance and repair. They include a wide variety of products ranging from cleaning solvents and degreasers to lubricants and protective sprays for rubber, plastic and vinyl.

● **Contact point/spark plug cleaner** is a solvent used to clean oily film and dirt from points, grime from electrical connectors and oil deposits from spark plugs. It is oil free and leaves no residue. It can also be used to remove gum and varnish from carburettor jets and other orifices.

● **Carburettor cleaner** is similar to contact point/spark plug cleaner but it usually has a stronger solvent and may leave a slight oily reside. It is not recommended for cleaning electrical components or connections.

● **Brake system cleaner** is used to remove grease or brake fluid from brake system components (where clean surfaces are absolutely necessary and petroleum-based solvents cannot be used); it also leaves no residue.

● **Silicone-based lubricants** are used to protect rubber parts such as hoses and grommets, and are used as lubricants for hinges and locks.

● **Multi-purpose grease** is an all purpose lubricant used wherever grease is more practical than a liquid lubricant such as oil. Some multi-purpose grease is coloured white and specially formulated to be more resistant to water than ordinary grease.

● **Gear oil** (sometimes called gear lube) is a specially designed oil used in transmissions and final drive units, as well as other areas where high friction, high temperature lubrication is required. It is available in a number of viscosities (weights) for various applications.

● **Motor oil**, of course, is the lubricant specially formulated for use in the engine. It normally contains a wide variety of additives to prevent corrosion and reduce foaming and wear. Motor oil comes in various weights (viscosity ratings) of from 5 to 80. The recommended weight of the oil depends on the seasonal temperature and the demands on the engine. Light oil is used in cold climates and under light load conditions; heavy oil is used in hot climates and where high loads are encountered. Multi-viscosity oils are designed to have characteristics of both light and heavy oils and are available in a number of weights from 5W-20 to 20W-50.

● **Petrol additives** perform several functions, depending on their chemical makeup. They usually contain solvents that help dissolve gum and varnish that build up on carburettor and inlet parts. They also serve to break down carbon deposits that form on the inside surfaces of the combustion chambers. Some additives contain upper cylinder lubricants for valves and piston rings.

● **Brake and clutch fluid** is a specially formulated hydraulic fluid that can withstand the heat and pressure encountered in brake/clutch systems. Care must be taken that this fluid does not come in contact with painted surfaces or plastics. An opened container should always be resealed to prevent contamination by water or dirt.

● **Chain lubricants** are formulated especially for use on motorcycle final drive chains. A good chain lube should adhere well and have good penetrating qualities to be effective as a lubricant inside the chain and on the side plates, pins and rollers. Most chain lubes are either the foaming type or quick drying type and are usually marketed as sprays. Take care to use a lubricant marked as being suitable for O-ring chains.

● **Degreasers** are heavy duty solvents used to remove grease and grime that may accumulate on engine and frame components. They can be sprayed or brushed on and, depending on the type, are rinsed with either water or solvent.

● **Solvents** are used alone or in combination with degreasers to clean parts and assemblies during repair and overhaul. The home mechanic should use only solvents that are non-flammable and that do not produce irritating fumes.

● **Gasket sealing compounds** may be used in conjunction with gaskets, to improve their sealing capabilities, or alone, to seal metal-to-metal joints. Many gasket sealers can withstand extreme heat, some are impervious to petrol and lubricants, while others are capable of filling and sealing large cavities. Depending on the intended use, gasket sealers either dry hard or stay relatively soft and pliable. They are usually applied by hand, with a brush, or are sprayed on the gasket sealing surfaces.

● **Thread locking compound** is an adhesive locking compound that prevents threaded fasteners from loosening because of vibration. It is available in a variety of types for different applications.

● **Moisture dispersants** are usually sprays that can be used to dry out electrical components such as the fuse block and wiring connectors. Some types can also be used as treatment for rubber and as a lubricant for hinges, cables and locks.

● **Waxes and polishes** are used to help protect painted and plated surfaces from the weather. Different types of paint may require the use of different types of wax polish. Some polishes utilise a chemical or abrasive cleaner to help remove the top layer of oxidised (dull) paint on older vehicles. In recent years, many non-wax polishes (that contain a wide variety of chemicals such as polymers and silicones) have been introduced. These non-wax polishes are usually easier to apply and last longer than conventional waxes and polishes.

About the MOT Test

In the UK, all vehicles more than three years old are subject to an annual test to ensure that they meet minimum safety requirements. A current test certificate must be issued before a machine can be used on public roads, and is required before a road fund licence can be issued. Riding without a current test certificate will also invalidate your insurance.

For most owners, the MOT test is an annual cause for anxiety, and this is largely due to owners not being sure what needs to be checked prior to submitting the motorcycle for testing. The simple answer is that a fully roadworthy motorcycle will have no difficulty in passing the test.

This is a guide to getting your motorcycle through the MOT test. Obviously it will not be possible to examine the motorcycle to the same standard as the professional MOT tester, particularly in view of the equipment required for some of the checks. However, working through the following procedures will enable you to identify any problem areas before submitting the motorcycle for the test.

It has only been possible to summarise the test requirements here, based on the regulations in force at the time of printing. Test standards are becoming increasingly stringent, although there are some exemptions for older vehicles. More information about the MOT test can be obtained from the HMSO publications, *How Safe is your Motorcycle* and *The MOT Inspection Manual for Motorcycle Testing.*

Many of the checks require that one of the wheels is raised off the ground. If the motorcycle doesn't have a centre stand, note that an auxiliary stand will be required. Additionally, the help of an assistant may prove useful.

Certain exceptions apply to machines under 50 cc, machines without a lighting system, and Classic bikes - if in doubt about any of the requirements listed below seek confirmation from an MOT tester prior to submitting the motorcycle for the test.

Check that the frame number is clearly visible.

> **HAYNES HiNT** *If a component is in borderline condition, the tester has discretion in deciding whether to pass or fail it. If the motorcycle presented is clean and evidently well cared for, the tester may be more inclined to pass a borderline component than if the motorcycle is scruffy and apparently neglected.*

Electrical System

Lights, turn signals, horn and reflector

✔ With the ignition on, check the operation of the following electrical components. **Note:** *The electrical components on certain small-capacity machines are powered by the generator, requiring that the engine is run for this check.*

  a) Headlight and tail light. Check that both illuminate in the low and high beam switch positions.
  b) Position lights. Check that the front position (or sidelight) and tail light illuminate in this switch position.
  c) Turn signals. Check that all flash at the correct rate, and that the warning light(s) function correctly. Check that the turn signal switch works correctly.
  c) Hazard warning system (where fitted). Check that all four turn signals flash in this switch position.
  d) Brake stop light. Check that the light comes on when the front and rear brakes are independently applied. Models first used on or after 1st April 1986 must have a brake light switch on each brake.
  e) Horn. Check that the sound is continuous and of reasonable volume.

✔ Check that there is a red reflector on the rear of the machine, either mounted separately or as part of the tail light lens.
✔ Check the condition of the headlight, tail light and turn signal lenses.

Headlight beam height

✔ The MOT tester will perform a headlight beam height check using specialised beam setting equipment (see illustration 1). This equipment will not be available to the home mechanic, but if you suspect that the headlight is incorrectly set or may have been maladjusted in the past, you can perform a rough test as follows.
✔ Position the bike in a straight line facing a brick wall. The bike must be off its stand, upright and with a rider seated. Measure the height from the ground to the centre of the headlight and mark a horizontal line on the wall at this height. Position the motorcycle 3.8 metres from the wall and draw a vertical

Headlight beam height checking equipment

line up the wall central to the centreline of the motorcycle. Switch to dipped beam and check that the beam pattern falls slightly lower than the horizontal line and to the left of the vertical line (see illustration 2).

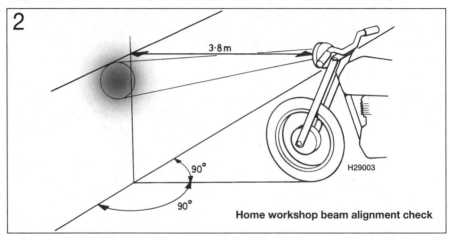

Home workshop beam alignment check

Exhaust System and Final Drive

Exhaust

✔ Check that the exhaust mountings are secure and that the system does not foul any of the rear suspension components.
✔ Start the motorcycle. When the revs are increased, check that the exhaust is neither holed nor leaking from any of its joints. On a linked system, check that the collector box is not leaking due to corrosion.

✔ Note that the exhaust decibel level ("loudness" of the exhaust) is assessed at the discretion of the tester. If the motorcycle was first used on or after 1st January 1985 the silencer must carry the BSAU 193 stamp, or a marking relating to its make and model, or be of OE (original equipment) manufacture. If the silencer is marked NOT FOR ROAD USE, RACING USE ONLY or similar, it will fail the MOT.

Final drive

✔ On chain or belt drive machines, check that the chain/belt is in good condition and does not have excessive slack. Also check that the sprocket is securely mounted on the rear wheel hub. Check that the chain/belt guard is in place.
✔ On shaft drive bikes, check for oil leaking from the drive unit and fouling the rear tyre.

Steering and Suspension

Steering

✔ With the front wheel raised off the ground, rotate the steering from lock to lock. The handlebar or switches must not contact the fuel tank or be close enough to trap the rider's hand. Problems can be caused by damaged lock stops on the lower yoke and frame, or by the fitting of non-standard handlebars.
✔ When performing the lock to lock check, also ensure that the steering moves freely without drag or notchiness. Steering movement can be impaired by poorly routed cables, or by overtight head bearings or worn bearings. The

tester will perform a check of the steering head bearing lower race by mounting the front wheel on a surface plate, then performing a lock to lock check with the weight of the machine on the lower bearing (see illustration 3).
✔ Grasp the fork sliders (lower legs) and attempt to push and pull on the forks (see illustration 4). Any play in the steering head bearings will be felt. Note that in extreme cases, wear of the front fork bushes can be misinterpreted for head bearing play.
✔ Check that the handlebars are securely mounted.
✔ Check that the handlebar grip rubbers are secure. They should by bonded to the bar left end and to the throttle cable pulley on the right end.

Front suspension

✔ With the motorcycle off the stand, hold the front brake on and pump the front forks up and down (see illustration 5). Check that they are adequately damped.
✔ Inspect the area above and around the front fork oil seals (see illustration 6). There should be no sign of oil on the fork tube (stanchion) nor leaking down the slider (lower leg). On models so equipped, check that there is no oil leaking from the anti-dive units.
✔ On models with swingarm front suspension, check that there is no freeplay in the linkage when moved from side to side.

Rear suspension

✔ With the motorcycle off the stand and an assistant supporting the motorcycle by its handlebars, bounce the rear suspension (see illustration 7). Check that the suspension components do not foul on any of the cycle parts and check that the shock absorber(s) provide adequate damping.

3 Front wheel mounted on a surface plate for steering head bearing lower race check

4 Checking the steering head bearings for freeplay

5 Hold the front brake on and pump the front forks up and down to check operation

6 Inspect the area around the fork dust seal for oil leakage (arrow)

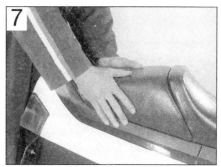

7 Bounce the rear of the motorcycle to check rear suspension operation

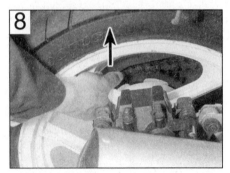

Checking for rear suspension linkage play

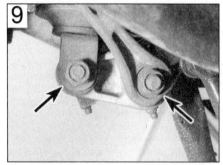

Worn suspension linkage pivots (arrows) are usually the cause of play in the rear suspension

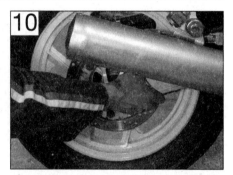

Grasp the swingarm at the ends to check for play in its pivot bearings

✔ Visually inspect the shock absorber(s) and check that there is no sign of oil leakage from its damper. This is somewhat restricted on certain single shock models due to the location of the shock absorber.

✔ With the rear wheel raised off the ground, grasp the wheel at the highest point

and attempt to pull it up **(see illustration 8)**. Any play in the swingarm pivot or suspension linkage bearings will be felt as movement. **Note:** *Do not confuse play with actual suspension movement.* Failure to lubricate suspension linkage bearings can lead to bearing failure **(see illustration 9)**.

✔ With the rear wheel raised off the ground, grasp the swingarm ends and attempt to move the swingarm from side to side and forwards and backwards - any play indicates wear of the swingarm pivot bearings **(see illustration 10)**.

Brakes, Wheels and Tyres

Brakes

✔ With the wheel raised off the ground, apply the brake then free it off, and check that the wheel is about to revolve freely without brake drag.

✔ On disc brakes, examine the disc itself. Check that it is securely mounted and not cracked.

✔ On disc brakes, view the pad material through the caliper mouth and check that the pads are not worn down beyond the limit **(see illustration 11)**.

✔ On drum brakes, check that when the brake is applied the angle between the operating lever and cable or rod is not too great **(see illustration 12)**. Check also that the operating lever doesn't foul any other components.

✔ On disc brakes, examine the flexible hoses from top to bottom. Have an assistant hold the brake on so that the fluid in the hose is under pressure, and check that there is no sign of fluid leakage, bulges or cracking. If there are any metal brake pipes or unions, check that these are free from corrosion and damage. Where a brake-linked anti-dive system is fitted, check the hoses to the anti-dive in a similar manner.

✔ Check that the rear brake torque arm is secure and that its fasteners are secured by self-locking nuts or castellated nuts with split-pins or R-pins **(see illustration 13)**.

✔ On models with ABS, check that the self-check warning light in the instrument panel works.

✔ The MOT tester will perform a test of the motorcycle's braking efficiency based on a calculation of rider and motorcycle weight. Although this cannot be carried out at home, you can at least ensure that the braking

systems are properly maintained. For hydraulic disc brakes, check the fluid level, lever/pedal feel (bleed of air if its spongy) and pad material. For drum brakes, check adjustment, cable or rod operation and shoe lining thickness.

Wheels and tyres

✔ Check the wheel condition. Cast wheels should be free from cracks and if of the built-up design, all fasteners should be secure. Spoked wheels should be checked for broken, corroded, loose or bent spokes.

✔ With the wheel raised off the ground, spin the wheel and visually check that the tyre and wheel run true. Check that the tyre does not foul the suspension or mudguards.

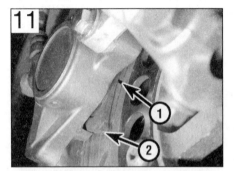

Brake pad wear can usually be viewed without removing the caliper. Most pads have wear indicator grooves (1) and some also have indicator tangs (2)

On drum brakes, check the angle of the operating lever with the brake fully applied. Most drum brakes have a wear indicator pointer and scale.

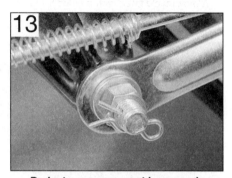

Brake torque arm must be properly secured at both ends

Check for wheel bearing play by trying to move the wheel about the axle (spindle)

Checking the tyre tread depth

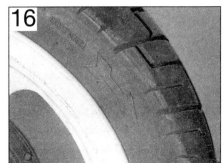

Tyre direction of rotation arrow can be found on tyre sidewall

Castellated type wheel axle (spindle) nut must be secured by a split pin or R-pin

Two straightedges are used to check wheel alignment

✔ With the wheel raised off the ground, grasp the wheel and attempt to move it about the axle (spindle) **(see illustration 14)**. Any play felt here indicates wheel bearing failure.
✔ Check the tyre tread depth, tread condition and sidewall condition **(see illustration 15)**.
✔ Check the tyre type. Front and rear tyre types must be compatible and be suitable for road use. Tyres marked NOT FOR ROAD USE, COMPETITION USE ONLY or similar, will fail the MOT.
✔ If the tyre sidewall carries a direction of rotation arrow, this must be pointing in the direction of normal wheel rotation **(see illustration 16)**.
✔ Check that the wheel axle (spindle) nuts (where applicable) are properly secured. A self-locking nut or castellated nut with a split-pin or R-pin can be used **(see illustration 17)**.
✔ Wheel alignment is checked with the motorcycle off the stand and a rider seated. With the front wheel pointing straight ahead, two perfectly straight lengths of metal or wood and placed against the sidewalls of both tyres **(see illustration 18)**. The gap each side of the front tyre must be equidistant on both sides. Incorrect wheel alignment may be due to a cocked rear wheel (often as the result of poor chain adjustment) or in extreme cases, a bent frame.

General checks and condition

✔ Check the security of all major fasteners, bodypanels, seat, fairings (where fitted) and mudguards.

✔ Check that the rider and pillion footrests, handlebar levers and brake pedal are securely mounted.

✔ Check for corrosion on the frame or any load-bearing components. If severe, this may affect the structure, particularly under stress.

Sidecars

A motorcycle fitted with a sidecar requires additional checks relating to the stability of the machine and security of attachment and swivel joints, plus specific wheel alignment (toe-in) requirements. Additionally, tyre and lighting requirements differ from conventional motorcycle use. Owners are advised to check MOT test requirements with an official test centre.

Preparing for storage

Before you start

If repairs or an overhaul is needed, see that this is carried out now rather than left until you want to ride the bike again.

Give the bike a good wash and scrub all dirt from its underside. Make sure the bike dries completely before preparing for storage.

Engine

● Remove the spark plug(s) and lubricate the cylinder bores with approximately a teaspoon of motor oil using a spout-type oil can **(see illustration 1)**. Reinstall the spark plug(s). Crank the engine over a couple of times to coat the piston rings and bores with oil. If the bike has a kickstart, use this to turn the engine over. If not, flick the kill switch to the OFF position and crank the engine over on the starter **(see illustration 2)**. If the nature on the ignition system prevents the starter operating with the kill switch in the OFF position,

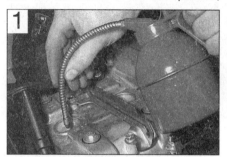

Squirt a drop of motor oil into each cylinder

Flick the kill switch to OFF . . .

. . . and ensure that the metal bodies of the plugs (arrows) are earthed against the cylinder head

remove the spark plugs and fit them back in their caps; ensure that the plugs are earthed (grounded) against the cylinder head when the starter is operated **(see illustration 3)**.

⚠ *Warning: It is important that the plugs are earthed (grounded) away from the spark plug holes otherwise there is a risk of atomised fuel from the cylinders igniting.*

HAYNES HiNT *On a single cylinder four-stroke engine, you can seal the combustion chamber completely by positioning the piston at TDC on the compression stroke.*

Connect a hose to the carburettor float chamber drain stub (arrow) and unscrew the drain screw

● Drain the carburettor(s) otherwise there is a risk of jets becoming blocked by gum deposits from the fuel **(see illustration 4)**.

● If the bike is going into long-term storage, consider adding a fuel stabiliser to the fuel in the tank. If the tank is drained completely, corrosion of its internal surfaces may occur if left unprotected for a long period. The tank can be treated with a rust preventative especially for this purpose. Alternatively, remove the tank and pour half a litre of motor oil into it, install the filler cap and shake the tank to coat its internals with oil before draining off the excess. The same effect can also be achieved by spraying WD40 or a similar water-dispersant around the inside of the tank via its flexible nozzle.

● Make sure the cooling system contains the correct mix of antifreeze. Antifreeze also contains important corrosion inhibitors.

● The air intakes and exhaust can be sealed off by covering or plugging the openings. Ensure that you do not seal in any condensation; run the engine until it is hot, then switch off and allow to cool. Tape a piece of thick plastic over the silencer end(s) **(see illustration 5)**. Note that some advocate pouring a tablespoon of motor oil into the silencer(s) before sealing them off.

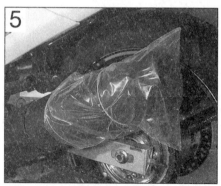

Exhausts can be sealed off with a plastic bag

Battery

● Remove it from the bike - in extreme cases of cold the battery may freeze and crack its case **(see illustration 6)**.

Disconnect the negative lead (A) first, followed by the positive lead (B)

● Check the electrolyte level and top up if necessary (conventional refillable batteries). Clean the terminals.

● Store the battery off the motorcycle and away from any sources of fire. Position a wooden block under the battery if it is to sit on the ground.

● Give the battery a trickle charge for a few hours every month **(see illustration 7)**.

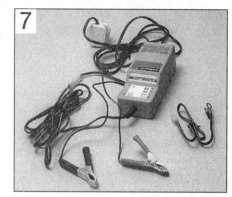

Use a suitable battery charger - this kit also assess battery condition

Tyres

● Place the bike on its centrestand or an auxiliary stand which will support the motorcycle in an upright position. Position wood blocks under the tyres to keep them off the ground and to provide insulation from damp. If the bike is being put into long-term storage, ideally both tyres should be off the ground; not only will this protect the tyres, but will also ensure that no load is placed on the steering head or wheel bearings.

● Deflate each tyre by 5 to 10 psi, no more or the beads may unseat from the rim, making subsequent inflation difficult on tubeless tyres.

Pivots and controls

● Lubricate all lever, pedal, stand and footrest pivot points. If grease nipples are fitted to the rear suspension components, apply lubricant to the pivots.

● Lubricate all control cables.

Cycle components

● Apply a wax protectant to all painted and plastic components. Wipe off any excess, but don't polish to a shine. Where fitted, clean the screen with soap and water.

● Coat metal parts with Vaseline (petroleum jelly). When applying this to the fork tubes, do not compress the forks otherwise the seals will rot from contact with the Vaseline.

● Apply a vinyl cleaner to the seat.

Storage conditions

● Aim to store the bike in a shed or garage which does not leak and is free from damp.

● Drape an old blanket or bedspread over the bike to protect it from dust and direct contact with sunlight (which will fade paint). This also hides the bike from prying eyes. Beware of tight-fitting plastic covers which may allow condensation to form and settle on the bike.

Getting back on the road

Engine and transmission

● Change the oil and replace the oil filter. If this was done prior to storage, check that the oil hasn't emulsified - a thick whitish substance which occurs through condensation.

● Remove the spark plugs. Using a spout-type oil can, squirt a few drops of oil into the cylinder(s). This will provide initial lubrication as the piston rings and bores comes back into contact. Service the spark plugs, or fit new ones, and install them in the engine.

● Check that the clutch isn't stuck on. The plates can stick together if left standing for some time, preventing clutch operation. Engage a gear and try rocking the bike back and forth with the clutch lever held against the handlebar. If this doesn't work on cable-operated clutches, hold the clutch lever back against the handlebar with a strong elastic band or cable tie for a couple of hours (see illustration 8).

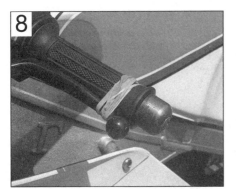

Hold clutch lever back against the handlebar with elastic bands or a cable tie

● If the air intakes or silencer end(s) were blocked off, remove the bung or cover used.

● If the fuel tank was coated with a rust preventative, oil or a stabiliser added to the fuel, drain and flush the tank and dispose of the fuel sensibly. If no action was taken with the fuel tank prior to storage, it is advised that the old fuel is disposed of since it will go off over a period of time. Refill the fuel tank with fresh fuel.

Frame and running gear

● Oil all pivot points and cables.

● Check the tyre pressures. They will definitely need inflating if pressures were reduced for storage.

● Lubricate the final drive chain (where applicable).

● Remove any protective coating applied to the fork tubes (stanchions) since this may well destroy the fork seals. If the fork tubes weren't protected and have picked up rust spots, remove them with very fine abrasive paper and refinish with metal polish.

● Check that both brakes operate correctly. Apply each brake hard and check that it's not possible to move the motorcycle forwards, then check that the brake frees off again once released. Brake caliper pistons can stick due to corrosion around the piston head, or on the sliding caliper types, due to corrosion of the slider pins. If the brake doesn't free after repeated operation, take the caliper off for examination. Similarly drum brakes can stick due to a seized operating cam, cable or rod linkage.

● If the motorcycle has been in long-term storage, renew the brake fluid and clutch fluid (where applicable).

● Depending on where the bike has been stored, the wiring, cables and hoses may have been nibbled by rodents. Make a visual check and investigate disturbed wiring loom tape.

Battery

● If the battery has been previously removal and given top up charges it can simply be reconnected. Remember to connect the positive cable first and the negative cable last.

● On conventional refillable batteries, if the battery has not received any attention, remove it from the motorcycle and check its electrolyte level. Top up if necessary then charge the battery. If the battery fails to hold a charge and a visual checks show heavy white sulphation of the plates, the battery is probably defective and must be renewed. This is particularly likely if the battery is old. Confirm battery condition with a specific gravity check.

● On sealed (MF) batteries, if the battery has not received any attention, remove it from the motorcycle and charge it according to the information on the battery case - if the battery fails to hold a charge it must be renewed.

Starting procedure

● If a kickstart is fitted, turn the engine over a couple of times with the ignition OFF to distribute oil around the engine. If no kickstart is fitted, flick the engine kill switch OFF and the ignition ON and crank the engine over a couple of times to work oil around the upper cylinder components. If the nature of the ignition system is such that the starter won't work with the kill switch OFF, remove the spark plugs, fit them back into their caps and earth (ground) their bodies on the cylinder head. Reinstall the spark plugs afterwards.

● Switch the kill switch to RUN, operate the choke and start the engine. If the engine won't start don't continue cranking the engine - not only will this flatten the battery, but the starter motor will overheat. Switch the ignition off and try again later. If the engine refuses to start, go through the fault finding procedures in this manual. **Note:** *If the bike has been in storage for a long time, old fuel or a carburettor blockage may be the problem. Gum deposits in carburettors can block jets - if a carburettor cleaner doesn't prove successful the carburettors must be dismantled for cleaning.*

● Once the engine has started, check that the lights, turn signals and horn work properly.

● Treat the bike gently for the first ride and check all fluid levels on completion. Settle the bike back into the maintenance schedule.

This Section provides an easy reference-guide to the more common faults that are likely to afflict your machine. Obviously, the opportunities are almost limitless for faults to occur as a result of obscure failures, and to try and cover all eventualities would require a book. Indeed, a number have been written on the subject.

Successful troubleshooting is not a mysterious 'black art' but the application of a bit of knowledge combined with a systematic and logical approach to the problem. Approach any troubleshooting by first accurately identifying the symptom and then checking through the list of possible causes, starting with the simplest or most obvious and progressing in stages to the most complex.

Take nothing for granted, but above all apply liberal quantities of common sense.

The main symptom of a fault is given in the text as a major heading below which are listed the various systems or areas which may contain the fault. Details of each possible cause for a fault and the remedial action to be taken are given, in brief, in the paragraphs below each heading. Further information should be sought in the relevant Chapter.

1 Engine doesn't start or is difficult to start

- [ ] Starter motor doesn't rotate
- [ ] Starter motor rotates but engine does not turn over
- [ ] Starter works but engine won't turn over (seized)
- [ ] No fuel flow
- [ ] Engine flooded
- [ ] No spark or weak spark
- [ ] Compression low
- [ ] Stalls after starting
- [ ] Rough idle

2 Poor running at low speed

- [ ] Spark weak
- [ ] Fuel/air mixture incorrect
- [ ] Compression low
- [ ] Poor acceleration

3 Poor running or no power at high speed

- [ ] Firing incorrect
- [ ] Fuel/air mixture incorrect
- [ ] Compression low
- [ ] Knocking or pinking
- [ ] Miscellaneous causes

4 Overheating

- [ ] Cooling system not operating properly
- [ ] Firing incorrect
- [ ] Fuel/air mixture incorrect
- [ ] Compression too high
- [ ] Engine load excessive
- [ ] Lubrication inadequate
- [ ] Miscellaneous causes

5 Clutch problems

- [ ] Clutch slipping
- [ ] Clutch not disengaging completely

6 Gearchanging problems

- [ ] Doesn't go into gear, or lever doesn't return
- [ ] Jumps out of gear
- [ ] Overselects

7 Abnormal engine noise

- [ ] Knocking or pinking
- [ ] Piston slap or rattling
- [ ] Valve noise
- [ ] Other noise

8 Abnormal driveline noise

- [ ] Clutch noise
- [ ] Transmission noise
- [ ] Chain or final drive noise

9 Abnormal frame and suspension noise

- [ ] Front end noise
- [ ] Shock absorber noise
- [ ] Brake noise

10 Oil pressure light comes on

- [ ] Engine lubrication system
- [ ] Electrical system

11 Excessive exhaust smoke

- [ ] White smoke
- [ ] Black smoke
- [ ] Brown smoke

12 Poor handling or stability

- [ ] Handlebar hard to turn
- [ ] Handlebar shakes or vibrates excessively
- [ ] Handlebar pulls to one side
- [ ] Poor shock absorbing qualities

13 Braking problems

- [ ] Brakes are spongy, don't hold
- [ ] Brake lever or pedal pulsates
- [ ] Brakes drag

14 Electrical problems

- [ ] Battery dead or weak
- [ ] Battery overcharged

1 Engine doesn't start or is difficult to start

Starter motor does'nt rotate

- [ ] Engine stop switch Off.
- [ ] Fuse blown. Check circuit fuses (Chapter 9).
- [ ] Battery voltage low. Check and recharge battery (Chapter 9).
- [ ] Starter motor defective. Make sure the wiring to the starter is secure. Make sure the starter solenoid (relay) clicks when the start button is pushed. If the solenoid clicks, then the fault is in the wiring or motor.
- [ ] Starter solenoid (relay) faulty. It is located behind the left side cover. Check it according to the procedure in Chapter 9.
- [ ] Starter button not contacting. The contacts could be wet, corroded or dirty. Disassemble and clean the switch (Chapter 9).
- [ ] Wiring open or shorted. Check all wiring connections and harnesses to make sure that they are dry, tight and not corroded. Also check for broken or frayed wires that can cause a short to ground (see wiring diagrams, Chapter 9).
- [ ] Ignition switch defective. Check the switch according to the procedure in Chapter 9. Replace the switch with a new one if it is defective.
- [ ] Engine stop switch defective. Check for wet, dirty or corroded contacts. Clean or replace the switch as necessary (Chapter 9).
- [ ] Faulty starter lockout switch. Check the wiring to the switch and the switch itself according to the procedures in Chapter 9.

Starter motor rotates but engine does not turn over

- [ ] Starter motor clutch defective. Inspect and repair or replace (Chapter 2).
- [ ] Damaged idler or starter gears. Inspect and replace the damaged parts (Chapter 2).

Starter works but engine won't turn over (seized)

- [ ] Seized engine caused by one or more internally damaged components. Failure due to wear, abuse or lack of lubrication. Damage can include seized valves, camshafts, pistons, crankshaft, connecting rod bearings, or transmission gears or bearings. Refer to Chapter 2 for engine disassembly.

No fuel flow

- [ ] No fuel in tank.
- [ ] Fuel tap vacuum hose broken or disconnected.
- [ ] Tank cap air vent obstructed. Usually caused by dirt or water. Remove it and clean the cap vent hole.
- [ ] Fuel tap clogged. Remove the tap and clean it and the filter (Chapter 4).
- [ ] Fuel line clogged. Pull the fuel line loose and carefully blow through it.
- [ ] Inlet needle valves clogged. For all of the valves to be clogged, either a very bad batch of fuel with an unusual additive has been used, or some other foreign object has entered the tank. Many times after a machine has been stored for many months without running, the fuel turns to a varnish-like liquid and forms deposits on the inlet needle valves and jets. The carburetors should be removed and overhauled if draining the float bowls does not alleviate the problem.

Engine flooded

- [ ] Float level too high. Check and adjust as described in Chapter 4.
- [ ] Inlet needle valve worn or stuck open. A piece of dirt, rust or other debris can cause the inlet needle to seat improperly, causing excess fuel to be admitted to the float bowl. In this case, the float chamber should be cleaned and the needle and seat inspected. If the needle and seat are worn, then the leaking will persist and the parts should be replaced with new ones (Chapter 4).
- [ ] Starting technique incorrect. Under normal circumstances (i.e., if all the carburetor functions are sound) the machine should start with little or no throttle. When the engine is cold, the choke should be operated and the engine started without opening the throttle. When the engine is at operating temperature, only a very slight amount of throttle should be necessary. If the engine is flooded, turn the fuel tap off and hold the throttle open while cranking the engine. This will allow additional air to reach the cylinders. Remember to turn the gas back on after the engine starts.

No spark or weak spark

- [ ] Ignition switch Off.
- [ ] Engine stop switch turned to the Off position.
- [ ] Battery voltage low. Check and recharge battery as necessary (Chapter 9).
- [ ] Spark plug dirty, defective or worn out. Locate reason for fouled plug(s) using spark plug condition chart on the inside rear cover and follow the plug maintenance procedures in Chapter 1.
- [ ] Spark plug cap or high-tension wiring faulty. Check condition. Replace either or both components if cracks or deterioration are evident (Chapter 5).
- [ ] Spark plug cap not making good contact. Make sure the plug cap fits snugly over the plug end.
- [ ] IC igniter defective. Check the unit. referring to Chapter 5 for details.
- [ ] Pickup coil defective. Check the unit, referring to Chapter 5 for details.
- [ ] Ignition coil(s) defective. Check the coils, referring to Chapter 5.
- [ ] Ignition or stop switch shorted. This is usually caused by water, corrosion, damage or excessive wear. The switches can be disassembled and cleaned with electrical contact cleaner. If cleaning does not help, replace the switches (Chapter 9).
- [ ] Wiring shorted or broken between:
  - a) Ignition switch and engine stop switch
  - b) IC igniter and engine stop switch
  - c) IC igniter and ignition coil
  - d) Ignition coil and plug
  - e) IC igniter and pickup coils
- [ ] Make sure that all wiring connections are clean, dry and tight. Look for chafed and broken wires (Chapters 5 and 9).

Compression low

- [ ] Spark plug loose. Remove the plug and inspect the threads. Reinstall and tighten to the specified torque (Chapter 1).
- [ ] Cylinder head not sufficiently tightened down. If the cylinder head is suspected of being loose, then there's a chance that the gasket or head is damaged if the problem has persisted for any length of time. The head nuts/bolts should be tightened to the proper torque in the correct sequence (Chapter 2).
- [ ] Improper valve clearance. This means that the valve is not closing completely and compression pressure is leaking past the valve. Check and adjust the valve clearances (Chapter 1).
- [ ] Cylinder and/or piston worn. Excessive wear will cause compression pressure to leak past the rings. This is usually accompanied by worn rings as well. A top end overhaul is necessary (Chapter 2).
- [ ] Piston rings worn, weak, broken, or sticking. Broken or sticking piston rings usually indicate a lubrication or carburetion problem that causes excess carbon deposits or seizures to form on the pistons and rings. Top end overhaul is necessary (Chapter 2).
- [ ] Piston ring-to-groove clearance excessive. This is caused by excessive wear of the piston ring lands. Piston replacement is necessary (Chapter 2).
- [ ] Cylinder head gasket damaged. If the head is allowed to become loose, or if excessive carbon build-up on the piston crown and combustion chamber causes extremely high compression, the head gasket may leak. Retorquing the head is not always sufficient to restore the seal, so gasket replacement is necessary (Chapter 2).

1 Engine doesn't start or is difficult to start (continued)

- [ ] Cylinder head warped. This is caused by overheating or improperly tightened head nuts/bolts. Machine shop resurfacing or head replacement is necessary (Chapter 2).
- [ ] Valve spring broken or weak. Caused by component failure or wear; the spring(s) must be replaced (Chapter 2).
- [ ] Valve not seating properly. This is caused by a bent valve (from over-revving or improper valve adjustment), burned valve or seat (improper carburetion) or an accumulation of carbon deposits on the seat (from carburetion, lubrication problems). The valves must be cleaned and/or replaced and the seats serviced if possible (Chapter 2).

Stalls after starting

- [ ] Improper choke action. Make sure the choke rod is getting a full stroke and staying in the "out" position. Adjustment of the cable slack is covered in Chapter 4.
- [ ] Ignition malfunction. See Chapter 5.
- [ ] Carburetor malfunction. See Chapter 4.
- [ ] Fuel contaminated. The fuel can be contaminated with either dirt or water, or can change chemically if the machine is allowed to sit

for several months or more. Drain the tank and float bowls (Chapter 4).
- [ ] Intake air leak. Check for loose carburetor-to-intake manifold connections, loose or missing vacuum gauge access port cap or hose, or loose carburetor top (Chapter 4).
- [ ] Idle speed incorrect. Turn idle speed adjuster screw until the engine idles at the specified rpm (Chapters 1 and 4).

Rough idle

- [ ] Ignition malfunction. See Chapter 5.
- [ ] Idle speed incorrect. See Chapter 1.
- [ ] Carburetors not synchronised. Adjust carburetors with vacuum gauge set or manometer as outlined in Chapter 1.
- [ ] Carburetor malfunction. See Chapter 4.
- [ ] Fuel contaminated. The fuel can be contaminated with either dirt or water, or can change chemically if the machine is allowed to sit for several months or more. Drain the tank and float bowls. If the problem is severe, a carburetor overhaul may be necessary (Chapters 1 and 4).
- [ ] Intake air leak (Chapter 4).
- [ ] Air cleaner clogged. Service or replace air filter element (Chapter 1).

2 Poor running at low speeds

Spark weak

- [ ] Battery voltage low. Check and recharge battery (Chapter 9).
- [ ] Spark plug fouled, defective or worn out. Refer to Chapter 1 for spark plug maintenance.
- [ ] Spark plug cap or high tension wiring defective. Refer to Chapters 1 and 5 for details on the ignition system.
- [ ] Spark plug cap not making contact.
- [ ] Incorrect spark plug. Wrong type, heat range or cap configuration. Check and install correct plugs listed in Chapter 1. A cold plug or one with a recessed firing electrode will not operate at low speeds without fouling.
- [ ] IC igniter defective. See Chapter 5.
- [ ] Pickup coil defective. See Chapter 5.
- [ ] Ignition coil(s) defective. See Chapter 5.

Fuel/air mixture incorrect

- [ ] Pilot screw(s) out of adjustment (Chapters 1 and 4).
- [ ] Pilot jet or air passage clogged. Remove and overhaul the carburetors (Chapter 4).
- [ ] Air bleed holes clogged. Remove carburetor and blow out all passages (Chapter 4).
- [ ] Air cleaner clogged, poorly sealed or missing.
- [ ] Air cleaner-to-carburetor boot poorly sealed. Look for cracks, holes or loose clamps and replace or repair defective parts.
- [ ] Fuel level too high or too low. Adjust the floats (Chapter 4).
- [ ] Fuel tank air vent obstructed. Make sure that the air vent passage in the filler cap is open.
- [ ] Carburetor intake manifolds loose. Check for cracks, breaks, tears or loose clamps or bolts. Repair or replace the rubber boots.

Compression low

- [ ] Spark plug loose. Remove the plug and inspect the threads. Reinstall and tighten to the specified torque (Chapter 1).
- [ ] Cylinder head not sufficiently tightened down. If the cylinder head is suspected of being loose, then there's a chance that the gasket and head are damaged if the problem has persisted for any length of time. The head nuts/bolts should be tightened to the proper torque in the correct sequence (Chapter 2).
- [ ] Improper valve clearance. This means that the valve is not closing completely and compression pressure is leaking past the valve. Check and adjust the valve clearances (Chapter 1).

- [ ] Cylinder and/or piston worn. Excessive wear will cause compression pressure to leak past the rings. This is usually accompanied by worn rings as well. A top end overhaul is necessary (Chapter 2).
- [ ] Piston rings worn, weak, broken, or sticking. Broken or sticking piston rings usually indicate a lubrication or carburetion problem that causes excess carbon deposits or seizures to form on the pistons and rings. Top end overhaul is necessary (Chapter 2).
- [ ] Piston ring-to-groove clearance excessive. This is caused by excessive wear of the piston ring lands. Piston replacement is necessary (Chapter 2).
- [ ] Cylinder head gasket damaged. If the head is allowed to become loose, or if excessive carbon build-up on the piston crown and combustion chamber causes extremely high compression, the head gasket may leak. Retorquing the head is not always sufficient to restore the seal, so gasket replacement is necessary (Chapter 2).
- [ ] Cylinder head warped. This is caused by overheating or improperly tightened head nuts/bolts. Machine shop resurfacing or head replacement is necessary (Chapter 2).
- [ ] Valve spring broken or weak. Caused by component failure or wear; the spring(s) must be replaced (Chapter 2).
- [ ] Valve not seating properly. This is caused by a bent valve (from over-revving or improper valve adjustment), burned valve or seat (improper carburetion) or an accumulation of carbon deposits on the seat (from carburetion, lubrication problems). The valves must be cleaned and/or replaced and the seats serviced if possible (Chapter 2).

Poor acceleration

- [ ] Carburetors leaking or dirty. Overhaul the carburetors (Chapter 4).
- [ ] Timing not advancing. The pickup coil unit or the IC igniter may be defective. If so, they must be replaced with new ones, as they cannot be repaired.
- [ ] Carburetors not synchronised. Adjust them with a vacuum gauge set or manometer (Chapter 1).
- [ ] Engine oil viscosity too high. Using a heavier oil than that recommended in Chapter 1 can damage the oil pump or lubrication system and cause drag on the engine.
- [ ] Brakes dragging. Usually caused by debris which has entered the brake piston sealing boot, or from a warped disc or bent axle. Repair as necessary (Chapter 7).

3 Poor running or no power at high speed

Firing incorrect

- [ ] Air filter restricted. Clean or replace filter (Chapter 1).
- [ ] Spark plug fouled, defective or worn out. See Chapter 1 for spark plug maintenance.
- [ ] Spark plug cap or high tension wiring defective. See Chapters 1 and 5 for details on the ignition system.
- [ ] Spark plug cap not in good contact. See Chapter 5.
- [ ] Incorrect spark plug. Wrong type, heat range or cap configuration. Check and install correct plugs listed in Chapter 1. A cold plug or one with a recessed firing electrode will not operate at low speeds without fouling.
- [ ] IC igniter defective. See Chapter 5.
- [ ] Ignition coil(s) defective. See Chapter 5.

Fuel/air mixture incorrect

- [ ] Main jet clogged. Dirt, water and other contaminants can clog the main jets. Clean the fuel tap filter, the float bowl area, and the jets and carburetor orifices (Chapter 4).
- [ ] Main jet wrong size. The standard jetting is for sea level atmospheric pressure and oxygen content.
- [ ] Throttle shaft-to-carburetor body clearance excessive. Refer to Chapter 4 for inspection and part replacement procedures.
- [ ] Air bleed holes clogged. Remove and overhaul carburetors (Chapter 4).
- [ ] Air cleaner clogged, poorly sealed or missing.
- [ ] Air cleaner-to-carburetor boot poorly sealed. Look for cracks, holes or loose clamps, and replace or repair defective parts.
- [ ] Fuel level too high or too low. Adjust the float(s) (Chapter 4).
- [ ] Fuel tank air vent obstructed. Make sure the air vent passage in the filler cap is open.
- [ ] Carburetor intake manifolds loose. Check for cracks, breaks, tears or loose clamps or bolts. Repair or replace the rubber boots (Chapter 2).
- [ ] Fuel tap clogged. Remove the tap and clean it and the filter (Chapter 1).
- [ ] Fuel line clogged. Pull the fuel line loose and carefully blow through it.

Compression low

- [ ] Spark plug loose. Remove the plug and inspect the threads. Reinstall and tighten to the specified torque (Chapter 1).
- [ ] Cylinder head not sufficiently tightened down. If the cylinder head is suspected of being loose, then there's a chance that the gasket and head are damaged if the problem has persisted for any length of time. The head nuts/bolts should be tightened to the proper torque in the correct sequence (Chapter 2).
- [ ] Improper valve clearance. This means that the valve is not closing completely and compression pressure is leaking past the valve. Check and adjust the valve clearances (Chapter 1).
- [ ] Cylinder and/or piston worn. Excessive wear will cause compression pressure to leak past the rings. This is usually accompanied by worn rings as well. A top end overhaul is necessary (Chapter 2).

- [ ] Piston rings worn, weak, broken, or sticking. Broken or sticking piston rings usually indicate a lubrication or carburetion problem that causes excess carbon deposits or seizures to form on the pistons and rings. Top end overhaul is necessary (Chapter 2).
- [ ] Piston ring-to-groove clearance excessive. This is caused by excessive wear of the piston ring lands. Piston replacement is necessary (Chapter 2).
- [ ] Cylinder head gasket damaged. If the head is allowed to become loose, or if excessive carbon build-up on the piston crown and combustion chamber causes extremely high compression, the head gasket may leak. Retorquing the head is not always sufficient to restore the seal, so gasket replacement is necessary (Chapter 2).
- [ ] Cylinder head warped. This is caused by overheating or improperly tightened head nuts/bolts. Machine shop resurfacing or head replacement is necessary (Chapter 2).
- [ ] Valve spring broken or weak. Caused by component failure or wear; the spring(s) must be replaced (Chapter 2).
- [ ] Valve not seating properly. This is caused by a bent valve (from over-revving or improper valve adjustment), burned valve or seat (improper carburetion) or an accumulation of carbon deposits on the seat (from carburetion, lubrication problems). The valves must be cleaned and/or replaced and the seats serviced if possible (Chapter 2).

Knocking or pinging

- [ ] Carbon build-up in combustion chamber. Use of a fuel additive that will dissolve the adhesive bonding the carbon particles to the crown and chamber is the easiest way to remove the build-up. Otherwise, the cylinder head will have to be removed and decarbonized (Chapter 2).
- [ ] Incorrect or poor quality fuel. Old or improper grades of gasoline (petrol) can cause detonation. This causes the piston to rattle, thus the knocking or pinging sound. Drain old gas and always use the recommended fuel grade.
- [ ] Spark plug heat range incorrect. Uncontrolled detonation indicates the plug heat range is too hot. The plug in effect becomes a glow plug, raising cylinder temperatures. Install the proper heat range plug (Chapter 1).
- [ ] Improper air/fuel mixture. This will cause the cylinder to run hot, which leads to detonation. Clogged jets or an air leak can cause this imbalance. See Chapter 4.

Miscellaneous causes

- [ ] Throttle valve doesn't open fully. Adjust the cable slack (Chapter 1).
- [ ] Clutch slipping. Caused by a cable that is improperly adjusted or snagging or damaged, loose or worn clutch components. Refer to Chapters 1 and 2 for adjustment and overhaul procedures.
- [ ] Timing not advancing.
- [ ] Engine oil viscosity too high. Using a heavier oil than the one recommended in Chapter 1 can damage the oil pump or lubrication system and cause drag on the engine.
- [ ] Brakes dragging. Usually caused by debris which has entered the brake piston sealing boot, or from a warped disc or bent axle. Repair as necessary.

4 Overheating

Cooling system not operating properly

☐ Coolant level low. Check coolant level as described in Daily (pre-ride) checks. If coolant level is low, the engine will overheat.

☐ Leak in cooling system. Check cooling system hoses and radiator for leaks and other damage. Repair or replace parts as necessary (Chapter 3).

☐ Thermostat sticking open or closed. Check and replace as described in Chapter 3.

☐ Faulty radiator cap. Remove the cap and have it pressure checked at a service station.

☐ Coolant passages clogged. Have the entire system drained and flushed, then refill with new coolant.

☐ Water pump defective. Remove the pump and check the components.

☐ Clogged radiator fins. Clean them by blowing compressed air through the fins from the back side.

Firing incorrect

☐ Spark plug fouled, defective or worn out. See Chapter 1 for spark plug maintenance.

☐ Incorrect spark plug.

☐ Faulty ignition coil(s) (Chapter 5).

Fuel/air mixture incorrect

☐ Main jet clogged. Dirt, water and other contaminants can clog the main jets. Clean the fuel tap filter, the float bowl area and the jets and carburetor orifices (Chapter 4).

☐ Main jet wrong size. The standard jetting is for sea level atmospheric pressure and oxygen content.

☐ Air cleaner poorly sealed or missing.

☐ Air cleaner-to-carburetor boot poorly sealed. Look for cracks, holes or loose clamps and replace or repair.

☐ Fuel level too low. Adjust the float(s) (Chapter 4).

☐ Fuel tank air vent obstructed. Make sure the air vent passage in the filler cap is open.

☐ Carburetor intake manifolds loose. Check for cracks, breaks, tears or loose clamps or bolts. Repair or replace the rubber boots (Chapter 2).

Compression too high

☐ Carbon build-up in combustion chamber. Use of a fuel additive that will dissolve the adhesive bonding the carbon particles to the piston crown and chamber is the easiest way to remove the build-up. Otherwise, the cylinder head will have to be removed and decarbonized (Chapter 2).

☐ Improperly machined head surface or installation of incorrect gasket during engine assembly. Check Specifications (Chapter 2).

Engine load excessive

☐ Clutch slipping - 600 models. Caused by an out of adjustment or snagging cable or damaged, loose or worn clutch components. Refer to Chapters 1 and 2 for adjustment and overhaul procedures.

☐ Clutch slipping - 750 models. Caused by overfilled fluid reservoir or worn clutch plates or springs.

☐ Engine oil level too high. The addition of too much oil will cause pressurization of the crankcase and inefficient engine operation. Check Specifications and drain to proper level (Daily (pre-ride) checks).

☐ Engine oil viscosity too high. Using a heavier oil than the one recommended in Chapter 1 can damage the oil pump or lubrication system as well as cause drag on the engine.

☐ Brakes dragging. Usually caused by debris which has entered the brake piston sealing boot, or from a warped disc or bent axle. Repair as necessary.

Lubrication inadequate

☐ Engine oil level too low. Friction caused by intermittent lack of lubrication or from oil that is "overworked" can cause overheating. The oil provides a definite cooling function in the engine. Check the oil level (Daily (pre-ride) checks).

☐ Poor quality engine oil or incorrect viscosity or type. Oil is rated not only according to viscosity but also according to type. Some oils are not rated high enough for use in this engine. Check the Specifications section and change to the correct oil (Chapter 1).

Miscellaneous causes

☐ Modification to exhaust system. Most aftermarket exhaust systems cause the engine to run leaner, which makes them run hotter. When installing an accessory exhaust system, always rejet the carburetors.

5 Clutch problems

Clutch slipping

☐ No clutch lever play - 600 models. Adjust clutch lever free play according to the procedure in Chapter 1.

☐ Friction plates worn or warped. Overhaul the clutch assembly (Chapter 2).

☐ Steel plates worn or warped (Chapter 2).

☐ Excess fluid in clutch reservoir - 750 models (Daily (pre-ride) checks).

☐ Clutch springs broken or weak. Old or heat-damaged (from slipping clutch) springs should be replaced with new ones (Chapter 2).

☐ Clutch release not adjusted properly - 600 models. See Chapter 1.

☐ Clutch inner cable hanging up - 600 models. Caused by a rusty or frayed cable or kinked outer cable. Replace the cable (Chapter 2A). Repair of a frayed cable is not advised.

☐ Clutch release mechanism defective - 600 models. Check the shaft, cam, actuating arm and pivot. Replace any defective parts (Chapter 2A).

☐ Clutch hub or housing unevenly worn. This causes improper engagement of the discs. Replace the damaged or worn parts (Chapter 2).

Clutch not disengaging completely

☐ Clutch lever play excessive - 600 models. Adjust at bars or at engine (Chapter 1).

☐ Insufficient fluid in master cylinder reservoir - 750 models (Daily (pre-ride) checks).

☐ Air in hydraulic line - 750 models. Bleed clutch line of air (Chapter 2B).

☐ Clutch plates warped or damaged. This will cause clutch drag, which in turn causes the machine to creep. Overhaul the clutch assembly (Chapter 2).

☐ Clutch spring tension uneven. Usually caused by a sagged or broken spring. Check and replace the springs (Chapter 2).

☐ Engine oil deteriorated. Old, thin, worn out oil will not provide proper lubrication for the discs, causing the clutch to drag. Replace the oil and filter (Chapter 1).

☐ Engine oil viscosity too high. Using a heavier oil than recommended in Chapter 1 can cause the plates to stick together, putting a drag on the engine. Change to the correct weight oil (Chapter 1).

☐ Clutch housing seized on shaft. Lack of lubrication, severe wear or damage can cause the housing to seize on the shaft. Overhaul of the clutch, and perhaps transmission, may be necessary to repair damage (Chapter 2).

5 Clutch problems (continued)

☐ Clutch release mechanism defective - 600 models. Worn or damaged release mechanism parts can stick and fail to apply force to the pressure plate. Overhaul the clutch cover components (Chapter 2).

☐ Loose clutch hub nut. Causes housing and hub misalignment putting a drag on the engine. Engagement adjustment continually varies. Overhaul the clutch assembly (Chapter 2).
☐ Master cylinder or release cylinder seals defective - 750 models (Chapter 2B).

6 Gear shifting problems

Doesn't go into gear or lever doesn't return

☐ Clutch not disengaging. See Section 5.
☐ Shift fork(s) bent or seized. Often caused by dropping the machine or from lack of lubrication. Overhaul the transmission (Chapter 2).
☐ Gear(s) stuck on shaft. Most often caused by a lack of lubrication or excessive wear in transmission bearings and bushings. Overhaul the transmission (Chapter 2).
☐ Shift drum binding. Caused by lubrication failure or excessive wear. Replace the drum and bearings (Chapter 2).
☐ Shift lever return spring weak or broken (Chapter 2).
☐ Shift lever broken. Splines stripped out of lever or shaft, caused by allowing the lever to get loose or from dropping the machine. Replace necessary parts (Chapter 2).
☐ Shift mechanism pawl broken or worn. Full engagement and rotary movement of shift drum results. Replace shaft assembly (Chapter 2).

☐ Pawl spring broken. Allows pawl to "float", causing sporadic shift operation. Replace spring (Chapter 2).

Jumps out of gear

☐ Shift fork(s) worn. Overhaul the transmission (Chapter 2).
☐ Gear groove(s) worn. Overhaul the transmission (Chapter 2).
☐ Gear dogs or dog slots worn or damaged. The gears should be inspected and replaced. No attempt should be made to service the worn parts.

Overshifts

☐ Pawl spring weak or broken (Chapter 2).
☐ Shift drum stopper lever not functioning (Chapter 2).
☐ Overshift limiter broken or distorted (Chapter 2).

7 Abnormal engine noise

Knocking or pinging

☐ Carbon build-up in combustion chamber. Use of a fuel additive that will dissolve the adhesive bonding the carbon particles to the piston crown and chamber is the easiest way to remove the build-up. Otherwise, the cylinder head will have to be removed and decarbonized (Chapter 2).
☐ Incorrect or poor quality fuel. Old or improper fuel can cause detonation. This causes the piston to rattle, thus the knocking or pinging sound. Drain the old gas and always use the recommended grade fuel (Chapter 4).
☐ Spark plug heat range incorrect. Uncontrolled detonation indicates that the plug heat range is too hot. The plug in effect becomes a glow plug, raising cylinder temperatures. Install the proper heat range plug (Chapter 1).
☐ Improper air/fuel mixture. This will cause the cylinder to run hot and lead to detonation. Clogged jets or an air leak can cause this imbalance. See Chapter 4.

Piston slap or rattling

☐ Cylinder-to-piston clearance excessive. Caused by improper assembly. Inspect and overhaul top end parts (Chapter 2).
☐ Connecting rod bent. Caused by over-revving, trying to start a badly flooded engine or from ingesting a foreign object into the combustion chamber. Replace the damaged parts (Chapter 2).
☐ Piston pin or piston pin bore worn or seized from wear or lack of lubrication. Replace damaged parts (Chapter 2).
☐ Piston ring(s) worn, broken or sticking. Overhaul the top end (Chapter 2).
☐ Piston seizure damage. Usually from lack of lubrication or overheating. Replace the pistons and bore the cylinders, as necessary (Chapter 2).
☐ Connecting rod bearing and/or piston pin-end clearance excessive. Caused by excessive wear or lack of lubrication. Replace worn parts.

Valve noise

☐ Incorrect valve clearances. Adjust the clearances by referring to Chapter 1.
☐ Valve spring broken or weak. Check and replace weak valve springs (Chapter 2).
☐ Camshaft or cylinder head worn or damaged. Lack of lubrication at high rpm is usually the cause of damage. Insufficient oil or failure to change the oil at the recommended intervals are the chief causes. Since there are no replaceable bearings in the head, the head itself will have to be replaced if there is excessive wear or damage (Chapter 2).

Other noise

☐ Cylinder head gasket leaking. This will cause compression leakage into the cooling system (which may show up as air bubbles in the coolant in the radiator). Also, coolant may get into the oil (which will turn the oil gray). In either case, have the cooling system checked by a dealer service department.
☐ Exhaust pipe leaking at cylinder head connection. Caused by improper fit of pipe(s) or loose exhaust flange. All exhaust fasteners should be tightened evenly and carefully. Failure to do this will lead to a leak.
☐ Crankshaft runout excessive. Caused by a bent crankshaft (from over-revving) or damage from an upper cylinder component failure. Can also be attributed to dropping the machine on either of the crankshaft ends.
☐ Engine mounting bolts loose. Tighten all engine mount bolts to the specified torque (Chapter 2).
☐ Crankshaft bearings worn (Chapter 2).
☐ Camshaft chain tensioner defective. Replace according to the procedure in Chapter 2.
☐ Camshaft chain, sprockets or guides worn (Chapter 2).
☐ Loose alternator rotor - 600 models. Tighten the mounting bolt to the specified torque (Chapter 2A).

8 Abnormal driveline noise

Clutch noise

☐ Clutch housing/friction plate clearance excessive (Chapter 2).
☐ Loose or damaged clutch pressure plate and/or bolts (Chapter 2).

Transmission noise

☐ Bearings worn. Also includes the possibility that the shafts are worn. Overhaul the transmission (Chapter 2).
☐ Gears worn or chipped (Chapter 2).
☐ Metal chips jammed in gear teeth. Probably pieces from a broken clutch, gear or shift mechanism that were picked up by the gears. This will cause early bearing failure (Chapter 2).

☐ Engine oil level too low. Causes a howl from transmission. Also affects engine power and clutch operation (Daily (pre-ride) checks).

Chain or final drive noise

☐ Chain not adjusted properly (Chapter 1).
☐ Sprocket (primary sprocket or rear sprocket) loose. Tighten fasteners (Chapter 6).
☐ Sprocket(s) worn. Replace sprocket(s) (Chapter 6).
☐ Rear sprocket warped. Replace (Chapter 6).
☐ Wheel coupling worn. Replace coupling (Chapter 6).

9 Abnormal frame and suspension noise

Front end noise

☐ Low fluid level or improper viscosity oil in forks. This can sound like "spurting" and is usually accompanied by irregular fork action (Chapter 6).
☐ Spring weak or broken. Makes a clicking or scraping sound. Fork oil, when drained, will have a lot of metal particles in it (Chapter 6).
☐ Steering head bearings loose or damaged. Clicks when braking. Check and adjust or replace as necessary (Chapter 6).
☐ Fork clamps loose. Make sure all fork clamp pinch bolts are tight (Chapter 6).
☐ Fork tube bent. Good possibility if machine has been dropped. Replace tube with a new one (Chapter 6).
☐ Front axle or axle clamp bolt loose. Tighten them to the specified torque (Chapter 7).

Shock absorber noise

☐ Fluid level incorrect. Indicates a leak caused by defective seal. Shock will be covered with oil. Replace shock (Chapter 6).
☐ Defective shock absorber with internal damage. This is in the body of the shock and cannot be remedied. The shock must be replaced with a new one (Chapter 6).

☐ Bent or damaged shock body. Replace the shock with a new one (Chapter 6).

Disc brake noise

☐ Squeal caused by shim not installed or positioned correctly (Chapter 7).
☐ Squeal caused by dust on brake pads. Usually found in combination with glazed pads. Clean using brake cleaning solvent (Chapter 7).
☐ Contamination of brake pads. Oil, brake fluid or dirt causing brake to chatter or squeal. Clean or replace pads (Chapter 7).
☐ Pads glazed. Caused by excessive heat from prolonged use or from contamination. Do not use sandpaper, emery cloth, carborundum cloth or any other abrasive to roughen the pad surfaces as abrasives will stay in the pad material and damage the disc. A very fine flat file can be used, but pad replacement is suggested as a cure (Chapter 7).
☐ Disc warped. Can cause a chattering, clicking or intermittent squeal. Usually accompanied by a pulsating lever and uneven braking. Replace the disc (Chapter 7).

10 Oil pressure warning light comes on

Engine lubrication system

☐ Engine oil pump defective (Chapter 2).
☐ Engine oil level low. Inspect for leak or other problem causing low oil level and add recommended lubricant (Daily (pre-ride) checks).
☐ Engine oil viscosity too low. Very old, thin oil or an improper weight of oil used in engine. Change to correct lubricant (Chapter 1).
☐ Camshaft or journals worn. Excessive wear causing drop in oil pressure. Replace cam and/or head. Abnormal wear could be caused by oil starvation at high rpm from low oil level or improper oil weight or type (Chapter 1).

☐ Crankshaft and/or bearings worn. Same problems as previous item. Check and replace crankshaft and/or bearings (Chapter 2).

Electrical system

☐ Oil pressure switch defective. Check the switch according to the procedure in Chapter 9. Replace it if it is defective.
☐ Oil pressure warning light circuit defective. Check for pinched, shorted, disconnected or damaged wiring (Chapter 9).

11 Excessive exhaust smoke

White smoke

☐ Piston oil ring worn. The ring may be broken or damaged, causing oil from the crankcase to be pulled past the piston into the combustion chamber. Replace the rings with new ones (Chapter 2).
☐ Cylinders worn, cracked, or scored. Caused by overheating or oil starvation. The cylinders will have to be rebored and new pistons installed.
☐ Valve oil seal damaged or worn. Replace oil seals with new ones (Chapter 2).
☐ Valve guide worn. Perform a complete valve job (Chapter 2).
☐ Engine oil level too high, which causes oil to be forced past the rings. Drain oil to the proper level (Daily (pre-ride) checks).
☐ Head gasket broken between oil return and cylinder. Causes oil to be pulled into combustion chamber. Replace the head gasket and check the head for warpage (Chapter 2).
☐ Abnormal crankcase pressurization, which forces oil past the rings. Clogged breather or hoses usually the cause (Chapter 4).

Black smoke

☐ Air cleaner clogged. Clean or replace the element (Chapter 1).
☐ Main jet too large or loose. Compare the jet size to the Specifications (Chapter 4).
☐ Choke stuck, causing fuel to be pulled through choke circuit (Chapter 4).
☐ Fuel level too high. Check and adjust the float level as necessary (Chapter 4).
☐ Inlet needle held off needle seat. Clean float bowl and fuel line and replace needle and seat if necessary (Chapter 4).

Brown smoke

☐ Main jet too small or clogged. Lean condition caused by wrong size main jet or by a restricted orifice. Clean float bowl and jets and compare jet size to Specifications (Chapter 4).
☐ Fuel flow insufficient. Fuel inlet needle valve stuck closed due to chemical reaction with old gas. Float level incorrect. Restricted fuel line. Clean line and float bowl and adjust floats if necessary (Chapter 4).
☐ Carburetor intake manifolds loose (Chapter 4).
☐ Air cleaner poorly sealed or not installed (Chapter 1).

12 Poor handling or stability

Handlebar hard to turn

☐ Steering stem locknut too tight (Chapter 6).
☐ Bearings damaged. Roughness can be felt as the bars are turned from side-to-side. Replace bearings and races (Chapter 6).
☐ Races dented or worn. Denting results from wear in only one position (i.e., straight ahead) from impacting an immovable object or hole or from dropping the machine. Replace races and bearings (Chapter 6).
☐ Steering stem lubrication inadequate. Causes are grease getting hard from age or being washed out by high pressure car washes. Disassemble steering head and repack bearings (Chapter 6).
☐ Steering stem bent. Caused by hitting a curb or hole or from dropping the machine. Replace damaged part. Do not try to straighten stem (Chapter 6).
☐ Front tire air pressure too low (Daily (pre-ride) checks).

Handlebar shakes or vibrates excessively

☐ Tires worn or out of balance.
☐ Swingarm bearings worn. Replace worn bearings by referring to Chapter 6.
☐ Rim(s) warped or damaged. Inspect wheels for runout (Chapter 7).
☐ Wheel bearings worn. Worn front or rear wheel bearings can cause poor tracking. Worn front bearings will cause wobble (Chapter 7).
☐ Handlebar clamp bolts loose (Chapter 6).
☐ Steering stem or fork clamps loose. Tighten them to the specified torque (Chapter 6).
☐ Engine mounting bolts loose. Will cause excessive vibration with increased engine rpm (Chapter 2).

Handlebar pulls to one side

☐ Frame bent. Definitely suspect this if the machine has been dropped. May or may not be accompanied by cracking near the bend. Replace the frame (Chapter 6).

☐ Wheels out of alignment. Caused by improper location of axle spacers or from bent steering stem or frame (Chapter 6).
☐ Swingarm bent or twisted. Caused by age (metal fatigue) or impact damage. Replace the arm (Chapter 6).
☐ Steering stem bent. Caused by impact damage or from dropping the motorcycle. Replace the steering stem (Chapter 6).
☐ Fork leg bent. Disassemble the forks and replace the damaged parts (Chapter 6).
☐ Fork oil level uneven (Chapter 1).
☐ Uneven setting of AVDS (600 A and B models) or ESCS (early 600 C models). Ensure that each fork leg is set to the same position.

Poor shock absorbing qualities

☐ Too hard:
 a) Fork oil level excessive (Chapter 6).
 b) Fork air pressure excessive - 600 A and B models.
 c) Incorrect setting of AVDS (600 A and B models) or ESCS (early 600 C models).
 d) Fork oil viscosity too high. Use a lighter oil, (see the Specifications in Chapter 6).
 e) Fork tube bent. Causes a harsh, sticking feeling (Chapter 6).
 f) Fork internal damage (Chapter 6).
 g) Shock shaft or body bent or damaged (Chapter 6).
 h) Shock internal damage.
 i) Shock air pressure and damping too high (Chapter 1).
 j) Tire pressure too high (Daily (pre-ride) checks).
☐ Too soft:
 a) Fork or shock oil insufficient and/or leaking (Chapter 6).
 b) Fork oil level too low (Chapter 6).
 c) Fork oil viscosity too light (Chapter 6).
 d) Fork springs weak or broken (Chapter 6).
 e) Fork air pressure too low - 600 A and B models.
 f) Incorrect setting of AVDS (600 A and B models) or ESCS (early 600 C models).
 g) Shock air pressure and damping too low (Chapter 1).

13 Braking problems

Brakes are spongy, don't hold

☐ Air in brake line. Caused by inattention to master cylinder fluid level or by leakage. Locate problem and bleed brakes (Chapter 7).
☐ Pad or disc worn (Chapters 1 and 7).
☐ Brake fluid leak. See item 1.
☐ Contaminated pads. Caused by contamination with oil, grease, brake fluid, etc. Clean or replace pads. Clean disc thoroughly with brake cleaner (Chapter 7).
☐ Brake fluid deteriorated. Fluid is old or contaminated. Drain system, replenish with new fluid and bleed the system (Chapter 7).
☐ Master cylinder internal parts worn or damaged causing fluid to bypass (Chapter 7).
☐ Master cylinder bore scratched. From ingestion of foreign material or broken spring. Repair or replace master cylinder (Chapter 7).
☐ Disc warped. Replace disc (Chapter 7).

Brake lever pulsates

☐ Disc warped. Replace disc (Chapter 7).
☐ Axle bent. Replace axle (Chapter 6).
☐ Brake caliper bolts loose (Chapter 7).

☐ Brake caliper shafts damaged or sticking, causing caliper to bind. Lube the shafts and/or replace them if they are corroded or bent (Chapter 7).
☐ Wheel warped or otherwise damaged (Chapter 7).
☐ Wheel bearings damaged or worn (Chapter 7).

Brakes drag

☐ Master cylinder piston seized. Caused by wear or damage to piston or cylinder bore (Chapter 7).
☐ Lever balky or stuck. Check pivot and lubricate (Chapter 7).
☐ Brake caliper binds. Caused by inadequate lubrication, corrosion or damage to caliper shafts (Chapter 7).
☐ Brake caliper piston seized in bore. Caused by wear or ingestion of dirt past deteriorated seal (Chapter 7).
☐ Brake pad damaged. Pad material separating from backing plate. Usually caused by faulty manufacturing process or from contact with chemicals. Replace pads (Chapter 7).
☐ Pads improperly installed (Chapter 7).
☐ Rear brake pedal free play insufficient (Chapter 1).

14 Electrical problems

Battery dead or weak

☐ Battery faulty. Caused by sulphated plates which are shorted through the sedimentation or low electrolyte level. Also, broken battery terminal making only occasional contact. Check fluid level and specific gravity (Chapters 1 and 9).
☐ Battery cables making poor contact (Chapter 9).
☐ Load excessive. Caused by addition of high wattage lights or other electrical accessories.
☐ Ignition switch defective. Switch either grounds internally or fails to shut off system. Replace the switch (Chapter 9).
☐ Regulator/rectifier defective (Chapter 9).

☐ Stator coil open or shorted (Chapter 9).
☐ Wiring faulty. Wiring grounded or connections loose in ignition, charging or lighting circuits (Chapter 9).

Battery overcharged

☐ Regulator/rectifier defective. Overcharging is noticed when battery gets excessively warm or "boils" over (Chapter 9).
☐ Battery defective. Replace battery with a new one (Chapter 9).
☐ Battery amperage too low, wrong type or size. Install manufacturer's specified amp-hour battery to handle charging load (Chapter 9).

Fault Finding Equipment

Checking engine compression

● Low compression will result in exhaust smoke, heavy oil consumption, poor starting and poor performance. A compression test will provide useful information about an engine's condition and if performed regularly, can give warning of trouble before any other symptoms become apparent.
● A compression gauge will be required, along with an adapter to suit the spark plug hole thread size. Note that the screw-in type gauge/adapter set up is preferable to the rubber cone type.

● Before carrying out the test, first check the valve clearances as described in Chapter 1.
1 Run the engine until it reaches normal operating temperature, then stop it and remove the spark plug(s), taking care not to scald your hands on the hot components.
2 Install the gauge adapter and compression gauge in No. 1 cylinder spark plug hole (see illustration 1).
3 On kickstart-equipped motorcycles, make sure the ignition switch is OFF, then open the throttle fully and kick the engine over a couple of times until the gauge reading stabilises.
4 On motorcycles with electric start only, the procedure will differ depending on the nature of the ignition system. Flick the engine kill switch (engine stop switch) to OFF and turn

Screw the compression gauge adapter into the spark plug hole, then screw the gauge into the adapter

the ignition switch ON; open the throttle fully and crank the engine over on the starter motor for a couple of revolutions until the gauge reading stabilises. If the starter will not operate with the kill switch OFF, turn the ignition switch OFF and refer to the next paragraph.

5 Install the spark plugs back into their suppressor caps and arrange the plug electrodes so that their metal bodies are earthed (grounded) against the cylinder head; this is essential to prevent damage to the ignition system as the engine is spun over **(see illustration 2)**. Position the plugs well away from the plug holes otherwise there is a risk of atomised fuel escaping from the combustion chambers and igniting. As a safety precaution, cover the top of the valve cover with rag. Now turn the ignition switch ON and kill switch ON, open the throttle fully and crank the engine over on the starter motor for a couple of revolutions until the gauge reading stabilises.

All spark plugs must be earthed (grounded) against the cylinder head

6 After one or two revolutions the pressure should build up to a maximum figure and then stabilise. Take a note of this reading and on multi-cylinder engines repeat the test on the remaining cylinders.

7 The correct pressures are given in Chapter 2 Specifications. If the results fall within the specified range and on multi-cylinder engines all are relatively equal, the engine is in good condition. If there is a marked difference between the readings, or if the readings are lower than specified, inspection of the top-end components will be required.

8 Low compression pressure may be due to worn cylinder bores, pistons or rings, failure of the cylinder head gasket, worn valve seals, or poor valve seating.

9 To distinguish between cylinder/piston wear and valve leakage, pour a small quantity of oil into the bore to temporarily seal the piston rings, then repeat the compression tests **(see illustration 3)**. If the readings show a noticeable increase in pressure this confirms that the cylinder bore, piston, or rings are worn. If, however, no change is indicated, the cylinder head gasket or valves should be examined.

Bores can be temporarily sealed with a squirt of motor oil

10 High compression pressure indicates excessive carbon build-up in the combustion chamber and on the piston crown. If this is the case the cylinder head should be removed and the deposits removed. Note that excessive carbon build-up is less likely with the used on modern fuels.

Checking battery open-circuit voltage

⚠ **Warning: The gases produced by the battery are explosive - never smoke or create any sparks in the vicinity of the battery. Never allow the electrolyte to contact your skin or clothing - if it does, wash it off and seek immediate medical attention.**

● Before any electrical fault is investigated the battery should be checked.

● You'll need a dc voltmeter or multimeter to check battery voltage. Check that the leads are inserted in the correct terminals on the meter, red lead to positive (+ve), black lead to negative (-ve). Incorrect connections can damage the meter.

● A sound fully-charged 12 volt battery should produce between 12.3 and 12.6 volts across its terminals (12.8 volts for a maintenance-free battery). On machines with a 6 volt battery, voltage should be between 6.1 and 6.3 volts.

1 Set a multimeter to the 0 to 20 volts dc range and connect its probes across the

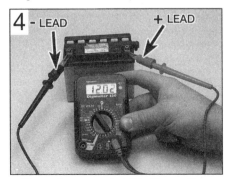

Measuring open-circuit battery voltage

battery terminals. Connect the meter's positive (+ve) probe, usually red, to the battery positive (+ve) terminal, followed by the meter's negative (-ve) probe, usually black, to the battery negative terminal (-ve) **(see illustration 4)**.

2 If battery voltage is low (below 10 volts on a 12 volt battery or below 4 volts on a six volt battery), charge the battery and test the voltage again. If the battery repeatedly goes flat, investigate the motorcycle's charging system.

Checking battery specific gravity (SG)

⚠ **Warning: The gases produced by the battery are explosive - never smoke or create any sparks in the vicinity of the battery. Never allow the electrolyte to contact your skin or clothing - if it does, wash it off and seek immediate medical attention.**

● The specific gravity check gives an indication of a battery's state of charge.

● A hydrometer is used for measuring specific gravity. Make sure you purchase one which has a small enough hose to insert in the aperture of a motorcycle battery.

● Specific gravity is simply a measure of the electrolyte's density compared with that of water. Water has an SG of 1.000 and fully-charged battery electrolyte is about 26% heavier, at 1.260.

● Specific gravity checks are not possible on maintenance-free batteries. Testing the open-circuit voltage is the only means of determining their state of charge.

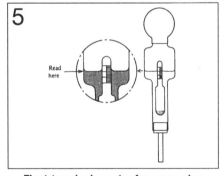

Float-type hydrometer for measuring battery specific gravity

1 To measure SG, remove the battery from the motorcycle and remove the first cell cap. Draw some electrolyte into the hydrometer and note the reading **(see illustration 5)**. Return the electrolyte to the cell and install the cap.

2 The reading should be in the region of 1.260 to 1.280. If SG is below 1.200 the battery needs charging. Note that SG will vary with temperature; it should be measured at 20°C (68°F). Add 0.007 to the reading for every 10°C above 20°C, and subtract 0.007 from the reading for every 10°C below 20°C. Add 0.004 to the reading for every 10°F above 68°F, and subtract 0.004 from the reading for every 10°F below 68°F.

3 When the check is complete, rinse the hydrometer thoroughly with clean water.

Checking for continuity

● The term continuity describes the uninterrupted flow of electricity through an electrical circuit. A continuity check will determine whether an **open-circuit** situation exists.

● Continuity can be checked with an ohmmeter, multimeter, continuity tester or battery and bulb test circuit **(see illustrations 6, 7 and 8)**.

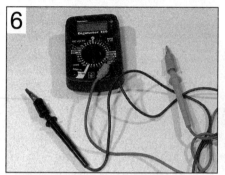

Digital multimeter can be used for all electrical tests

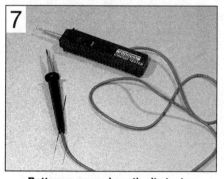

Battery-powered continuity tester

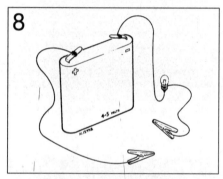

Battery and bulb test circuit

● All of these instruments are self-powered by a battery, therefore the checks are made with the ignition OFF.

● As a safety precaution, always disconnect the battery negative (-ve) lead before making checks, particularly if ignition switch checks are being made.

● If using a meter, select the appropriate ohms scale and check that the meter reads infinity (∞). Touch the meter probes together and check that meter reads zero; where necessary adjust the meter so that it reads zero.

● After using a meter, always switch it OFF to conserve its battery.

Switch checks

1 If a switch is at fault, trace its wiring up to the wiring connectors. Separate the wire connectors and inspect them for security and condition. A build-up of dirt or corrosion here will most likely be the cause of the problem - clean up and apply a water dispersant such as WD40.

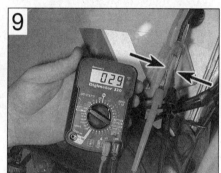

Continuity check of front brake light switch using a meter - note split pins used to access connector terminals

2 If using a test meter, set the meter to the ohms x 10 scale and connect its probes across the wires from the switch **(see illustration 9)**. Simple ON/OFF type switches, such as brake light switches, only have two wires whereas combination switches, like the ignition switch, have many internal links. Study the wiring diagram to ensure that you are connecting across the correct pair of wires. Continuity (low or no measurable resistance - 0 ohms) should be indicated with the switch ON and no continuity (high resistance) with it OFF.

3 Note that the polarity of the test probes doesn't matter for continuity checks, although care should be taken to follow specific test procedures if a diode or solid-state component is being checked.

4 A continuity tester or battery and bulb circuit can be used in the same way. Connect its probes as described above **(see illustration 10)**. The light should come on to indicate continuity in the ON switch position, but should extinguish in the OFF position.

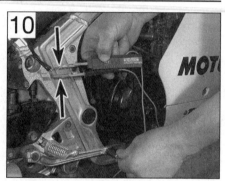

Continuity check of rear brake light switch using a continuity tester

Wiring checks

● Many electrical faults are caused by damaged wiring, often due to incorrect routing or chaffing on frame components.

● Loose, wet or corroded wire connectors can also be the cause of electrical problems, especially in exposed locations.

1 A continuity check can be made on a single length of wire by disconnecting it at each end and connecting a meter or continuity tester across both ends of the wire **(see illustration 11)**.

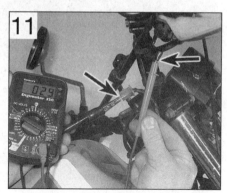

Continuity check of front brake light switch sub-harness

2 Continuity (low or no resistance - 0 ohms) should be indicated if the wire is good. If no continuity (high resistance) is shown, suspect a broken wire.

Checking for voltage

● A voltage check can determine whether current is reaching a component.

● Voltage can be checked with a dc voltmeter, multimeter set on the dc volts scale, test light or buzzer **(see illustrations 12 and 13)**. A meter has the advantage of being able to measure actual voltage.

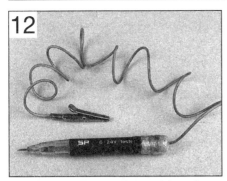

A simple test light can be used for voltage checks

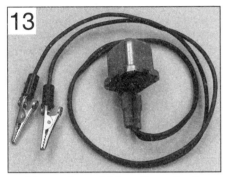

A buzzer is useful for voltage checks

● When using a meter, check that its leads are inserted in the correct terminals on the meter, red to positive (+ve), black to negative (-ve). Incorrect connections can damage the meter.
● A voltmeter (or multimeter set to the dc volts scale) should always be connected in parallel (across the load). Connecting it in series will destroy the meter.
● Voltage checks are made with the ignition ON.

1 First identify the relevant wiring circuit by referring to the wiring diagram at the end of this manual. If other electrical components share the same power supply (ie are fed from the same fuse), take note whether they are working correctly - this is useful information in deciding where to start checking the circuit.

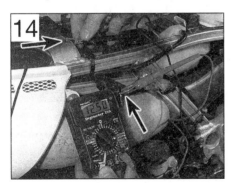

Checking for voltage at the rear brake light power supply wire using a meter . . .

2 If using a meter, check first that the meter leads are plugged into the correct terminals on the meter (see above). Set the meter to the dc volts function, at a range suitable for the battery voltage. Connect the meter red probe (+ve) to the power supply wire and the black probe to a good metal earth (ground) on the motorcycle's frame or directly to the battery negative (-ve) terminal **(see illustration 14)**. Battery voltage should be shown on the meter with the ignition switched ON.

3 If using a test light or buzzer, connect its positive (+ve) probe to the power supply terminal and its negative (-ve) probe to a good earth (ground) on the motorcycle's frame or directly to the battery negative (-ve) terminal **(see illustration 15)**. With the ignition ON, the test light should illuminate or the buzzer sound.

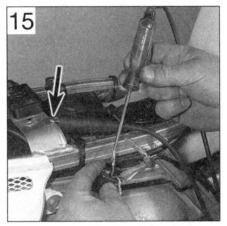

. . . or a test light - note the earth connection to the frame (arrow)

4 If no voltage is indicated, work back towards the fuse continuing to check for voltage. When you reach a point where there is voltage, you know the problem lies between that point and your last check point.

Checking the earth (ground)

● Earth connections are made either directly to the engine or frame (such as sensors, neutral switch etc. which only have a positive feed) or by a separate wire into the earth circuit of the wiring harness. Alternatively a short earth wire is sometimes run directly from the component to the motorcycle's frame.
● Corrosion is often the cause of a poor earth connection.
● If total failure is experienced, check the security of the main earth lead from the negative (-ve) terminal of the battery and also the main earth (ground) point on the wiring harness. If corroded, dismantle the connection and clean all surfaces back to bare metal.

1 To check the earth on a component, use an insulated jumper wire to temporarily bypass its earth connection **(see illustration 16)**. Connect one end of the jumper wire between the earth terminal or metal body of the component and the other end to the motorcycle's frame.

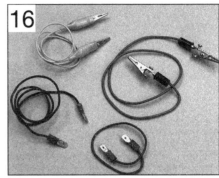

A selection of jumper wires for making earth (ground) checks

2 If the circuit works with the jumper wire installed, the original earth circuit is faulty. Check the wiring for open-circuits or poor connections. Clean up direct earth connections, removing all traces of corrosion and remake the joint. Apply petroleum jelly to the joint to prevent future corrosion.

Tracing a short-circuit

● A short-circuit occurs where current shorts to earth (ground) bypassing the circuit components. This usually results in a blown fuse.

● A short-circuit is most likely to occur where the insulation has worn through due to wiring chafing on a component, allowing a direct path to earth (ground) on the frame.

1 Remove any bodypanels necessary to access the circuit wiring.

2 Check that all electrical switches in the circuit are OFF, then remove the circuit fuse and connect a test light, buzzer or voltmeter (set to the dc scale) across the fuse terminals. No voltage should be shown.

3 Move the wiring from side to side whilst observing the test light or meter. When the test light comes on, buzzer sounds or meter shows voltage, you have found the cause of the short. It will usually shown up as damaged or burned insulation.

4 Note that the same test can be performed on each component in the circuit, even the switch.

A

ABS (Anti-lock braking system) A system, usually electronically controlled, that senses incipient wheel lockup during braking and relieves hydraulic pressure at wheel which is about to skid.

Aftermarket Components suitable for the motorcycle, but not produced by the motorcycle manufacturer.

Allen key A hexagonal wrench which fits into a recessed hexagonal hole.

Alternating current (ac) Current produced by an alternator. Requires converting to direct current by a rectifier for charging purposes.

Alternator Converts mechanical energy from the engine into electrical energy to charge the battery and power the electrical system.

Ampere (amp) A unit of measurement for the flow of electrical current. Current = Volts ÷ Ohms.

Ampere-hour (Ah) Measure of battery capacity.

Angle-tightening A torque expressed in degrees. Often follows a conventional tightening torque for cylinder head or main bearing fasteners **(see illustration)**.

Angle-tightening cylinder head bolts

Antifreeze A substance (usually ethylene glycol) mixed with water, and added to the cooling system, to prevent freezing of the coolant in winter. Antifreeze also contains chemicals to inhibit corrosion and the formation of rust and other deposits that would tend to clog the radiator and coolant passages and reduce cooling efficiency.

Anti-dive System attached to the fork lower leg (slider) to prevent fork dive when braking hard.

Anti-seize compound A coating that reduces the risk of seizing on fasteners that are subjected to high temperatures, such as exhaust clamp bolts and nuts.

API American Petroleum Institute. A quality standard for 4-stroke motor oils.

Asbestos A natural fibrous mineral with great heat resistance, commonly used in the composition of brake friction materials. Asbestos is a health hazard and the dust created by brake systems should never be inhaled or ingested.

ATF Automatic Transmission Fluid. Often used in front forks.

ATU Automatic Timing Unit. Mechanical device for advancing the ignition timing on early engines.

ATV All Terrain Vehicle. Often called a Quad.

Axial play Side-to-side movement.

Axle A shaft on which a wheel revolves. Also known as a spindle.

B

Backlash The amount of movement between meshed components when one component is held still. Usually applies to gear teeth.

Ball bearing A bearing consisting of a hardened inner and outer race with hardened steel balls between the two races.

Bearings Used between two working surfaces to prevent wear of the components and a build-up of heat. Four types of bearing are commonly used on motorcycles: plain shell bearings, ball bearings, tapered roller bearings and needle roller bearings.

Bevel gears Used to turn the drive through 90°. Typical applications are shaft final drive and camshaft drive **(see illustration)**.

Bevel gears are used to turn the drive
through 90°

BHP Brake Horsepower. The British measurement for engine power output. Power output is now usually expressed in kilowatts (kW).

Bias-belted tyre Similar construction to radial tyre, but with outer belt running at an angle to the wheel rim.

Big-end bearing The bearing in the end of the connecting rod that's attached to the crankshaft.

Bleeding The process of removing air from an hydraulic system via a bleed nipple or bleed screw.

Bottom-end A description of an engine's crankcase components and all components contained there-in.

BTDC Before Top Dead Centre in terms of piston position. Ignition timing is often expressed in terms of degrees or millimetres BTDC.

Bush A cylindrical metal or rubber component used between two moving parts.

Burr Rough edge left on a component after machining or as a result of excessive wear.

C

Cam chain The chain which takes drive from the crankshaft to the camshaft(s).

Canister The main component in an evaporative emission control system (California market only); contains activated charcoal granules to trap vapours from the fuel system rather than allowing them to vent to the atmosphere.

Castellated Resembling the parapets along the top of a castle wall. For example, a castellated wheel axle or spindle nut.

Catalytic converter A device in the exhaust system of some machines which converts certain pollutants in the exhaust gases into less harmful substances.

Charging system Description of the components which charge the battery, ie the alternator, rectifer and regulator.

Circlip A ring-shaped clip used to prevent endwise movement of cylindrical parts and shafts. An internal circlip is installed in a groove in a housing; an external circlip fits into a groove on the outside of a cylindrical piece such as a shaft. Also known as a snap-ring.

Clearance The amount of space between two parts. For example, between a piston and a cylinder, between a bearing and a journal, etc.

Coil spring A spiral of elastic steel found in various sizes throughout a vehicle, for example as a springing medium in the suspension and in the valve train.

Compression Reduction in volume, and increase in pressure and temperature, of a gas, caused by squeezing it into a smaller space.

Compression damping Controls the speed the suspension compresses when hitting a bump.

Compression ratio The relationship between cylinder volume when the piston is at top dead centre and cylinder volume when the piston is at bottom dead centre.

Continuity The uninterrupted path in the flow of electricity. Little or no measurable resistance.

Continuity tester Self-powered bleeper or test light which indicates continuity.

Cp Candlepower. Bulb rating common found on US motorcycles.

Crossply tyre Tyre plies arranged in a criss-cross pattern. Usually four or six plies used, hence 4PR or 6PR in tyre size codes.

Cush drive Rubber damper segments fitted between the rear wheel and final drive sprocket to absorb transmission shocks **(see illustration)**.

Cush drive rubbers dampen out
transmission shocks

D

Degree disc Calibrated disc for measuring piston position. Expressed in degrees.

Dial gauge Clock-type gauge with adapters for measuring runout and piston position. Expressed in mm or inches.

Diaphragm The rubber membrane in a master cylinder or carburettor which seals the upper chamber.

Diaphragm spring A single sprung plate often used in clutches.

Direct current (dc) Current produced by a dc generator.

Decarbonisation The process of removing carbon deposits - typically from the combustion chamber, valves and exhaust port/system.

Detonation Destructive and damaging explosion of fuel/air mixture in combustion chamber instead of controlled burning.

Diode An electrical valve which only allows current to flow in one direction. Commonly used in rectifiers and starter interlock systems.

Disc valve (or rotary valve) A induction system used on some two-stroke engines.

Double-overhead camshaft (DOHC) An engine that uses two overhead camshafts, one for the intake valves and one for the exhaust valves.

Drivebelt A toothed belt used to transmit drive to the rear wheel on some motorcycles. A drivebelt has also been used to drive the camshafts. Drivebelts are usually made of Kevlar.

Driveshaft Any shaft used to transmit motion. Commonly used when referring to the final driveshaft on shaft drive motorcycles.

E

Earth return The return path of an electrical circuit, utilising the motorcycle's frame.

ECU (Electronic Control Unit) A computer which controls (for instance) an ignition system, or an anti-lock braking system.

EGO Exhaust Gas Oxygen sensor. Sometimes called a Lambda sensor.

Electrolyte The fluid in a lead-acid battery.

EMS (Engine Management System) A computer controlled system which manages the fuel injection and the ignition systems in an integrated fashion.

Endfloat The amount of lengthways movement between two parts. As applied to a crankshaft, the distance that the crankshaft can move side-to-side in the crankcase.

Endless chain A chain having no joining link. Common use for cam chains and final drive chains.

EP (Extreme Pressure) Oil type used in locations where high loads are applied, such as between gear teeth.

Evaporative emission control system Describes a charcoal filled canister which stores fuel vapours from the tank rather than allowing them to vent to the atmosphere. Usually only fitted to California models and referred to as an EVAP system.

Expansion chamber Section of two-stroke engine exhaust system so designed to improve engine efficiency and boost power.

F

Feeler blade or gauge A thin strip or blade of hardened steel, ground to an exact thickness, used to check or measure clearances between parts.

Final drive Description of the drive from the transmission to the rear wheel. Usually by chain or shaft, but sometimes by belt.

Firing order The order in which the engine cylinders fire, or deliver their power strokes, beginning with the number one cylinder.

Flooding Term used to describe a high fuel level in the carburettor float chambers, leading to fuel overflow. Also refers to excess fuel in the combustion chamber due to incorrect starting technique.

Free length The no-load state of a component when measured. Clutch, valve and fork spring lengths are measured at rest, without any preload.

Freeplay The amount of travel before any action takes place. The looseness in a linkage, or an assembly of parts, between the initial application of force and actual movement. For example, the distance the rear brake pedal moves before the rear brake is actuated.

Fuel injection The fuel/air mixture is metered electronically and directed into the engine intake ports (indirect injection) or into the cylinders (direct injection). Sensors supply information on engine speed and conditions.

Fuel/air mixture The charge of fuel and air going into the engine. See **Stoichiometric ratio**.

Fuse An electrical device which protects a circuit against accidental overload. The typical fuse contains a soft piece of metal which is calibrated to melt at a predetermined current flow (expressed as amps) and break the circuit.

G

Gap The distance the spark must travel in jumping from the centre electrode to the side electrode in a spark plug. Also refers to the distance between the ignition rotor and the pickup coil in an electronic ignition system.

Gasket Any thin, soft material - usually cork, cardboard, asbestos or soft metal - installed between two metal surfaces to ensure a good seal. For instance, the cylinder head gasket seals the joint between the block and the cylinder head.

Gauge An instrument panel display used to monitor engine conditions. A gauge with a movable pointer on a dial or a fixed scale is an analogue gauge. A gauge with a numerical readout is called a digital gauge.

Gear ratios The drive ratio of a pair of gears in a gearbox, calculated on their number of teeth.

Glaze-busting see **Honing**

Grinding Process for renovating the valve face and valve seat contact area in the cylinder head.

Gudgeon pin The shaft which connects the connecting rod small-end with the piston. Often called a piston pin or wrist pin.

H

Helical gears Gear teeth are slightly curved and produce less gear noise that straight-cut gears. Often used for primary drives.

Installing a Helicoil thread insert in a cylinder head

Helicoil A thread insert repair system. Commonly used as a repair for stripped spark plug threads **(see illustration)**.

Honing A process used to break down the glaze on a cylinder bore (also called glaze-busting). Can also be carried out to roughen a rebored cylinder to aid ring bedding-in.

HT High Tension Description of the electrical circuit from the secondary winding of the ignition coil to the spark plug.

Hydraulic A liquid filled system used to transmit pressure from one component to another. Common uses on motorcycles are brakes and clutches.

Hydrometer An instrument for measuring the specific gravity of a lead-acid battery.

Hygroscopic Water absorbing. In motorcycle applications, braking efficiency will be reduced if DOT 3 or 4 hydraulic fluid absorbs water from the air - care must be taken to keep new brake fluid in tightly sealed containers.

I

lbf ft Pounds-force feet. An imperial unit of torque. Sometimes written as ft-lbs.

lbf in Pound-force inch. An imperial unit of torque, applied to components where a very low torque is required. Sometimes written as in-lbs.

IC Abbreviation for Integrated Circuit.

Ignition advance Means of increasing the timing of the spark at higher engine speeds. Done by mechanical means (ATU) on early engines or electronically by the ignition control unit on later engines.

Ignition timing The moment at which the spark plug fires, expressed in the number of crankshaft degrees before the piston reaches the top of its stroke, or in the number of millimetres before the piston reaches the top of its stroke.

Infinity (∞) Description of an open-circuit electrical state, where no continuity exists.

Inverted forks (upside down forks) The sliders or lower legs are held in the yokes and the fork tubes or stanchions are connected to the wheel axle (spindle). Less unsprung weight and stiffer construction than conventional forks.

J

JASO Quality standard for 2-stroke oils.

Joule The unit of electrical energy.

Journal The bearing surface of a shaft.

K

Kickstart Mechanical means of turning the engine over for starting purposes. Only usually fitted to mopeds, small capacity motorcycles and off-road motorcycles.

Kill switch Handebar-mounted switch for emergency ignition cut-out. Cuts the ignition circuit on all models, and additionally prevent starter motor operation on others.

km Symbol for kilometre.

kph Abbreviation for kilometres per hour.

L

Lambda (λ) sensor A sensor fitted in the exhaust system to measure the exhaust gas oxygen content (excess air factor).

Lapping see **Grinding**.
LCD Abbreviation for Liquid Crystal Display.
LED Abbreviation for Light Emitting Diode.
Liner A steel cylinder liner inserted in a aluminium alloy cylinder block.
Locknut A nut used to lock an adjustment nut, or other threaded component, in place.
Lockstops The lugs on the lower triple clamp (yoke) which abut those on the frame, preventing handlebar-to-fuel tank contact.
Lockwasher A form of washer designed to prevent an attaching nut from working loose.
LT Low Tension Description of the electrical circuit from the power supply to the primary winding of the ignition coil.

M

Main bearings The bearings between the crankshaft and crankcase.
Maintenance-free (MF) battery A sealed battery which cannot be topped up.
Manometer Mercury-filled calibrated tubes used to measure intake tract vacuum. Used to synchronise carburettors on multi-cylinder engines.
Micrometer A precision measuring instrument that measures component outside diameters **(see illustration)**.

Tappet shims are measured with a micrometer

MON (Motor Octane Number) A measure of a fuel's resistance to knock.
Monograde oil An oil with a single viscosity, eg SAE80W.
Monoshock A single suspension unit linking the swingarm or suspension linkage to the frame.
mph Abbreviation for miles per hour.
Multigrade oil Having a wide viscosity range (eg 10W40). The W stands for Winter, thus the viscosity ranges from SAE10 when cold to SAE40 when hot.
Multimeter An electrical test instrument with the capability to measure voltage, current and resistance. Some meters also incorporate a continuity tester and buzzer.

N

Needle roller bearing Inner race of caged needle rollers and hardened outer race. Examples of uncaged needle rollers can be found on some engines. Commonly used in rear suspension applications and in two-stroke engines.
Nm Newton metres.
NOx Oxides of Nitrogen. A common toxic pollutant emitted by petrol engines at higher temperatures.

O

Octane The measure of a fuel's resistance to knock.
OE (Original Equipment) Relates to components fitted to a motorcycle as standard or replacement parts supplied by the motorcycle manufacturer.
Ohm The unit of electrical resistance. Ohms = Volts ÷ Current.
Ohmmeter An instrument for measuring electrical resistance.
Oil cooler System for diverting engine oil outside of the engine to a radiator for cooling purposes.
Oil injection A system of two-stroke engine lubrication where oil is pump-fed to the engine in accordance with throttle position.
Open-circuit An electrical condition where there is a break in the flow of electricity - no continuity (high resistance).
O-ring A type of sealing ring made of a special rubber-like material; in use, the O-ring is compressed into a groove to provide the **Oversize (OS)** Term used for piston and ring size options fitted to a rebored cylinder.
Overhead cam (sohc) engine An engine with single camshaft located on top of the cylinder head.
Overhead valve (ohv) engine An engine with the valves located in the cylinder head, but with the camshaft located in the engine block or crankcase.
Oxygen sensor A device installed in the exhaust system which senses the oxygen content in the exhaust and converts this information into an electric current. Also called a Lambda sensor.

P

Plastigauge A thin strip of plastic thread, available in different sizes, used for measuring clearances. For example, a strip of Plastigauge is laid across a bearing journal. The parts are assembled and dismantled; the width of the crushed strip indicates the clearance between journal and bearing.
Polarity Either negative or positive earth (ground), determined by which battery lead is connected to the frame (earth return). Modern motorcycles are usually negative earth.
Pre-ignition A situation where the fuel/air mixture ignites before the spark plug fires. Often due to a hot spot in the combustion chamber caused by carbon build-up. Engine has a tendency to 'run-on'.
Pre-load (suspension) The amount a spring is compressed when in the unloaded state. Preload can be applied by gas, spacer or mechanical adjuster.
Premix The method of engine lubrication on older two-stroke engines. Engine oil is mixed with the petrol in the fuel tank in a specific ratio. The fuel/oil mix is sometimes referred to as "petroil".
Primary drive Description of the drive from the crankshaft to the clutch. Usually by gear or chain.
PS Pfedestärke - a German interpretation of BHP.
PSI Pounds-force per square inch. Imperial measurement of tyre pressure and cylinder pressure measurement.
PTFE Polytetrafluoroethylene. A low friction substance.

Pulse secondary air injection system A process of promoting the burning of excess fuel present in the exhaust gases by routing fresh air into the exhaust ports.

Q

Quartz halogen bulb Tungsten filament surrounded by a halogen gas. Typically used for the headlight **(see illustration)**.

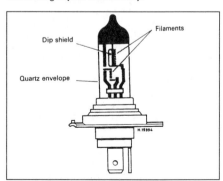

Quartz halogen headlight bulb construction

R

Rack-and-pinion A pinion gear on the end of a shaft that mates with a rack (think of a geared wheel opened up and laid flat). Sometimes used in clutch operating systems.
Radial play Up and down movement about a shaft.
Radial ply tyres Tyre plies run across the tyre (from bead to bead) and around the circumference of the tyre. Less resistant to tread distortion than other tyre types.
Radiator A liquid-to-air heat transfer device designed to reduce the temperature of the coolant in a liquid cooled engine.
Rake A feature of steering geometry - the angle of the steering head in relation to the vertical **(see illustration)**.

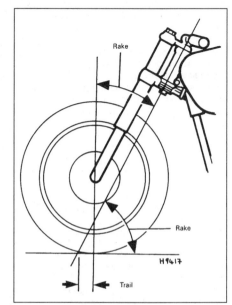

Steering geometry

Rebore Providing a new working surface to the cylinder bore by boring out the old surface. Necessitates the use of oversize piston and rings.

Rebound damping A means of controlling the oscillation of a suspension unit spring after it has been compressed. Resists the spring's natural tendency to bounce back after being compressed.

Rectifier Device for converting the ac output of an alternator into dc for battery charging.

Reed valve An induction system commonly used on two-stroke engines.

Regulator Device for maintaining the charging voltage from the generator or alternator within a specified range.

Relay A electrical device used to switch heavy current on and off by using a low current auxiliary circuit.

Resistance Measured in ohms. An electrical component's ability to pass electrical current.

RON (Research Octane Number) A measure of a fuel's resistance to knock.

rpm revolutions per minute.

Runout The amount of wobble (in-and-out movement) of a wheel or shaft as it's rotated. The amount a shaft rotates 'out-of-true'. The out-of-round condition of a rotating part.

S

SAE (Society of Automotive Engineers) A standard for the viscosity of a fluid.

Sealant A liquid or paste used to prevent leakage at a joint. Sometimes used in conjunction with a gasket.

Service limit Term for the point where a component is no longer useable and must be renewed.

Shaft drive A method of transmitting drive from the transmission to the rear wheel.

Shell bearings Plain bearings consisting of two shell halves. Most often used as big-end and main bearings in a four-stroke engine. Often called bearing inserts.

Shim Thin spacer, commonly used to adjust the clearance or relative positions between two parts. For example, shims inserted into or under tappets or followers to control valve clearances. Clearance is adjusted by changing the thickness of the shim.

Short-circuit An electrical condition where current shorts to earth (ground) bypassing the circuit components.

Skimming Process to correct warpage or repair a damaged surface, eg on brake discs or drums.

Slide-hammer A special puller that screws into or hooks onto a component such as a shaft or bearing; a heavy sliding handle on the shaft bottoms against the end of the shaft to knock the component free.

Small-end bearing The bearing in the upper end of the connecting rod at its joint with the gudgeon pin.

Spalling Damage to camshaft lobes or bearing journals shown as pitting of the working surface.

Specific gravity (SG) The state of charge of the electrolyte in a lead-acid battery. A measure of the electrolyte's density compared with water.

Straight-cut gears Common type gear used on gearbox shafts and for oil pump and water pump drives.

Stanchion The inner sliding part of the front forks, held by the yokes. Often called a fork tube.

Stoichiometric ratio The optimum chemical air/fuel ratio for a petrol engine, said to be 14.7 parts of air to 1 part of fuel.

Sulphuric acid The liquid (electrolyte) used in a lead-acid battery. Poisonous and extremely corrosive.

Surface grinding (lapping) Process to correct a warped gasket face, commonly used on cylinder heads.

T

Tapered-roller bearing Tapered inner race of caged needle rollers and separate tapered outer race. Examples of taper roller bearings can be found on steering heads.

Tappet A cylindrical component which transmits motion from the cam to the valve stem, either directly or via a pushrod and rocker arm. Also called a cam follower.

TCS Traction Control System. An electronically-controlled system which senses wheel spin and reduces engine speed accordingly.

TDC Top Dead Centre denotes that the piston is at its highest point in the cylinder.

Thread-locking compound Solution applied to fastener threads to prevent slackening. Select type to suit application.

Thrust washer A washer positioned between two moving components on a shaft. For example, between gear pinions on gearshaft.

Timing chain See **Cam Chain.**

Timing light Stroboscopic lamp for carrying out ignition timing checks with the engine running.

Top-end A description of an engine's cylinder block, head and valve gear components.

Torque Turning or twisting force about a shaft.

Torque setting A prescribed tightness specified by the motorcycle manufacturer to ensure that the bolt or nut is secured correctly. Undertightening can result in the bolt or nut coming loose or a surface not being sealed. Overtightening can result in stripped threads, distortion or damage to the component being retained.

Torx key A six-point wrench.

Tracer A stripe of a second colour applied to a wire insulator to distinguish that wire from another one with the same colour insulator. For example, Br/W is often used to denote a brown insulator with a white tracer.

Trail A feature of steering geometry. Distance from the steering head axis to the tyre's central contact point.

Triple clamps The cast components which extend from the steering head and support the fork stanchions or tubes. Often called fork yokes.

Turbocharger A centrifugal device, driven by exhaust gases, that pressurises the intake air. Normally used to increase the power output from a given engine displacement.

TWI Abbreviation for Tyre Wear Indicator. Indicates the location of the tread depth indicator bars on tyres.

U

Universal joint or U-joint (UJ) A double-pivoted connection for transmitting power from a driving to a driven shaft through an angle. Typically found in shaft drive assemblies.

Unsprung weight Anything not supported by the bike's suspension (ie the wheel, tyres, brakes, final drive and bottom (moving) part of the suspension).

V

Vacuum gauges Clock-type gauges for measuring intake tract vacuum. Used for carburettor synchronisation on multi-cylinder engines.

Valve A device through which the flow of liquid, gas or vacuum may be stopped, started or regulated by a moveable part that opens, shuts or partially obstructs one or more ports or passageways. The intake and exhaust valves in the cylinder head are of the poppet type.

Valve clearance The clearance between the valve tip (the end of the valve stem) and the rocker arm or tappet/follower. The valve clearance is measured when the valve is closed. The correct clearance is important - if too small the valve won't close fully and will burn out, whereas if too large noisy operation will result.

Valve lift The amount a valve is lifted off its seat by the camshaft lobe.

Valve timing The exact setting for the opening and closing of the valves in relation to piston position.

Vernier caliper A precision measuring instrument that measures inside and outside dimensions. Not quite as accurate as a micrometer, but more convenient.

VIN Vehicle Identification Number. Term for the bike's engine and frame numbers.

Viscosity The thickness of a liquid or its resistance to flow.

Volt A unit for expressing electrical "pressure" in a circuit. Volts = current x ohms.

W

Water pump A mechanically-driven device for moving coolant around the engine.

Watt A unit for expressing electrical power. Watts = volts x current.

Wear limit see **Service limit**

Wet liner A liquid-cooled engine design where the pistons run in liners which are directly surrounded by coolant **(see illustration).**

Wet liner arrangement

Wheelbase Distance from the centre of the front wheel to the centre of the rear wheel.

Wiring harness or loom Describes the electrical wires running the length of the motorcycle and enclosed in tape or plastic sheathing. Wiring coming off the main harness is usually referred to as a sub harness.

Woodruff key A key of semi-circular or square section used to locate a gear to a shaft. Often used to locate the alternator rotor on the crankshaft.

Wrist pin Another name for gudgeon or piston pin.

Haynes Motorcycle Manuals – The Complete List

Title	Book No
APRILIA RS50 (99 - 06) & RS125 (93 - 06)	4298
Aprilia RSV1000 Mille (98 - 03)	♦ 4255
Aprilia SR50	4755
BMW 2-valve Twins (70 - 96)	♦ 0249
BMW F650	♦ 4761
BMW K100 & 75 2-valve Models (83 - 96)	♦ 1373
BMW R850, 1100 & 1150 4-valve Twins (93 - 04)	♦ 3466
BMW R1200 (04 - 06)	♦ 4598
BSA Bantam (48 - 71)	0117
BSA Unit Singles (58 - 72)	0127
BSA Pre-unit Singles (54 - 61)	0326
BSA A7 & A10 Twins (47 - 62)	0121
BSA A50 & A65 Twins (62 - 73)	0155
Chinese Scooters	4768
DUCATI 600, 620, 750 and 900 2-valve V-Twins (91 - 05)	♦ 3290
Ducati MK III & Desmo Singles (69 - 76)	◊ 0445
Ducati 748, 916 & 996 4-valve V-Twins (94 - 01)	♦ 3756
GILERA Runner, DNA, Ice & SKP/Stalker (97 - 07)	4163
HARLEY-DAVIDSON Sportsters (70 - 08)	♦ 2534
Harley-Davidson Shovelhead and Evolution Big Twins (70 - 99)	♦ 2536
Harley-Davidson Twin Cam 88 (99 - 03)	♦ 2478
HONDA NB, ND, NP & NS50 Melody (81 - 85)	◊ 0622
Honda NE/NB50 Vision & SA50 Vision Met-in (85 - 95)	◊ 1278
Honda MB, MBX, MT & MTX50 (80 - 93)	0731
Honda C50, C70 & C90 (67 - 03)	0324
Honda XR80/100R & CRF80/100F (85 - 04)	2218
Honda XL/XR 80, 100, 125, 185 & 200 2-valve Models (78 - 87)	0566
Honda H100 & H100S Singles (80 - 92)	◊ 0734
Honda CB/CD125T & CM125C Twins (77 - 88)	◊ 0571
Honda CG125 (76 - 07)	◊ 0433
Honda NS125 (86 - 93)	◊ 3056
Honda CBR125R (04 - 07)	4620
Honda MBX/MTX125 & MTX200 (83 - 93)	◊ 1132
Honda CD/CM185 200T & CM250C 2-valve Twins (77 - 85)	0572
Honda XL/XR 250 & 500 (78 - 84)	0567
Honda XR250L, XR250R & XR400R (86 - 03)	2219
Honda CB250 & CB400N Super Dreams (78 - 84)	◊ 0540
Honda CR Motocross Bikes (86 - 01)	2222
Honda CRF250 & CRF450 (02 - 06)	2630
Honda CBR400RR Fours (88 - 99)	◊ ♦ 3552
Honda VFR400 (NC30) & RVF400 (NC35) V-Fours (89 - 98)	◊ ♦ 3496
Honda CB500 (93 - 02) & CBF500 03 - 08	◊ 3753
Honda CB400 & CB550 Fours (73 - 77)	0262
Honda CX/GL500 & 650 V-Twins (78 - 86)	0442
Honda CBX550 Four (82 - 86)	◊ 0940
Honda XL600R & XR600R (83 - 08)	♦ 2183
Honda XL600/650V Transalp & XRV750 Africa Twin (87 to 07)	♦ 3919
Honda CBR600F1 & 1000F Fours (87 - 96)	♦ 1730
Honda CBR600F2 & F3 Fours (91 - 98)	♦ 2070
Honda CBR600F4 (99 - 06)	♦ 3911
Honda CB600F Hornet & CBF600 (98 - 06)	◊ ♦ 3915
Honda CBR600RR (03 - 06)	♦ 4590
Honda CB650 sohc Fours (78 - 84)	0665
Honda NTV600 Revere, NTV650 and NT650V Deauville (88 - 05)	◊ ♦ 3243
Honda Shadow VT600 & 750 (USA) (88 - 03)	2312
Honda CB750 sohc Four (69 - 79)	0131
Honda V45/65 Sabre & Magna (82 - 88)	0820
Honda VFR750 & 700 V-Fours (86 - 97)	♦ 2101
Honda VFR800 V-Fours (97 - 01)	♦ 3703
Honda VFR800 V-Tec V-Fours (02 - 05)	♦ 4196
Honda CB750 & CB900 dohc Fours (78 - 84)	0535
Honda VTR1000 (FireStorm, Super Hawk) & XL1000V (Varadero) (97 - 08)	♦ 3744
Honda CBR900RR FireBlade (92 - 99)	♦ 2161
Honda CBR900RR FireBlade (00 - 03)	♦ 4060
Honda CBR1000RR Fireblade (04 - 07)	♦ 4604
Honda CBR1100XX Super Blackbird (97 - 07)	♦ 3901
Honda ST1100 Pan European V-Fours (90 - 02)	♦ 3384
Honda Shadow VT1100 (USA) (85 - 98)	2313
Honda GL1000 Gold Wing (75 - 79)	0309

Title	Book No
Honda GL1100 Gold Wing (79 - 81)	0669
Honda Gold Wing 1200 (USA) (84 - 87)	2199
Honda Gold Wing 1500 (USA) (88 - 00)	2225
KAWASAKI AE/AR 50 & 80 (81 - 95)	1007
Kawasaki KC, KE & KH100 (75 - 99)	1371
Kawasaki KMX125 & 200 (86 - 02)	◊ 3046
Kawasaki 250, 350 & 400 Triples (72 - 79)	0134
Kawasaki 400 & 440 Twins (74 - 81)	0281
Kawasaki 400, 500 & 550 Fours (79 - 91)	0910
Kawasaki EN450 & 500 Twins (Ltd/Vulcan) (85 - 07)	2053
Kawasaki EX500 (GPZ500S) & ER500 (ER-5) (87 - 08)	♦ 2052
Kawasaki ZX600 (ZZ-R600 & Ninja ZX-6) (90 - 06)	♦ 2146
Kawasaki ZX-6R Ninja Fours (95 - 02)	♦ 3541
Kawasaki ZX-6R (03 - 06)	♦ 4742
Kawasaki ZX600 (GPZ600R, GPX600R, Ninja 600R & RX) & ZX750 (GPX750R, Ninja 750R)	♦ 1780
Kawasaki 650 Four (76 - 78)	0373
Kawasaki Vulcan 700/750 & 800 (85 - 04)	♦ 2457
Kawasaki 750 Air-cooled Fours (80 - 91)	0574
Kawasaki ZR550 & 750 Zephyr Fours (90 - 97)	♦ 3382
Kawasaki Z750 & Z1000 (03 - 08)	♦ 4762
Kawasaki ZX750 (Ninja ZX-7 & ZXR750) Fours (89 - 96)	♦ 2054
Kawasaki Ninja ZX-7R & ZX-9R (94 - 04)	♦ 3721
Kawasaki 900 & 1000 Fours (73 - 77)	0222
Kawasaki ZX900, 1000 & 1100 Liquid-cooled Fours (83 - 97)	♦ 1681
KTM EXC Enduro & SX Motocross (00 - 07)	♦ 4629
MOTO GUZZI 750, 850 & 1000 V-Twins (74 - 78)	0339
MZ ETZ Models (81 - 95)	◊ 1680
NORTON 500, 600 & 750 Twins (57 - 70)	0187
Norton Commando (68 - 77)	0125
PEUGEOT Speedfight, Trekker & Vivacity Scooters (96 - 08)	◊ 3920
PIAGGIO (Vespa) Scooters (91 - 06)	◊ 3492
SUZUKI GT, ZR & TS50 (77 - 90)	◊ 0799
Suzuki TS50X (84 - 00)	◊ 1599
Suzuki 100, 125, 185 & 250 Air-cooled Trail bikes (79 - 89)	0797
Suzuki GP100 & 125 Singles (78 - 93)	◊ 0576
Suzuki GS, GN, GZ & DR125 Singles (82 - 05)	◊ 0888
Suzuki GSX-R600/750 (06 - 09)	♦ 4790
Suzuki 250 & 350 Twins (68 - 78)	0120
Suzuki GT250X7, GT200X5 & SB200 Twins (78 - 83)	◊ 0469
Suzuki GS/GSX250, 400 & 450 Twins (79 - 85)	0736
Suzuki GS500 Twin (89 - 06)	♦ 3238
Suzuki GS550 (77 - 82) & GS750 Fours (76 - 79)	0363
Suzuki GS/GSX550 4-valve Fours (83 - 88)	1133
Suzuki SV650 & SV650S (99 - 08)	♦ 3912
Suzuki GSX-R600 & 750 (96 - 00)	♦ 3553
Suzuki GSX-R600 (01 - 03), GSX-R750 (00 - 03) & GSX-R1000 (01 - 02)	♦ 3986
Suzuki GSX-R600/750 (04 - 05) & GSX-R1000 (03 - 06)	♦ 4382
Suzuki GSF600, 650 & 1200 Bandit Fours (95 - 06)	♦ 3367
Suzuki Intruder, Marauder, Volusia & Boulevard (85 - 06)	♦ 2618
Suzuki GS850 Fours (78 - 88)	0536
Suzuki GS1000 Four (77 - 79)	0484
Suzuki GSX-R750, GSX-R1100 (85 - 92), GSX600F, GSX750F, GSX1100F (Katana) Fours	♦ 2055
Suzuki GSX600/750F & GSX750 (98 - 02)	♦ 3987
Suzuki GS/GSX1000, 1100 & 1150 4-valve Fours (79 - 88)	0737
Suzuki TL1000S/R & DL1000 V-Strom (97 - 04)	♦ 4083
Suzuki GSF650/1250 (05 - 09)	♦ 4798
Suzuki GSX1300R Hayabusa (99 - 04)	♦ 4184
Suzuki GSX1400 (02 - 07)	♦ 4758
TRIUMPH Tiger Cub & Terrier (52 - 68)	0414
Triumph 350 & 500 Unit Twins (58 - 73)	0137
Triumph Pre-Unit Twins (47 - 62)	0251
Triumph 650 & 750 2-valve Unit Twins (63 - 83)	0122
Triumph Trident & BSA Rocket 3 (69 - 75)	0136
Triumph Bonneville (01 - 07)	♦ 4364
Triumph Daytona, Speed Triple, Sprint & Tiger (97 - 05)	♦ 3755
Triumph Triples and Fours (carburettor engines) (91 - 04)	♦ 2162
VESPA P/PX125, 150 & 200 Scooters (78 - 06)	0707
Vespa Scooters (59 - 78)	0126
YAMAHA DT50 & 80 Trail Bikes (78 - 95)	◊ 0800
Yamaha T50 & 80 Townmate (83 - 95)	◊ 1247

Title	Book No
Yamaha YB100 Singles (73 - 91)	◊ 0474
Yamaha RS/RXS100 & 125 Singles (74 - 95)	0331
Yamaha RD & DT125LC (82 - 95)	◊ 0887
Yamaha TZR125 (87 - 93) & DT125R (88 - 07)	◊ 1655
Yamaha TY50, 80, 125 & 175 (74 - 84)	◊ 0464
Yamaha XT & SR125 (82 - 03)	◊ 1021
Yamaha YBR125	4797
Yamaha Trail Bikes (81 - 00)	2350
Yamaha 2-stroke Motocross Bikes 1986 - 2006	2662
Yamaha YZ & WR 4-stroke Motocross Bikes (98 - 08)	2689
Yamaha 250 & 350 Twins (70 - 79)	0040
Yamaha XS250, 360 & 400 sohc Twins (75 - 84)	0378
Yamaha RD250 & 350LC Twins (80 - 82)	0803
Yamaha RD350 YPVS Twins (83 - 95)	1158
Yamaha RD400 Twin (75 - 79)	0333
Yamaha XT, TT & SR500 Singles (75 - 83)	0342
Yamaha XZ550 Vision V-Twins (82 - 85)	0821
Yamaha FJ, FZ, XJ & YX600 Radian (84 - 92)	2100
Yamaha XJ600S (Diversion, Seca II) & XJ600N Fours (92 - 03)	♦ 2145
Yamaha YZF600R Thundercat & FZS600 Fazer (96 - 03)	♦ 3702
Yamaha FZ-6 Fazer (04 - 07)	♦ 4751
Yamaha YZF-R6 (99 - 02)	♦ 3900
Yamaha YZF-R6 (03 - 05)	♦ 4601
Yamaha 650 Twins (70 - 83)	0341
Yamaha XJ650 & 750 Fours (80 - 84)	0738
Yamaha XS750 & 850 Triples (76 - 85)	0340
Yamaha TDM850, TRX850 & XTZ750 (89 - 99)	◊ ♦ 3540
Yamaha YZF750R & YZF1000R Thunderace (93 - 00)	♦ 3720
Yamaha FZR600, 750 & 1000 Fours (87 - 96)	♦ 2056
Yamaha XV (Virago) V-Twins (81 - 03)	♦ 0802
Yamaha XVS650 & 1100 Drag Star/V-Star (97 - 05)	♦ 4195
Yamaha XJ900F Fours (83 - 94)	♦ 3239
Yamaha XJ900S Diversion (94 - 01)	♦ 3739
Yamaha YZF-R1 (98 - 03)	♦ 3754
Yamaha YZF-R1 (04 - 06)	♦ 4605
Yamaha FZS1000 Fazer (01 - 05)	♦ 4287
Yamaha FJ1100 & 1200 Fours (84 - 96)	♦ 2057
Yamaha XJR1200 & 1300 (95 - 06)	♦ 3981
Yamaha V-Max (85 - 03)	♦ 4072

ATVs

Title	Book No
Honda ATC70, 90, 110, 185 & 200 (71 - 85)	0565
Honda Rancher, Recon & TRX250EX ATVs	2553
Honda TRX300 Shaft Drive ATVs (88 - 00)	2125
Honda Foreman (95 - 07)	2465
Honda TRX300EX, TRX400EX & TRX450R/ER ATVs (93 - 06)	2318
Kawasaki Bayou 220/250/300 & Prairie 300 ATVs (86 - 03)	2351
Polaris ATVs (85 - 97)	2302
Polaris ATVs (98 - 06)	2508
Yamaha YFS200 Blaster ATV (88 - 06)	2317
Yamaha YFB250 Timberwolf ATVs (92 - 00)	2217
Yamaha YFM350 & YFM400 (ER and Big Bear) ATVs (87 - 03)	2126
Yamaha Banshee and Warrior ATVs (87 - 03)	2314
Yamaha Kodiak and Grizzly ATVs (93 - 05)	2567
ATV Basics	10450

TECHBOOK SERIES

Title	Book No
Twist and Go (automatic transmission) Scooters Service and Repair Manual	4082
Motorcycle Basics TechBook (2nd Edition)	3515
Motorcycle Electrical TechBook (3rd Edition)	3471
Motorcycle Fuel Systems TechBook	3514
Motorcycle Maintenance TechBook	4071
Motorcycle Modifying	4272
Motorcycle Workshop Practice TechBook (2nd Edition)	3470

◊ = not available in the USA   ♦ = Superbike

The manuals on this page are available through good motorcycle dealers and accessory shops.
In case of difficulty, contact: **Haynes Publishing**
(UK) +44 1963 442030   (USA) +1 805 498 6703
(SV) +46 18 124016
(Australia/New Zealand) +61 3 9763 8100

MCL24.08/09

Preserving Our Motoring Heritage

< The Model J Duesenberg Derham Tourster. Only eight of these magnificent cars were ever built – this is the only example to be found outside the United States of America

Almost every car you've ever loved, loathed or desired is gathered under one roof at the Haynes Motor Museum. Over 300 immaculately presented cars and motorbikes represent every aspect of our motoring heritage, from elegant reminders of bygone days, such as the superb Model J Duesenberg to curiosities like the bug-eyed BMW Isetta. There are also many old friends and flames. Perhaps you remember the 1959 Ford Popular that you did your courting in? The magnificent 'Red Collection' is a spectacle of classic sports cars including AC, Alfa Romeo, Austin Healey, Ferrari, Lamborghini, Maserati, MG, Riley, Porsche and Triumph.

A Perfect Day Out

Each and every vehicle at the Haynes Motor Museum has played its part in the history and culture of Motoring. Today, they make a wonderful spectacle and a great day out for all the family. Bring the kids, bring Mum and Dad, but above all bring your camera to capture those golden memories for ever. You will also find an impressive array of motoring memorabilia, a comfortable 70 seat video cinema and one of the most extensive transport book shops in Britain. The Pit Stop Cafe serves everything from a cup of tea to wholesome, home-made meals or, if you prefer, you can enjoy the large picnic area nestled in the beautiful rural surroundings of Somerset.

> John Haynes O.B.E., Founder and Chairman of the museum at the wheel of a Haynes Light 12.

< The 1936 490cc sohc-engined International Norton – well known for its racing success

The Museum is situated on the A359 Yeovil to Frome road at Sparkford, just off the A303 in Somerset. It is about 40 miles south of Bristol, and 25 minutes drive from the M5 intersection at Taunton.
Open 9.30am - 5.30pm (10.00am - 4.00pm Winter) 7 days a week, *except Christmas Day, Boxing Day and New Years Day*
Special rates available for schools, coach parties and outings Charitable Trust No. 292048